Smart Low-Carbon Development of Cities in China

Editors: Pan Jiahua, Wang Hanqing, Liang Benfan, Zhou Yueyun

Paths International Ltd

中国社会科学出版社
CHINA SOCIAL SCIENCES PRESS

Acknowledgements

We sincerely appreciate the input of China Social Sciences Press's President Zhao Jianying's support for this project and Xia Xia, Yang Yang and Zhao Shixiong in the department of international Cooperation & Publishing at CSSP served as the editors, making the proof-reading. They have done much work in arranging the preparation of the English version.

We also sincerely appreciate the hard work of the translator, Miss Jin Jun. Without her diligent work, this English version would not exist.

Finally we especially appreciate the financial support in the publication of this book from CASS Innovation Translation Fund.

January 23, 2016

A Smart Low-Carbon City: New Hope for the Development of Human Civilization (Preface)

Social progress and city development not only bring unlimited innovations, production vitality, material wealth and new hopes to human beings, but also present huge problems, puzzles and challenges. With environmental pollution, ecological unbalance complicated by various natural and man-made disasters and problems, people have realized that city development in modern society entails higher, greater intelligence.

Smart and low carbon represent the future of human city development. Low carbon imposes constraints on global sustainable development and is the task and mission, which must be accomplished in city development. Sustainable development of the earth and human beings cannot be fostered without low-carbon development of cities. Intelligence serves as the best way for coping with and addressing all issues, achieving low-carbon and sustainable development in cities. Intricate city development contradictions and problems can't be conquered without intelligence or efficiency.

Building a smart low-carbon city is the latest trend in city development since the outbreak of the financial crisis in 2008 and also the core project for a smart planet. With wide applications of satellite navigation, wireless Internet access and the voice phone, people are overwhelmed by ubiquitous intelligence anytime, anywhere in city life. The development of the Internet of Things and cloud computing quickly leads our cities to the stage of smart low-carbon development. Building a smart low-carbon city has become one bottom-to-top campaign for China's city development since 2011. The China Association for Science and Technology has initiated Smart City Forum, while declarations about building a smart city have been made in many cities; national, provincial and municipal "12th Five-year" plans for the development of the Internet of Things and cloud computing have been officially published.

How to plan and design a smart low-carbon city has become the issue which urgently needs to be addressed by us and also the hot spot which attracts attention from scientific research, industry, teaching circles and city management departments. As the 'think bank' and 'think tank' for the Party Central Committee and the State Council, the Chinese Academy of Social Sciences has more necessity to seize this golden historic opportunity for city development.

This report is prepared by the Institute for Urban and Environmental Studies, Chinese Academy of Social Sciences, in conjunction with bodies including the Joint Research Center of a Global Low Carbon City at Hunan University of Technology, to make a preliminary attempt to introduce, research and explore theories and modes concerning a smart low-carbon city. The highlighted bright spot of this report lies in defining a smart city from the perspective of an urban political and economic cultural complex and developing a relatively ideal Chinese urban low-carbon development evaluation indicator system and data which can provide a reference for cities to promote energy conservation and emission reduction.

Authors of this report made painstaking efforts, collected a mass of data and conducted in-depth, insightful analysis of some issues. The smart city and low-carbon development constitute one key and important issue for today's world and involve the responsibilities and obligations of developed countries and developing countries for mankind and the earth; developed countries possess lots of advanced technologies and should offer more assistance to developing countries; a smart low-carbon city can be built and low-carbon operation can be performed to bring about the harmonious coexistence of human beings and nature only when input into information technology is intensified, a cloud service mode based on cloud computing is relied on and Internet of Things technologies are extensively used for improving efficiency and reducing energy consumption. Rapid development of information communication technology provides one efficient, pollution-free, livable, non-complicated technological and material base for building the smart city. A smart city brings good news to human beings and is a new hope for the development of mankind.

I hope that people from all walks of life pay attention to the development of the smart city, apply information technologies to achieve the goal of a smart low-carbon city so that China can create a bright future along the scientific development path.

Wu Jichuan

Minister of the Ministry of the Information Industry, China

September 4, 2012

Contents

Chapter 1 Low-Carbon Building Development Policies and Trends in China 1

Chapter 2 Comprehensive Evaluation of Low-Carbon Development of 110 Cities in China 14

Chapter 3 The Low-Carbon Development Situation in Chinese Cities and the Main Objectives during the 12th Five-Year Plan Period 26

Chapter 4 Technical Methods for Overall Energy Planning for Low-Carbon Cities and their Applications 38

Chapter 5 Integrated Technology for Efficient Coal Water Slurry Preparation and Combustion and Applications——Take Zhuzhou Lanyu Thermal Energy Technology Development Co., Ltd. as an example 57

Chapter 6 Applied Research on Urban Low-Carbon Transportation Technologies in China 81

Chapter 7 Research on Urban Waste Recycling Technology Applications 106

Chapter 8 Research on Low-carbon Technologies for Packaging Waste Disposal 122

Chapter 9 Carbon Reduction Technologies for the Packaging Product Manufacturing Process and their Application Research 137

Chapter 10 Exploration and Discussions of Carbon Reduction in Paper Packaging Material Production 160

Chapter 11 Practice and Countermeasures of Low-Carbon Transportation Development of Zhuzhou City in the Urban Transformation Process 173

Chapter 12 Discussion of Typical Foreign and Domestic Urban Low-Carbon Development Modes ... 190

Chapter 13 Transformation Modes and Reference for Domestic and Foreign Industrial Low-carbon Development ... 211

Appendix I The Top-ten Cities of Low-carbon Development for 110 Selected Chinese Cities in 2009 ... 222

Appendix II The Bottom-ten Cities of Low-carbon Development for 110 Selected Chinese Cities in 2009 ... 226

Appendix III The Data of Low-carbon Development Indicators for 110 Selected Chinese Cities in 2008 ... 232

Appendix IV The Order of 110 Selected Chinese Cities for Low-carbon Development in 2008 ... 239

Appendix V Dimensional Data of Low-carbon Development Indicators for 110 Selected Chinese Cities in 2009 ... 260

Appendix VI Dimensional and Dimensionless Data of Low-carbon Development Indicators for 110 Selected Chinese Cities in 2009 ... 274

Appendix VII Chronicle of Events about Low-Carbon Development Indicators in China (2011-2012) ... 295

Postscript ... 309

Chapter 1
Low-Carbon Building Development Policies and Trends in China

Yuan Lu Liang Benfan

Abstract: The development of both smart building and low-carbon building embodies people's pursuit of a higher living quality and also represents the main direction for the current development of the building industry. This chapter focuses on discussions about the development mode of low-carbon smart buildings. It starts with reviewing relevant released Chinese policies for low-carbon buildings and smart buildings and then summarizes the low-carbon smart building development mode in China, followed by further examination of low-carbon smart building development trends in the light of corporate achievements in the development of low-carbon smart buildings.
Keywords: Low-Carbon Building; Smart Building

Rapid urbanization in China presents urgent requirements for the development of low-carbon smart buildings; however, currently a well-proven low-carbon smart building development mode has not yet been made available in China. Against such a background, there is an urgent need for methodically clarifying low-carbon smart building development trends and relevant domestic policies so as to promote the popularization of low-carbon smart buildings.

1.1 Chinese Policies for Low-Carbon Smart Building Development

1.1.1 Chinese policies for low-carbon building development

Building a green environment with a reduction of carbon emissions from buildings is mainly achieved by the following two means: firstly, enhance energy efficiency in buildings and reduce energy consumption; secondly, popularize renewable energy and improve the energy structure. Current national policies for promoting building energy conservation and a green environment have fully taken into account both aspects.

1) Building energy conservation

China's urbanization started accelerating in the mid-1990s when the State also began attaching importance to building energy conservation and released series policies for building energy conservation, which mainly covered national energy conservation standards and intensified efforts in energy conservation.

In 1986, the State promulgated a *Design Standard for Residential Energy Conservation in Northern Heating Regions JGJ26-86* and carried out editing and revision twice. Design standards JGJ134-2001 and JGJ75-2003 for residential energy conservation in hot-summer and cold-winter regions, hot-summer and warm-winter regions were published in 2001 and 2003 respectively. A *Design Standard for Energy Conservation of Public Buildings GB50189-2005* was released in 2005. These three residential energy conservation standards and one public building energy conservation standard established 50% as the energy conservation goal to determine the performance standard for the external envelope structure of buildings and energy efficiency of relevant heating and air-conditioning equipment, which are commonly known as "50% energy conservation standards". In August 2010, the Ministry of Housing and Urban-Rural Development officially promulgated and implemented a new national standard for 60% residential energy conservation, covering northern cold and severe cold regions and hot-summer and cold-winter regions. Currently, efforts for preparing a 75% energy conversation standard are underway in some regions such as Beijing and Chongqing. As there is a restriction on air-conditioning and heating equipment efficiency improvement, the energy conservation rate increase in public buildings is lower than that in residential buildings; the 50% energy conservation standard is still prevailing nationwide at present; however, a regional 65% public building energy conservation standard has been promulgated or prepares to be implemented in some regions such as Jiangsu Province, Anhui Province and Chongqing etc.

On October 24,2007, the Ministry of Finance released *Provisional Measures for Administration of Special Funds for Energy Conservation in Office Buildings of State Organs and Large Public Buildings (C. J. (2007) No. 558)*, which specifies fund support for energy conservation renovation of public buildings and energy conservation supervision system. In 2011, the Ministry of Finance also issued a *Circular Concerning Further Promoting Energy Conservation in Public Buildings (C. J. (2011) No. 207)*, which presents specific subsidy standards and clarifies the working goal for public building energy conservation during the 12th Five-year Plan period: strive to reduce the energy consumption per unit area in public buildings by 10% and enable 15% reduction of energy conservation in large public buildings during the 12th Five-year Plan period. The State Council also successively issued No. 530 Order of the State Council of the People's Republic of China *Regulations for Energy Conservation in Civil Buildings* and No. 531 Order of the State Council *Regulations for Energy Conservation in Public Institutions* in 2008, with the aim of pushing forward building energy conservation and improving energy utilization efficiency.

The State has gradually stepped up efforts in propelling building energy conservation, as evidenced by the fact that in a Circular *Concerning Further Deepening Heating Metering and Energy Conservation Renovation of Existing Residential Buildings in Northern Heating Regions (C. J. (2011) No. 12)* as distributed in 2011, the Ministry of Finance and the Ministry of Housing and Urban-Rural Development vowed to further expand the building energy conservation renovation scale and basically complete heating metering and energy conservation renovation of old residential buildings with a renovation value in North China by 2020; make provinces (regions, municipalities) complete heating metering renovation of at least 35% of old residential buildings by the end of the 12th Five-year Plan period, encourage provinces (regions, municipalities) where conditions permit to increase the completion rate as well as stress that valuation and charging based on heat consumption and separately targeted at households must be concurrently conducted in projects in which heating metering renovation is completed.

2) Utilization of renewable energy

Subsidised policies for promoting renewable energy in buildings were mainly released by the Ministry of Finance and the Ministry of Construction. In 2006, the Ministry of Finance and the Ministry of Construction jointly announced *Provisional Measures for Administration of Special Funds for Application of Renewable Energy in Buildings (C. J. (2006) No. 460)*, which provides for special funds that are arranged from central finance to support the application of renewable energy in buildings. In 2009, a *Circular Concerning Printing and Distribution of Urban Demonstration Implementation Plan for Application of Renewable Energy in Buildings (C. J. (2009) No. 305) and a Circular Concerning Printing and Distribution of Implementation Plan for Accelerating Application of Renewable Energy in Buildings in Rural Areas (C. J. (2009) No. 306)* were successively published, which explicitly specify requirements for renewable energy utilization and calculation in demonstration cities. In 2010, a *Circular Concerning* Strengthening Administration of Urban Demonstration and Rural-area County-level Demonstration of Application of Renewable Energy in Buildings (C. J. (2010) No. 455) and a Circular Concerning Strengthening Administration of Follow-on Work on Demonstration of Application of Renewable Energy in Buildings and Budget Execution *(C. J. (2010) No. 484)* were released to revise and supplement existing policies. The goal for application of the promotion of renewable energy in buildings during the 12th Five-year Plan period has been clarified in the latest Circular Concerning Further Promoting Application of Renewable Energy in Buildings *(C. J. (2011) No. 61)*. The Circular stressed efforts in increasing proportions of renewable energy including solar energy, shallow geothermal energy, biomass energy etc. in buildings and making renewable energy consumption in buildings account for more than 15% of building energy consumption by 2020. During the 12th Five-year Plan period, both ministries will push ahead with centralized and massively contiguous application of renewable energy in buildings and further enrich the mode of application of renewable energy in buildings, and actively expand application fields, seek to add more than 2.5 billion m^2 involving application of renewable energy in buildings and bring about conventional energy replacement equivalent to 30-million-ton standard coal by the end of 2015.

Increasing proportions of renewable energy in building energy consumption produces a direct effect on and is of great significance in reducing pollution and carbon emissions. Solar energy development and utilization is particularly valued by the State thanks to its advantages, including suitability for vast areas, convenience for renovation of existing buildings and relatively mature technologies etc. *Implementation Opinions Concerning Quickening Solar Energy Photovoltaic Application in Buildings (C. J. (2009) No. 128), Provisional Measures for Administration of Fiscal Subsidy Funds for Energy Photovoltaic Application in Buildings (C. J. (2009) No. 129), Circular Concerning Implementation of Golden Sun Demonstration Project (C. J. (2009) No. 397), Circular Concerning Strengthening Construction Management of Golden Sun Demonstration Project and Demonstration Project for Solar Energy Photovoltaic Application in Buildings (C. J. (2010) No. 662)* etc. have been successively released since 2009. They have provided vigorous policy and fund support for solar energy photovoltaic application in buildings.

3) Green building and comprehensive policy

Evaluation Standard for Green Building was released later in China; it is a green building evaluation system established on the basis of fully analyzing and studying various domestic and foreign advanced green building evaluation systems in the light of the specific national conditions of China; thus this evaluation system is relatively comprehensive. Green building is defined in such a way that efforts are made to maximize saving of resources (energy, water, building materials, land), protect the environment and reduce pollution throughout the lifecycle of a building, so as to offer healthy, comfortable and efficient usable spaces for people and harmoniously coexist with nature. The first green building in China was completed in Beijing in 2004, which is an office building serving the purpose of a green building demonstration and is the result of Sino-US cooperation. The Ministry of Housing and Urban-Rural Development promulgated *Chinese Evaluation Standard for Green Building* in 2006, which is commonly known as a "three-star standard", while certification commenced accordingly in 2008. Currently, more than 30 buildings have passed certification based on this three-star standard.

On August 31,2011, a *Circular of the State Council Concerning Printing and Distribution of Comprehensive Work Plan for Energy Conservation and Emission Reduction during the 12th Five-year Plan Period (G. F.(2011) No. 26)* was issued, which calls for fully recognizing the importance, urgency and arduousness of energy conservation and emission reduction during the 12th Five-Year Plan period. With respect to building energy conservation, this circular states "develop and implement action plan for green buildings, push forward building energy conservation in an all-round way from the perspectives of planning, laws and regulations, technology, standard and design etc." *Implementation Opinions Concerning Pilot Demonstration of Key Small Green Towns (C. J. (2011) No. 341)* released in the same year emphasized actions to highlight green ecological development, guarantee key projects, and conscientiously enhance energy conservation and the emission reduction capability of pilot demonstration towns, and focus on construction of green ecological facilities.

It is shown that with accelerated urbanization and deepening work on building energy conservation and a green environment, the State has gradually reinforced efforts in building energy conservation and emission reduction in recent years. Release of relevant national policies also has reached a peak level in the past two years.

1.1.2 Policies for smart building development

Smart building policies were released earlier compared with low-carbon building ones in China and there was also no peak period during which smart building policies were released.

1) *Building Smart System Integration Design Atlas (03X801-1).*

The Building Smart System Integration Design Atlas which became effective as from May 1, 2003 includes preparation descriptions, smart system integration design descriptions, graphic symbols, letter symbols, structure diagrams of various control systems, an integration mode atlas of building smart subsystems. It is suitable for smart system integration of new, expansion and reconstruction projects.

2) *Configuration and Technical Requirements for Smart Systems in Residential Areas (CJ/T 174-2003).*

This standard specifies basic configuration and optional configuration of smart system construction in residential areas and serves as the basis for developers to choose smart systems. This standard sets forth configuration and technical requirements for smart systems in residential areas etc., mainly including definitions, technical classification, construction requirements, technical requirements, safety protection subsystems, management and monitoring subsystems, communication network subsystems etc. It is suitable for construction of smart systems in new residential areas and can also serve as a technical basis for real estate developers to choose systems and subsystems in the construction of smart residential areas.

3) *Key Points and Technical Guidelines for Construction of Smart Systems in Residential Quarters (2003).*

Key Points and Technical Guidelines for Construction of Smart Systems in Residential Quarters prepared by the Housing Industrialization Promotion Center of the Ministry of Construction entrusted by Project Quality Safety Supervision & Industry Management Department of the Ministry of Construction have played an important role in the construction of smart systems in residential quarters in China and have served as the important basis for China's real estate developers, design institutes, system integrators and smart product manufacturers to build and manage smart systems in residential quarters. These guidelines are designed to adapt to a life style in the information society in the 21st century and improve the functional quality of residential buildings. The overall objective for smart systems in residential quarters is to adopt modern information transmission technology, network technology and information integration technology to carry out precise design, optimal integration and elaborate construction, increase the content of high and new technologies in residential buildings and the level of the living environment so as to meet the modern living needs of residents.

4) *Design Standard for Smart Buildings (GB/T 50314-2006).*

Design Standard for Smart Buildings (GB/T 50314-2000) was released as early as 2000 and was revised in 2006. Smart building is a big concept and encompasses an information facility system, information-based application system, building equipment management system, public security system and machine room works. It incorporates original security protection, fire protection, building automation and control, telephone/ TV/computers, network system, and addresses such technical fields as information communication, computers, automatic control and building electrical development etc. and covers the engineering design of smart systems in civil and industrial buildings including new, expanded and reconstructed offices, commercial, cultural, media, sports, hospital, school, transport and residential buildings etc. This standard is a recommended, non-mandatory national standard.

5) *Acceptance Specification for Engineering Quality of Smart Building (GB 50339• 2003).*

Acceptance Specification for Engineering Quality of Smart Building jointly released by the Ministry of Construction and General Administration of Quality Supervision, Inspection and Quarantine is mainly applicable to the acceptance check of engineering quality of smart buildings in new, expanded and reconstruction building projects. It has been implemented as from October 1 2003, in which part of the provisions (Subparagraphs 5.5.2, 5.5.3, 7.2.6, 7.2.9, 7.2.11, 11.1.7) are mandatory provisions.

Acceptance Specification for Engineering Quality of Smart Building is based on *Design Standard for Smart Buildings (GB/T 50314-2000)* and follows preparation principles specified in *Unified Acceptance Standard for Construction Quality of Building Project (GB 50300 - 2001)*. It mainly specifies provisions for quality control, system tests and completion acceptance of smart building works including communication network systems, information network systems, building equipment monitoring system, automatic fire alarm and fire protection linkage system, security protection system, integrated wiring systems, smart system integration, power supply and grounding, environment and residential building (residential quarters) intelligentization etc.

6) *The 12th Five-year Development Plan for Internet of Things Industry*

According to the National 12th Five-year Plan, smart power grids, intelligent transport, intelligent logistics, smart homes, environmental and safety tests, industrial and automatic control, medical treatment and health, precision agriculture and animal husbandry, finance and services, the national defense and military industry are ten key development fields in the Internet of Things industry. *The 12th Five-Year Development Plan for the Internet of Things Industry* provides targeted special planning for these fields.

China's Internet of Things is still at the stage of demonstration application during the 12th Five-Year Plan period during which industrial development thinking is seeking breakthroughs in parts of fields and then driving the development of the upstream and downstream industry chain, gradually extending to more fields. Three large fields including transport, environmental protection, agriculture and animal husbandry present a huge market demand and relevant applications show good market prospects and relatively large opportunities for substantial development; furthermore, these three fields involve many enterprises and technologies in upstream and downstream sections of the industry chain and can produce a very good driving effect; thus they will be given priority in the development of the Internet of Things industry.

7) Evaluation Standard for Digital Systems in Residential Areas (CJ/T 376-2011)

Evaluation Standard for Digital Systems in Residential Areas was published and distributed by Standards Press of China organized by the Standard Rating Research Institute of the Ministry of Construction; it is a product standard for the urban construction industry and became effective as from February 1, 2012.

The bright spot of this standard lies in making provisions for functions requirements for digital systems in residential areas, rules for testing, evaluation grading of various products. It can serve as a reference basis for equipment manufacturers to develop products and for real estate development enterprises and system engineering organizations to build digital systems in residential areas and choose digital system functions for residential areas. With regard to content, this standard also embodies the philosophy of a green environment, energy conservation and low carbon. On the principle of "putting people first", with household safety, comfort, convenience and economy as goals, applications in ecology, energy conservation, environmental protection and sustainable development are considered as key points for smart construction in the future.

1.2 Low-Carbon Smart Building Development Mode in China

1.2.1 Internet of Things development mode: IGRS

IGRS started with active involvement in the digital home plan of Guangdong Province and has delivered a lot of results in promoting digital home consumption and 3C fusion. IGRS outlines a roadmap for realizing a digital home by firstly enabling local interconnection, and then fostering full-scene interconnection and finally interconnecting with the entire 3C network. QUERY is this 3C or 3G?

The IGRS standard has been included in a White Paper as a national standard for system-related general requirements in *Digital Technology Applications in Buildings and Residential Areas* and has been officially implemented as from December, 2006. This White Paper was jointly initiated by the Ministry of Construction and the Ministry of Information Industry and was released by the General Administration of Quality

Supervision and Standardization Administration of the People's Republic of China. In this national standard text, the IGRS standard has, as system-related general requirements, defined the specification for realizing the application layer functions of a home electronic system, while IGRS has also become the only standard for this field. IGRS is actively involved in a National Digital Demonstration Community Project nationwide; the IGRS standard has been adopted to build five smart residential quarters and many digital communities in Qingdao and Ji'nan since 2008.

1.2.2 Home appliance product integration mode: Haier

Intelligentization rapidly enhances the quality of life, while the U-home system launched by Haier makes an smart home accessible to the public and takes the lead in future trends.

A U-home is a brand-new network home platform for displaying a new lifestyle to people in the age of computer networks. It helps share information between people and home appliances, between home appliances, between home appliances and the external network, between home appliances and the after-sales system. In a U-home-enabled home, people can watch various programs from the network by means of high-definition TV and monitor the inside and outside of the room with TV, and people even freely go online for shopping. People outside the home can control the home appliances and learn about every corner of the home by mobile phone from anywhere in the world.

A U-home brings new life experience to people. For example, sending short messages to open and close a water heater, air conditioner, washing machine and microwave oven etc. became mature applications long ago; receiving images from a home video monitoring system through a network to immediately learn about the home also has been widely applied; automatic online diagnosis of home appliances and delivery of information to Haier's service center, automatic upgrading of software for home appliances, also have gone smoothly; an ultrawideband wireless high definition chip can facilitate wireless transmission of a high definition signal, while users can freely receive programs from various resources including networks and the PC etc. via TV, and can control other home appliances by way of the TV at any time.

1.2.3 Brand real estate development mode: Vanke

Vanke believes that the true standard for measuring community intelligentization consists in its individual functions close to consumer needs. Various modern technologies are deployed to collect, process, transmit, display and highly integrate information within the community, while a security protection system and property management system bring together security, convenience and comfort.

In order to offer a safe and reassuring living environment for owners, Vanke adopts a closed management mode under which a home anti-theft alarm system, boundary precaution alarm system, closed circuit television system and electronic guard tour system etc. constitute one all-around, three-dimensional integrated security protection system. A smart card system consisting of an automatic management system for the parking lot, access control system, clubhouse consumption system etc. is put in place, bringing great convenience for community property management and the life of residents.

In order to standardize design, construction and operation of smart systems in residential projects of the group, Vanke has developed "Smart System Standards of Vanke Group", including Configuration Standard, Design Standard, Component Standard, Construction Acceptance Standard and Operation Standard etc. They are mainly applied to overall planning for a smart system subarea in new residential areas, development of a smart system plan, preliminary design of a smart system and design of construction drawings for smart systems etc.

1.2.4 Green energy-saving technology cluster - Zhongguancun Exhibition Center

Zhongguancun Exhibition Center is designed in strict compliance with the highest three-star standard in the *Evaluation Standard for Green Buildings (GB50378-06)* issued by the Ministry of Construction in terms of conservation of land, energy, water, materials, indoor environment and operation management etc., and satisfies more than 50 indicators specified by it. This center incorporates multiple green energy saving technologies including high performance permeable ground, a siphonic roof rainwater discharge collection system, a water source heat pump system and optical lighting system etc. 100% of domestic water is recycled. A solar power system can be connected to the power grid and nearly 50,000 kWh power can be annually generated by 720 amorphous thin film modules installed on the roof of this center. Heat consumption in its composite vacuum insulated glass on a winter night is 83% and 70% lower than that in single white glass and common hollow glass respectively. Its heat recovery system can annually save nearly 570,000 kWh power equivalent to 122t standard coal. This center integrates currently the latest green building technologies in China and is expected to become the highest three-star green building; however, as it is a demonstration under test, success on a large-scale is still uncertain.

1.3 Development Trends of Low-Carbon and Smart Buildings in China

1.3.1 Low-carbon building development trends in China

1) Great achievements were made in China's building energy conservation during the 11th Five-year Plan period

Marked effects were produced on executing mandatory standards for energy conservation in new buildings. As of the end of 2010, new buildings in which mandatory standards for energy conservation were executed at the design stage in cities and towns nationwide accounted for 99.5% and that at the construction stage made up 95.4%, up 42 percentage points and 71 percentage points compared with 2005 respectively, which finished the work goal set by the State Council that "new buildings in which mandatory standards for energy conservation were executed at the construction stage should account for more than 95%". In 2010, the newly added energy-saving building area was 1.22 billion m^2, and an energy saving capability equivalent to 11.5 million tons of standard coal could be achieved. During the 11th Five-Year Plan period, the cumulative completed energy-saving building area was 4,857 million m^2 and an energy saving capability equivalent to 46 million tons of standard coal could be achieved. The energy-saving building area in cities and towns nationwide accounted for 23.1% of the existing building area; provinces, regions and municipalities where 30% was exceeded included Beijing, Tianjin, Shanghai, Chongqing, Hebei, Jilin, Liaoning, Jiangsu, Ningxia, Qinghai and Xinjiang.

Continued efforts were made to deepen the construction of an energy conservation monitoring system for office buildings of state organs and large public buildings. As of the end of 2010, energy consumption of 33,000 office buildings of state organs and large public buildings nationwide was statistically analyzed, while an energy audit of 4,850 buildings was completed, and energy consumption conditions of nearly 6,000 buildings was published and energy consumption of about 1,500 buildings was dynamically monitored. Pilot work on developing platforms for dynamic monitoring of energy consumption was carried out in nine provinces, regions and municipalities including Beijing, Tianjin, Shenzhen, Jiangsu, Chongqing, Inner Mongolia, Shanghai, Zhejiang and Guizhou. Pilot work on building 72 conservation-oriented campuses was initiated. Construction of the energy conservation monitoring system helped a comprehensive understanding of energy conservation levels and energy consumption characteristics of public buildings, and aroused enthusiasm for energy-saving operations and reconstruction and vigorously promoted conversion of the energy conservation potential into actual performance of energy conservation.

Renewable energy building applications showed a good momentum of rapid development. As of the end of 2010, the Ministry of Finance and the Ministry of Housing and Urban-Rural Development jointly implemented 371 renewable energy building application demonstration projects, 210 solar photoelectric building application demonstration projects, and completed a list of 47 renewable energy building application cities and 98 demonstration counties. Mandatory promotion of solar water heating systems was started in such provinces as Shandong, Jiangsu, Hainan and Hubei. The solar photo-thermal application area and shallow geothermal energy application area reached 1.48 billion m^2 and 227 million m^2 nationwide respectively, up 25.5% and 63.3% compared with 2009; installed capacity for photoelectric building applications which had been completed or was under construction hit 850.6 MW, a breakthrough has been made in the growth and annual capability for replacing traditional energy sources, equivalent to 20 million tons of standard coal.

Steady progress was made in the construction of green buildings and green ecological urban areas. As of the end of 2010, 112 projects were awarded the green building evaluation label nationwide, which covered a building area exceeding 13 million m^2, and there were more projects awarded such a label in Shanghai, Suzhou, Shenzhen, Hangzhou, Beijing and Tianjin. 217 green building demonstration projects were carried out nationwide, with a building area of more than 40 million m^2. Green ecological urban areas are being built in Binhai New Area in Tianjin, Guangming New Area in Shenzhen, Caofeidian New Area in Tangshan City, Hebei Province, the Industry Park in Suzhou City, Jiangsu Province, resource saving environmentally friendly supporting reform pilot areas in Changsha-Zhuzhou-Xiangtan, Hunan Province and Wuhan, Hubei Province.

There were some breakthroughs in rural building energy conservation work. Progress was made in building energy conservation work in rural areas in parts of provinces and municipalities; 13,829 rural households were organized to build new earthquake-resistant energy-saving houses, and 39,900 rural households conducted energy saving reconstruction of existing houses, and more than 400 rural centralized solar energy bath houses were completed, and energy conservation equivalent to more than 100,000 tons of standard coal was achieved. Rural housing and living conditions were significantly improved in Beijing during the 11th Five-year Plan period. With reconstruction of rural mud-thatch houses, farmers were guided to adopt new wall materials to build energy-saving houses in Harbin City. Popularization of new wall materials and the application of biomass, including straw, were taken as breakthrough points; beneficial progress was made in energy-saving house construction and new energy applications in rural areas in Shannxi and Gansu Provinces.

Positive achievements were made in wall material renovation. According to incomplete statistics, in 2010 the output of new wall materials surpassed 400 billion standard bricks nationwide, accounting for about 60% of total wall material output, while the actual use of new wall materials reached the equivalent of 350 billion standard bricks, accounting for about 70% of overall wall materials, which resulted in the completion of the development goal for wall material renovations as set by the State Council. Continuous actions have been taken to upgrade and transform new wall material technologies, enriching the product, improving product quality safety performance according to actual weather conditions and resource characteristics in various regions. An array of new materials and new products for building energy conservation, including integrated new building energy conservation systems with thermal insulation structure, and a light structure building system, have been introduced.

2) The 12th Five-Year goals and plans specific to various regions have been formulated for low-carbon and green development of buildings.

With the ecological city construction goal, Tianjin will become the first place to carry out four-step pilot work on energy conservation in China during the 12th Five-Year Plan period, and will see the implementation of five-year planning for newly developing energy-saving buildings covering 120 million m^2, making 30% of construction reach green building standards and enabling energy-saving reconstruction of the existing 15-million-square-meters of buildings.

At the end of 2009, Shandong Province vowed to energetically push ahead with solar energy and building integration, and focus on the popularization and application of solar photo-thermal systems while scientifically boosting solar photovoltaic applications, and take about three years to increase the popularization and application of urban solar water heaters from the current 20% to more than 40%, and elevate the rural popularization and application from less than 5% to more than 10%, seeking annual growth of about 25%. New, reconstructed and expanded residential buildings of 12 storeys and less and public buildings with a centralized supply of hot water within planning areas in cities above county level nationwide must be provided with solar photo-thermal systems, in which solar photo-thermal systems must be incorporated as part of building design and must be designed, constructed, checked and accepted synchronously with building works. Absence of synchronous design will result in failure to obtain approval from the planning department.

Wuxi Municipal Government has implemented a green building "4610 Plan" to promote low-carbon city construction since 2011. According to "4610 Plan", the following four support policies will be put into practice: a policy incentive for renewable energy development and utilization, policy support for reaching the national green building star standard, policy support for energy conservation reconstruction of existing buildings, and policy support for green energy-saving public buildings. The following six energy-saving technologies will be disseminated: ground source heat pump applications, solar energy applications, rainwater collection and water resource utilization, new wall material applications, energy-saving door and window applications, underground space utilization. Ten projects including public buildings, residential projects, existing building reconstruction projects will be selected and nurtured as leading municipal projects for building energy conservation.

1.3.2 China's smart building development trends

Enterprises are the main force for promoting smart building development; enterprises in various places have proposed their own new concepts and strive to become principal forces for leading future smart building development direction.

1) Haier Group's U-home

U represents Ubiquitous. As the biggest competitor for IGRS, led by Haier Group, Haier ITopHome was jointly initiated by Haier, Tsinghua Tongfang, China Netcom, SVA Group, Chunlan Group, Great Wall Group, Shanghai Belling to promote a home network system standard and platform industrialization. Haier Smarthome comprehensively launched a smart home U-home system in 2011, and smart home experience rooms have been extensively built nationwide. Haier Smarthome system focuses on intelligent control of home appliances such as smart air conditioners, smart refrigerators, smart TVs, etc; it is believed that in the next few years many home appliances will be equipped with a smart home communication module function for wired or wireless connection, by which a specific smart home remote control unit is used to exercise smart control of all home appliances in the entire home.

2) IGRS's smart building technology

IGRS was jointly initiated and established by Lenovo, TCL, Konka, Hisense and Great Wall, etc. under the leadership of the Ministry of Information Industry.

In March 2010, the International Organization for Standardization and the International Electrotechnical Commission (ISO/IEC) globally released official texts of IGRS's two international standards through their official websites. In 2011, IGRS and IGRS digital products including Great Wall smart home products, power carrier broadband communication products and IGRS Lingxi wireless connector, etc. were showcased at the China Hubei Industry-Academy-Research Cooperation Project Fair, Shanghai "100% Design Show" and China Hi-Tech Fair.

3) Large network technology mode

China's smart building development has shifted from a spontaneous practice stage to the stage of government-driven development. "Three-in-one Network" and "Internet of Things" industries vigorously supported by national policies will become two powerful engines for driving the smart home industry to develop by leaps and bounds. In 2010, the Chinese government explicitly stated efforts to energetically develop the "three-in-one network" and "Internet of Things" industries and started full implementation of a pilot program for "three-in-one network" cities. Rapid development of a "three-in-one network" will be conducive to building a faster and wider "information superhighway" and developing more information value-added service industries. As the home's intelligent information center, the smart home industry inevitably needs final fusion and information intercommunication with community intelligentization, city intelligentization, national intelligentization, even a transnational intelligent network to form a real "smart earth network". A smart home is the base and smallest unit for a "smart earth" and acts as the final service source for a "smart earth". We believe that robust development of a "three-in-one network" and "Internet of Things" will certainly drive the smart home industry to take a faster, more substantive industrial development path, and accelerate the large-scale, industrialized, normalized and standardized development of the smart home industry, so that smart home life really unfolds in numerous households.

References

[1] General Office of the Ministry of Housing and Rural-Urban Development: *Report on Special Energy Conservation and Emission Reduction Supervision Inspection and Building Energy Conservation Inspection in National Housing and the Rural-Urban Development Field in 2010* (JBK (2011) No. 25), December 6, 2010.

[2] The Ministry of Housing and Rural-Urban Development of the People's Republic of China: *IGRS has become the National Standard for Smart Building, China Quality Daily*, March 5, 2007.

Chapter 2
Comprehensive Evaluation of Low-Carbon Development of 110 Cities in China

Liang Benfan, Zhu Shouxian, Zhou Yueyun, Zhuang Guiyang

Abstract: This chapter describes the evaluation index system of urban low carbon development, including the indicators and their weights. It also uses this index system to calculate the low-carbon development of 110 selected Chinese cities and make some interpretation of this result.

Key words: urban low carbon development; evaluation index system; sort

Controlling greenhouse gas emissions is an important task for China to actively address global climate change and an important means for quickening the transformation of the economic development mode and stimulating sustainable economic and social development. In December 2011, the State Council issued a national 12th Five-Year work plan for controlling greenhouse gas emissions according to the goal, as set in the National 12th Five-Year Plan, for reducing carbon dioxide emissions per unit of GDP in 2015 by 17% compared with 2010, which made quantitative arrangements for indicators of carbon dioxide emission reduction per unit of GDP for provinces, regions and municipalities. Cities are the main centers for economic development and basic providers of low-carbon development at national, provincial and regional levels. In recent years, cities nationwide have witnessed remarkable achievements in energetically pushing forward energy conservation and emission reduction, and optimizing the energy structure, controlling greenhouse gas emissions, increasing the carbon sink, speeding up the formation of the industrial system and lifestyle characterized by low carbon. In this study 110 Chinese cities were chosen for a comprehensive ranking study of current low-carbon development levels in order to urge and test cities to take low-carbon development action and sum up their experience in the development of low-carbon cities and promote scientific emission reduction in cities.

2.1 Urban Low-Carbon Development Level Evaluation Indicator System

The urban low-carbon development evaluation indicator system is designed to guide practice and serve policy design and urban low-carbon development planning in addition to evaluating current urban low-carbon development. For example, the desired value, target value and current value are set for each evaluation indicator, and then the target value is determined according to the desired value, and current high-carbon development is improved according to the target value. The basis for selecting indicators and indicator values lies in considering resource endowment, the technical level and consumption mode in addition to the low-carbon development stage and its background factors. The potential for low-carbon development and the efforts of cities in transforming to a low-carbon economy are of realistic significance for evaluating the low-carbon development level of a city.

Development of a low-carbon economy must be based on the current economic development stage and resource endowment, and careful examination of the connotation and development trends of a low-carbon economy, as well as incorporation of a cleanness-oriented energy structure, industrial structure optimization and upgrading, technical level increase, consumption mode changes, carbon sink potential, into a strategic plan for economic and social development. Study shows that a cleaner energy structure can lower carbon emission intensity per unit of energy consumption, and optimization of the industrial structure can generally enhance carbon output efficiency (carbon productivity) of various sectors in the social economy, while advocating a green consumption mode can curb energy demand from terminals. However, the above route cannot work without supporting coordination from the institutional environment and motivation from policy tools. In 2010, the National Development and Reform Commission released a *Circular Concerning Pilot Work on Low-Carbon Provinces, Regions and Municipalities* (FGQH: 2010, No. 1587), in which formulation of a strategic plan for development of a low-carbon economy, establishment of carbon emission monitoring, a statistical and supervision system, awareness of the general public about a low-carbon economy, execution of building energy conservation standards, incentive measures for non-commercial energy including small biogas facilities, solar water heaters and biomass energy are deemed to reflect the degree of effort of a region or city in transforming to a low-carbon economy. Meanwhile, pilot work on a low-carbon city is required to focus on developing a low-carbon development plan and supporting policies for low-carbon green development, accelerating the establishment of the industrial system characterized by low-carbon emissions, and building greenhouse gas emission data statistics and management system, and actively advocating a low-carbon green lifestyle and consumption mode.

The urban low-carbon development evaluation indicator system used in this paper encompasses five first-grade indicators including economic transformation indicator, social transformation indicator, facility low carbon indicator, resource low carbon indicator, environmental low carbon indicator, among which the economic transformation indicator represents the low-carbon technology level and development stage; the social transformation indicator reflects the consumption mode and social development fairness degree; the facility low carbon indicator depicts low-carbon resource endowment and development utilization; the environmental low carbon indicator shows the urban green development level. Two component indicators are chosen under each first-grade indicator and are given the corresponding threshold values (see Table 2-1).

The indicator system is designed to evaluate current low-carbon development of domestic cities and stresses the relative evaluation of current low-carbon development. However, given differences between domestic cities in their economic development level and progress, evaluation merely based on relative criteria may be subject to restrictions. Thus evaluation criteria for absolute values and their weights are set according to the degree of importance of indicators in this paper. Indicators are calculated by actual value–minimum value/maximum value–minimum value. Five indicators including carbon economy intensity, per-capita carbon emission, low-carbon building, water body environment and atmospheric environment are reverse indicators, while five indicators including consumption structure, employment contribution, low-carbon transport, low-carbon energy and forest carbon sink are forward indicators.

Table 2-1 China's urban low-carbon development evaluation indicator system

<table>
<tr><th>First-grade indicators</th><th>Weights (%)</th><th>Second-grade indicators</th><th>Component indicators</th><th>Weights (%)</th></tr>
<tr><td rowspan="2">1. Economic transformation indicator</td><td rowspan="2">60</td><td>(1) Carbon economy intensity indicator</td><td>Carbon emission intensity per unit of GDP</td><td>45</td></tr>
<tr><td>(2) Per-capita carbon emission indicator</td><td>Per-capita carbon emission level</td><td>15</td></tr>
<tr><td rowspan="2">2. Social transformation indicator</td><td rowspan="2">10</td><td>(3) Consumption structure indicator</td><td>Proportion of low-carbon expenditure of urban residents</td><td>5</td></tr>
<tr><td>(4) Employment contribution indicator</td><td>Number of posts provided by unit of carbon emissions</td><td>5</td></tr>
<tr><td rowspan="2">3. Facility low-carbon indicator</td><td rowspan="2">15</td><td>(5) Low-carbon building indicator</td><td>Operation energy consumption per unit building area</td><td>8</td></tr>
<tr><td>(6) Low-carbon transport indicator</td><td>Public transport usage efficiency and preference for taking public transport</td><td>7</td></tr>
<tr><td rowspan="2">4. Resource low-carbon indicator</td><td rowspan="2">10</td><td>(7) Low-carbon energy indicator</td><td>Use of non-fossil energy</td><td>6</td></tr>
<tr><td>(8) Forest carbon sink indicator</td><td>Forest coverage rate</td><td>4</td></tr>
</table>

(Contd.)

First-grade indicators	Weights (%)	Second-grade indicators	Component indicators	Weights (%)
5. Environmental low-carbon indicator	5	(9) Water body environment indicator	COD emissions	3
		(10) Atmospheric environment indicator	SO_2 emissions	2

2.2 Determination of Overall Indicator Weights

2.2.1 Methods for determining indicator weight

Determination of various indicator weights is the core issue in multi-attribute decision-making. Weight determination methods are mainly divided into a subjective weighting method and objective weighting method, among which the subjective weighting method compares various indicators according to the degree of importance, and assigns values to them and calculates their weights on the basis of knowledge, experience or preference of the decision-maker, mainly including the Delphi method, AHP method, preference ratio method, DARE method, binomial coefficient method, comparative matrix method and importance ranking method. In this study, indicator weights are determined mainly by the Delphi method and AHP method.

2.2.2 First-grade indicator weight assignment

Here is a guiding thought for low-carbon indicator system weight assignment: give prominence to the importance of core indicators characterized by low-carbon development, and meanwhile fully reflect the low-carbon development process and transformation characteristics of various urban aspects, and the requirements of relevant national policies.

Operationally, a hierarchical control method is adopted. Weights are firstly assigned to five first-grade indicators. The relative importance, guiding force, relevance and of five first-grade indicators in urban low-carbon development and the difficulty degree of these indicators are weighed and compared through group discussion by experts so as to assign a weight to each first-grade indicator. For example, the economic transformation indicator is a basic indicator, which reflects the low-carbon development level, whose weight accounts for 60% of the weights of the whole indicator system.

2.2.3 Second-grade indicator weight assignment

1) Carbon economy intensity indicator

It refers to the carbon emission intensity per unit of GDP and is the core indicator for measuring low carbon. This indicator directly links carbon emissions resulting from energy consumption to GDP; it can objectively reflect the improvement in overall carbon resource utilization efficiency of the urban social economy and also measure the overall low-carbon technology level of a city in a specific period. In November 2009, the State Council held an executive meeting for studying and arranging work on climate change and decided to decrease China's carbon dioxide emission per unit of GDP in 2020 by 40%–45% compared with 2005, incorporate such a decrease, as a constraint indicator, into a national medium and long-term plan for economic and social development, and develop the corresponding domestic statistical, monitoring and evaluation measures. Therefore, this indicator has the biggest weight in ten second-grade indicators, accounting for 45%.

Economic transformation also covers the unit energy consumption indicator of key products such as overall energy consumption per ton of steel, overall energy consumption from cement, coal consumption from the power supply for thermal power, and can also involve the comparison of carbon emission indicators per unit industrial added value in key industries. Locations of cities in different climatic zones may affect heating energy consumption. However, given that different climatic conditions lead to different industrial competitiveness in different regions, climatic conditions are not considered here.

Lowering carbon intensity merely leads to decreasing the carbon dioxide emission per unit of GDP but does not result in reducing the total carbon dioxide. This is the weakness of the carbon intensity indicator. A remedial measure is the setting of a total carbon emission indicator. Total carbon emission means the total carbon emission in a country or region in a given period, while total emission reduction exerts a relatively large impact on economic development.

2) Per-capita carbon emission indicator

It reflects total carbon emission and total population and is one of the internationally accepted core indicators for comparing low-carbon development levels. Its value is described as follows: firstly, it contains a total carbon emission indicator and makes up for the deficiency of the carbon intensity indicator; secondly, it incorporates population into the indicator, which embodies equitable distribution requirements of human living rights for carbon space and carbon footprint differences in populations in different regions; thirdly, this is an internationally recognized indicator and can be used for international comparisons. Thus this indicator ranks only second to the carbon intensity indicator in terms of importance and enjoys 15% weight.

3) Consumption structure indicator

It refers to the proportion of low-carbon expenditure of urban residents. Generally, resident consumption is gradually shifting from traditional basic living consumption to development-oriented consumption; however, the internal consumption structure imbalance has a great impact on expansion of overall consumption. There is excessively rapid growth in educational and medical expenditure, for institutional reasons and due to the increasing hierarchy of needs, which make up a large proportion of consumption expenditure. Therefore we consider educational and medical expenditure as the core content of low-carbon consumption expenditure. Change in low-carbon consumption expenditure structure and preferences of urban residents serve as guidance and an indication of the direction of urban low-carbon development. This indicator's weight is 5%.

4) Employment contribution indicator

The number of posts provided by unit of carbon emissions refers to the posting output of carbon emissions and helps evaluate the quality of urban social low-carbon development. Employment is the foundation of China's social stability and also the precondition for fairness and freedom. Increasing the employment rate is an important goal of any country for macroeconomic regulation and control, thus this indicator's weight was also 5%.

5) Low-carbon building indicator

It refers to operation energy consumption per unit building area. Building energy consumption is divided into a broad one and a narrow one, among which broad building energy consumption involves energy consumption in the whole process from building materials manufacturing, building construction to building operation, while narrow building energy consumption is based on building operations energy consumption and is the leading part of building energy consumption. China's real estate industry is highly developed, in which there is huge space for building energy conservation. There is high motivation for implementing it at local level and it is a good indicator for evaluating the low-carbon development performance of a local city. Its weight is 8%.

6) Low-carbon transport indicator

Low-carbon transport can be evaluated by such indicators as energy use efficiency, public transport vehicle use efficiency, preference for taking public transport. Given that two indicators GDP per unit of carbon emissions and per-capita carbon emissions level contain an efficiency factor concerning urban transport energy use, only public transport usage efficiency and a preference for taking public transport are emphatically evaluated here. Like the low-carbon building indicator, low-carbon transport is also implemented in a highly motivated way at local level and is a good indicator capable of evaluating the low-carbon development performance of a local city. Its weight is 7%.

7) Low-carbon energy indicator

It is very difficult to complete adjustment of the energy consumption structure within the short term due to differences in resource endowments. Thus the proportion and development degree of non-fossil energy are chosen as the evaluation content of the low-carbon energy indicator. In November 2009, the Chinese Government clarified actions to vigorously develop renewable energy and actively push forward nuclear power construction, in order to make the proportion of non-fossil energy in primary energy consumption in China reach about 15% by 2020, which is an important reference indicator for the use of non-fossil energy in the city. Besides nuclear energy and large hydropower facilities construction, wind power, solar photo-thermal energy, earth energy and biomass energy, are also available for being massively developed and utilized in local cities. Thus this indicator's weight is relatively high (6%).

8) Forest carbon sink indicator

The forest carbon sinks means that forest plants absorb carbon dioxide from the atmosphere and fix it in vegetation or soil so as to lower carbon dioxide concentration in the atmosphere. The forest is the largest carbon pool in the terrestrial ecosystem and plays a very important unique role in decreasing greenhouse gas concentrations in the atmosphere and mitigating global warming. Expanding the forest covered area and increasing the forest coverage rate are economically feasible and important low-cost mitigation measures. The Chinese Government has emphasized efforts to carry out forest planting and strengthen forest management so as to increase, by 2020, the forest area and forest growing stock by 40 million hectares and 1.3 billion cubic meters compared with 2005 respectively, and the forest coverage rate to 21.66%. This indicator's weight is 4%.

9) Water body environment indicator

From the perspective of the current development situation in China, conceptually there is no necessary connection between low-carbon development and environmental pollution; however, the case where the environment is severely polluted, though total carbon emissions and intensity are very low, cannot be called scientific low-carbon development. Therefore, two indicators, water body environment and atmospheric environment, are introduced.

Chemical oxygen demand (COD) is oxygen consumption calculated according to the amount of residual chemical oxidants after chemical oxidants (such as potassium permanganate) oxidatively decompose oxidizable matter (such as organic matter, nitrites, ferrites, sulfides, etc.) in water and is the important indicator for measuring the degree of water pollution. COD's unit is ppm or mg/l; the higher COD value, the severer is the degree of water pollution. This indicator's weight is 3%.

10) Atmospheric environment indicator

Sulfur dioxide (SO_2) is mainly caused by combustion of sulphur-containing substances such as coal and fuel oil, and is a core indicator for evaluating the atmospheric environment. This indicator's weight is 2%.

2.3 Comprehensive Rankings of 110 Cities at the Low-Carbon Development Level

In this study, 110 cities were chosen for low-carbon development level analysis according to the above evaluation method and indicator system; comprehensive scores are shown in Table 2-2.

Table 2-2 Comprehensive scores of 110 cities at the low-carbon development level in 2009

Ranking	City	Comprehensive index score	Ranking	City	Comprehensive index score
1	Taizhou	80.53	56	Zhuzhou	47.36
2	Zhuhai	74.47	57	Kaifeng	47.06
3	Shenzhen	74.29	58	Mianyang	47.01
4	Xiamen	73.82	59	Xuzhou	46.62
5	Beijing	73.45	60	Shaoguan	46.47
6	Wenzhou	71.8	61	Tai'an	46.31
7	Haikou	70.62	61	Harbin	46.31
8	Zhanjiang	70.38	63	Nanjing	46.04
9	Yanan	69.79	64	Wuhu	45.76
10	Fuzhou	69.21	65	Mudanjiang	45.66
11	Sanya	67.44	66	Luzhou	45.5
12	Shantou	67.21	67	Zhengzhou	45.19
13	Hangzhou	66.05	68	Yueyang	45.01
14	Shanghai	65.88	69	Qinhuangdao	44.82
15	Zhongshan	65.54	70	Liuzhou	44.76
16	Guangzhou	65.23	71	Xianyang	44.46
17	Quanzhou	64.48	72	Luoyang	43.45
18	Chengdu	63.54	73	Guiyang	43.43
19	Changchun	61.78	74	Zunyi	42.97
20	Nanning	60.43	75	Jilin	42.88
21	Xi'an	60.15	76	Jingzhou	42.7
22	Ningbo	59.77	77	Qujing	42.41
23	Qingdao	59.64	78	Jining	42.12
24	Nantong	59.45	79	Qigihar	41.95
25	Changsha	58.56	80	Tongchuan	40.42
26	Yangzhou	57.9	81	Xiangtan	40.31
27	Foshan	57.49	82	Daqing	40.14
28	Zhangjiajie	57.35	83	Chifeng	39.5
29	Weihai	57.34	84	Xining	39.41
30	Kunming	57.32	85	Shijiazhuang	39.14
31	Yantai	56.92	86	Rizhao	39.13
32	Tianjin	56.78	87	Lanzhou	38.87
33	Shaoxing	56.25	88	Zaozhuang	38.31
34	Lianyungang	56.08	89	Hohhot	36.61
35	Guilin	55.41	90	Pingdingshan	36.53

(Contd.)

Ranking	City	Comprehensive index score	Ranking	City	Comprehensive index score
36	Jiujiang	55.06	91	Jinchang	36.46
37	Changde	54.97	92	Da Tong	36.09
38	Yichang	54.66	93	Zibo	35.78
39	Wuxi	53.99	94	Panzhihua	34.85
40	Huzhou	53.91	95	Anshan	34.72
41	Jiaxing	53.67	96	Yinchuan	34.37
42	Nanchang	53.29	97	Yangquan	33.88
43	Dalian	52.92	98	Taiyuan	33.62
44	Wuhan	52.27	99	Fushun	33.46
45	Hefei	51.42	100	Jiaozuo	33.37
46	Beihai	51.22	101	Changzhi	33.34
47	Suzhou	50.73	102	Handan	32.95
48	Baoding	49.95	103	Anyang	31.48
49	Baoji	49.72	104	Linfen	30.52
50	Changzhou	49.7	105	Urumqi	28.35
51	Ji'nan	48.63	106	Tangshan	26.95
52	Yibin	47.71	107	Ma'anshan	26.12
53	Chongqing	47.67	108	Baotou	24.51
54	Shenyang	47.56	109	Karamay	24.22
55	Weifang	47.54	110	Shizuishan	14.37

As shown by the analysis of comprehensive scores at the urban low-carbon development level, the top five cities in comprehensive rankings in 2009 were Taizhou, Zhuhai, Shenzhen, Xiamen and Beijing, while the last five cities in comprehensive rankings were Shizuishan, Karamay, Baotou, Ma'anshan and Tangshan, among which Taizhou, Zhuhai, Shenzhen, Xiamen and Beijing are located in the eastern region, while Shizuishan (coal), Karamay (oil), Baotou (iron and steel), Ma'anshan (iron and steel) and Tangshan (coal, iron and steel) are resource-based cities; furthermore, Shizuishan is among the first batch of resources-exhausted cities listed by the State.

2.4 Analysis of Low-Carbon Development Level Rankings of the Case Cities

A case study of ten cities with respect to core indicators of the low-carbon development level was conducted to discuss connotations of various elements in the low-carbon development indicator system, which can provide a decision-making basis for practices in urban low-carbon development.

Therefore, cities with relatively high scores in comprehensive rankings need to continue leading the low-carbon development trend, adopt a diversified development strategy, and boost a virtuous cycle for constantly upgrading the industrial structure while cultivating new economic growth points, enhancing international competitiveness and becoming the first to develop low-carbon demonstration cities.

As indicated by the analysis of energy consumption per unit of GDP and carbon emission intensity per unit of GDP, the energy consumption per unit of GDP in Taizhou, Shenzhen, Xiamen, Zhuhai and Beijing was 0.51-0.61, much lower than the national average –1.03, while that in Karamay, Baotou, Ma'anshan, Tangshan and Shizuishan was 1.6-6.8, among which that in Shizuishan was 6.5 times higher than the national average. Reducing energy consumption per unit of GDP, improving energy use efficiency and promoting industrial structure upgrading have become the important tasks for cities with high energy consumption such as Shizuishan during the 12th Five-Year Plan period, even beyond (Fig. 2-1, Fig. 2-5).

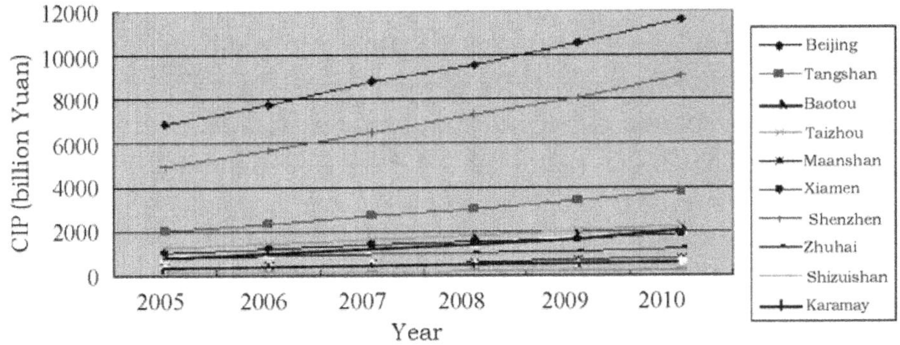

Fig. 2-1 Economic aggregate in case cities (2005-2010)

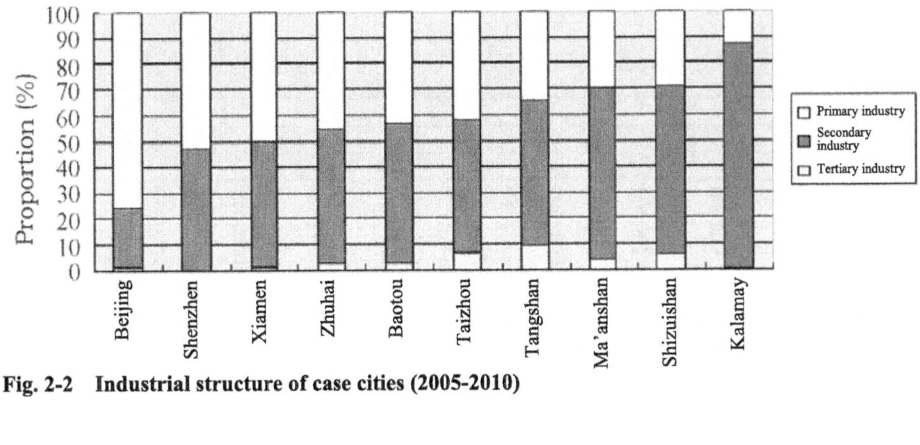

Fig. 2-2 Industrial structure of case cities (2005-2010)

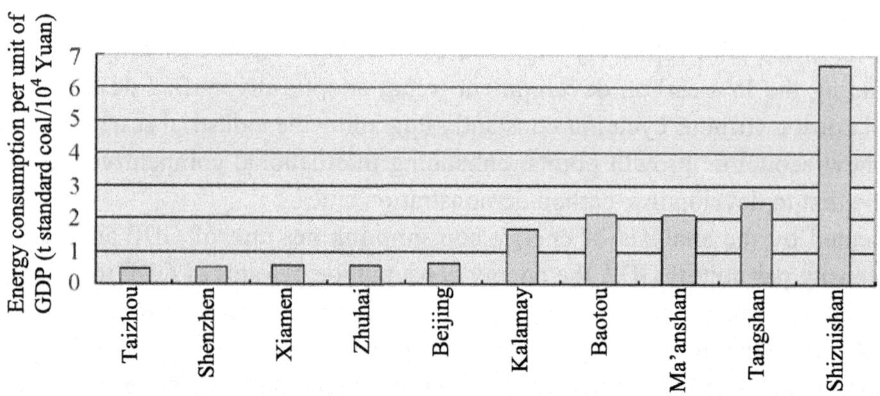

Fig. 2-3 Energy consumption per unit of GDP in case cities (2009)

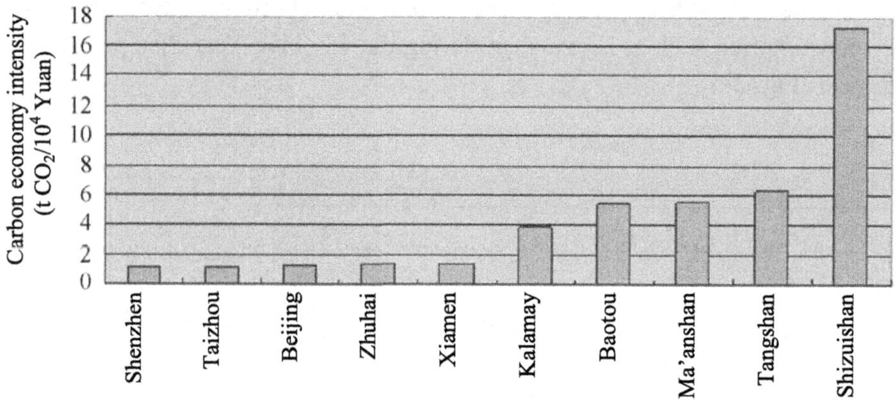

Fig. 2-4 Carbon economy intensity in case cities (2009)

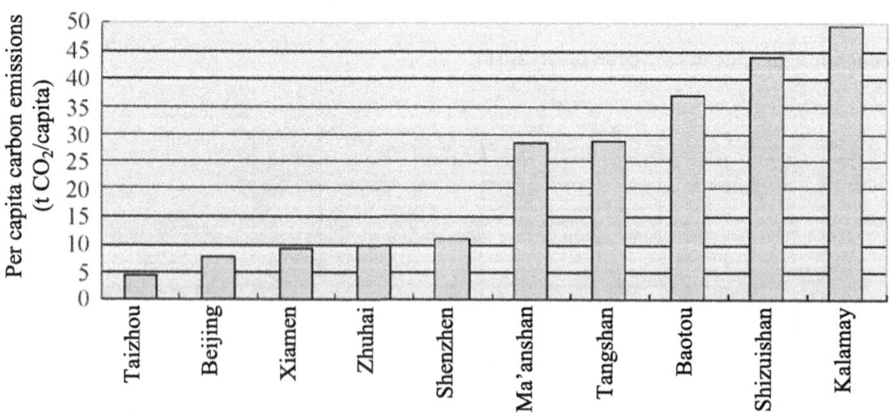

Fig. 2-5 Per-capita carbon emission levels in case cities (2009)

Therefore, cities with relatively high scores in comprehensive rankings need to continue leading the low-carbon development trend, adopt a diversified development strategy, and boost a virtuous cycle for constantly upgrading the industrial structure while cultivating new economic growth points, enhancing international competitiveness and becoming the first to develop low-carbon demonstration cities.

Cities with relatively low cores in comprehensive rankings, especially resource-based cities or resource-exhausted cities, need to open up a new route for urban low-carbon development. It is necessary to integrate resources, breaking the original resource layout structure, and conducting restructuring and reconstruction to integrate industrial enterprises with a relatively large impact on total carbon emissions so as to stimulate the upgrading of the industrial structure.

Overall, urban low-carbon development requires low-carbon development planning based on current situation evaluation. In particular, most of the cities are undergoing transformation in social and economic development, and it is particularly necessary to incorporate a low-carbon development goal into social and economic development planning. Currently, government documents for building low-carbon cities have been issued in some pilot cities for low-carbon development. Urban low-carbon development planning entails the consideration of their own energy conservation and emission reduction goals, an overview of the low-carbon development experience in different regions, formulation of a feasible low-carbon development roadmap and exploration of integrated solutions for low-carbon development.

Chapter 3
The Low-Carbon Development Situation in Chinese Cities

Liang Benfan, Zhou Yueyun

Abstract: 2011 was a year in which China's urban low-carbon development advanced rapidly. Government at all levels and departments put forward their own energy consumption and carbon dioxide emission reduction targets as well as the strategic layout of the energy saving work during the 12th Five-Year Plan period. This chapter has summarized the main achievements of the urban low-carbon development for the "12th Five-Year" on this basis. This chapter also proposes finally a few key tasks for China's "12th Five-Year".

Key words: urban low-carbon development; key tasks

In 2011 Chinese cities witnessed rapid low-carbon development, specifically as follows: Local cities and national departments specified the indicator of energy consumption per unit of GDP and the carbon dioxide emission reduction indicator, and made strategic arrangements for energy conservation, emission reduction and temperature control during the 12th Five-Year Plan period, and exercised total quantity control over energy consumption and carbon emissions; they further expanded development and utilization of clean and renewable energy including hydropower, wind power, optical energy, terrestrial heat and biomass, while the construction of waste and old materials recovery systems favorable to resource and energy conservation and carbon storage was consolidated and strengthened, and pilot projects for photoelectric integration in urban buildings, new energy city projects and pilot projects for low-carbon cities continued receiving support from the fiscal subsidy mechanism. Great headway was made in market-oriented energy conservation and emission reduction mechanism innovations such as a pilot program for carbon finance and carbon emission permit trading. Dual control, which covers both total quantity and intensity, has been exercised on fossil energy consumption and carbon emissions in a comprehensive way since 2012, which has become an important national policy for China to address climate change, promote transformation of China's economy and further reform and opening-up. During the 12th Five-Year Plan period, people in various cities shall clearly understand the State's determination to drive forward energy conservation, emissions reduction and temperature control, and seize opportunities for developing efficient new industries with low consumption and low emissions, as well as working on the following aspects: Firstly, proceed from a long-term and all-round development strategy to formulate comprehensive planning for low-carbon energy conservation, emission reduction, temperature control in cities; secondly, act on the goal for dual control involving both total carbon emissions and intensity, and comprehensively adjust the industrial development direction and goals, boost upgrading of the industrial structure and transformation of energy use; thirdly, utilize the carbon finance mechanism, the carbon emissions permit trading market and high technology to promote energy conservation and the rapid development of the new energy industry; fourthly, completely establish energy and carbon emission statistical data and an internal evaluation system, and gain the initiative in remission reduction and evaluation; fifthly, take full advantage of various policies introduced by the Central Government including a policy for building a pilot demonstration base and a fiscal subsidy policy to advance low-carbon development in regions and cities.

3.1 Main Achievements in Low-Carbon Development of Chinese Cities since 2011

China's energy intensity in 2011 decreased by only 2.01% compared with 2010, lower than the 3.2% planning goal for the average decrease during the 12th Five-Year Plan period; however, great achievements were still made, which laid a good foundation for smooth low-carbon energy conservation, emission reduction and temperature control over the next four years.

1. Overall strategic arrangements for energy conservation, emission reduction and temperature control have been made

2011 was the year for starting the implementation of the 12th Five-Year Plan; strategic arrangements for energy conservation, emission reduction and temperature control during the 12th Five-Year Plan period have a vital bearing on continuity, linkage and final success of low-carbon development at national, regional and municipal levels. In response to this, *the 12th Five-Year Plan for China's National Economic and Social Development* was adopted in the fourth session of the 11th National People's Congress in March, and the State Council issued an *Overall Work Plan for Energy Conservation and Emission Reduction during the 12th Five-Year Plan Period* in September, which broke down national energy conservation and emission reduction goals and assigned sub-goals to provinces, regions and municipalities, and required that national energy consumption per 10,000-yuan GDP would decrease to 0.869t standard coal (calculated according to price in 2005) by 2015, down 16% and 32% compared with 1.034t and 1.276t standard coal in 2010 and 2005 respectively. Thus a cumulative 0.67 billion tons standard coal would be saved in five years. In November, the State Council convened an executive meeting in which a *Work Plan for Control of Greenhouse Gas Emissions during the 12th Five-Year Plan Period* was discussed and adopted, which specified overall requirements and key tasks for controlling greenhouse gas emissions in China and stated that the carbon dioxide emission per unit of GDP would decline by 17% in China by 2015 compared with 2010. The above target has also been broken down and sub-targets have been assigned to provinces, regions and municipalities, while these sub-targets have been further broken down at provincial, regional and municipal levels and the corresponding sub-targets have been assigned to local cities and counties. These strategic arrangements constituted the biggest bright spot in China's urban low-carbon development in 2011.

2. Dual control, which covers both total quantity and intensity, was tested on energy consumption and carbon emissions in part of the provinces and municipalities

In the past, only energy consumption intensity control was exercised on energy conservation and emission reduction in China, which meant that only the energy quantity consumed by unit of GDP was evaluated. The disadvantages of energy intensity control emerged as follows: people in some regions were keen on developing high-carbon industries with high energy consumption so that total energy use and carbon dioxide emissions in China continued soaring with economic expansion. Though the energy consumption per unit of GDP passed evaluation, the total energy consumption and carbon dioxide emissions greatly increased, resulting in persistent tension in the national energy market and an overdraft of carbon space capacity. Therefore, dual control involving both total quantity and intensity was exercised on energy consumption and carbon dioxide emissions so as to promote economic transformation and industrial structure adjustment, and mitigate China's dependence on the world's energy market, which represents a new direction for China to manage energy conservation, emissions reduction and carbon emissions in the future. In 2011, the State was studying and developing a total quantity control plan for energy consumption and carbon emissions; the State had not yet required pilot work on total quantity control of energy consumption and carbon emissions in various regions, but in part of the provinces and municipalities, including Guangdong Province, total energy consumption and total carbon emissions had been calculated according to carbon intensity reduction sub-targets and GDP growth sub-goals for the 12th Five-Year Plan period, as assigned by the State so as to carry out dual control. The total carbon emission was 0.51 billion tons in Guangdong Province in 2010. As shown by estimations made according to the requirements that carbon intensity should decrease by 19.5%, total energy consumption and total carbon dioxide emissions in 2015 should be controlled at 0.35 billion tons standard coal and 0.66 billion tons respectively. The advantages of total quantity control over energy consumption and carbon emissions are as follows: It will initiate energy structure optimization and an industrial structure upgrading mechanism, and it will lead not only to a reduction in energy consumption per unit of GDP but also a dramatic decrease in the fossil energy use level and carbon dioxide emissions per unit of GDP, so as to better guarantee an overall sustained and balanced development of China's economy.

3. A series of new measures for energy conservation, emission reduction and temperature control were taken

Work Plan for Control of Greenhouse Gas Emissions during the 12th Five-Year Plan Period and Overall Work Plan for Energy Conservation and Emission Reductions during the 12th Five-Year Plan Period initiated as many as 26 and 48 new measures for energy conservation, emission reduction and temperature control respectively. The main contents include accelerating the adjustment of the industrial structure, giving great impetus to energy conservation and consumption reduction, and actively developing low-carbon energy, and endeavoring to increase carbon sinks, controlling greenhouse gas emissions from non-energy activities, and strengthening the saving and replacement of high emissions products; also pushing ahead with pilot work on low-carbon provinces, regions and municipalities, making pilot efforts in a low-carbon experimental industrial park, and developing pilot low-carbon communities, putting into practice a pilot low-carbon business and low-carbon products, and establishing a basic statistical system for greenhouse gas emissions. Finally, the contents include reinforcing greenhouse gas emission accounting, and building a voluntary emissions reduction trading mechanism, and carrying out pilot work on carbon emissions permit trading. *China's Roadmap for Phase-Out of Incandescent Lamps, Plan for Implementation of Energy Conservation and Low Carbon Action of a Myriad Enterprises and Administrative Measures for Fiscal Incentive Funds for Energy-Saving Technology Transformation.* are highlights of new measures for energy conservation, emissions reduction and temperature control as taken at national, provincial and municipal levels. The first one was jointly released by seven ministries and commissions including the National Development and Reform Commission, the Ministry of Commerce, General Administration of Customs, State Administration for Industry & Commerce and General Administration of Quality Supervision, Inspection and Quarantine; the second one was jointly issued by twelve ministries and commissions including the National Development and Reform Commission, the Ministry of Education, the Ministry of Industry and Information Technology, the Ministry of Finance, the Ministry of Housing and Rural-Urban Development and the Ministry of Transport; the third one was jointly promulgated by the Ministry of Finance and the National Development and Reform Commission. *Roadmap for Phase-Out of Incandescent Lamps* requires the gradual prohibition of import and sales of general lighting incandescent lamps in five stages from October 1, 2012. In order to phase out incandescent lamps and obtain fiscal incentives for energy-saving technology transformation, most of the provinces and municipalities have developed their own measures for implementation of the enterprise energy conservation low carbon action and work plan for the phase-out of incandescent lamps.

4. Development and utilization of clean and renewable energy were further expanded

The proportion of non-thermal power, including hydropower and nuclear power in China, increased from 19.7% to 23% from 2007 to March 2010. Cumulative wind power installed capacity and photovoltaic solar installed capacity were 25.8 million kW and 0.16 million kW respectively in 2009. Hydropower installed capacity, newly added and approved hydropower and capacity under construction were 211 million kW, 13.22 million kW and 77 million kW respectively in China in 2010. Nuclear power installed capacity and capacity under construction reached 10.82 million kW and 29.14 million kW (26 units) respectively. Full-year newly added wind power capacity connected to the power grid and total grid-connected capacity stood at 13.99 million kW and 31.07 million kW respectively. The proportion of clean and renewable energy installed capacity rose to 26.53%. In 2011, conventional hydropower installed capacity, nuclear power installed capacity, grid-connected wind power installed capacity were 0.23 billion kW, 12.57 million kW and 45.05 million kW respectively in China, accounting for 21.8%, 1.19% and 4.27% of total power installed capacity respectively. Grid-connected solar photovoltaic installed capacity, biomass power installed capacity, geothermal power installed capacity and ocean power installed capacity were 2.14 million kW, 4.36 million kW, 24,200 kW and 6000 kW respectively. As of the end of 2011, installed capacity based on clean and renewable energy including hydropower, nuclear power, wind power, solar power and biomass accounted for 27.5% of total power installed capacity, up 0.9 percentage points compared with 2010. Conventional fossil energy including coal, fuel oil and natural gas causes not only pollutant discharge but also carbon emissions. Though nuclear power does not lead to carbon emissions, it is not very safe. Renewable energy is clean energy without carbon emissions, thus it represents the general direction for future urban energy development. Exploration and development of clean energy and renewable energy is an historic responsibility of cities.

5. Increasing importance was attached to the construction of a waste and old materials recovery system and the development of an "urban mine"

According to *China Construction News*, 3.5 million tons of electronic waste was generated in China in 2011. Nearly 0.18 billion tons of municipal waste is generated annually in China, while municipal waste piled up over the years hit 10 billion tons. Waste is a misplaced resource; utilization and recovery of waste is the development of an "urban mine". As the *Circular Economy Promotion Law* took effect from January 1, 2009, the State has encouraged and stimulated the construction of a waste recovery system. *Opinions of General Office of the State Council Concerning the Establishment of Complete Advanced Waste and an Old Goods Recovery System,* released in November 2011, state that by 2015 a modern waste and discarded goods recovery system with a complete network, advanced technology, excellent sorting and standardized management will be preliminarily built, and the rate of recovery of the main varieties of waste and discarded goods will reach 70%. Many cities showed keen interest in the development of an "urban mine". In 2011, 75 new wall materials enterprises in Jingmen City, Hubei Province digested not only all 1.2 million annually accumulated coal ash but also more than one million tons accumulated over the years; city-wide cement enterprises utilized 2,426,600t of solid wastes, and building material enterprises made use of 1,637,000t of solid wastes, while forestry enterprises used 489,000t of sawdust, barks and crotches, and saved 420,000m^3 of timber. As a national pilot unit for the circular economy, Jingmen Gelinmei New Materials Co., Ltd. has disposed of 120,000t of hazardous wastes containing nickel, cobalt and zinc, and has recovered and disposed of more than 6,000t of metals, and has saved 580,000t of standard coal as well as reducing carbon dioxide emissions by more than 100,000t since 2008.

6. Pilot demonstration projects were further supported by fiscal, tax and financial policies

China has launched a series of pilot demonstration projects for propelling energy conservation and emission reduction since 2010. Specifically, they are the New Energy City Project, New Energy Demonstration District Project, New Energy Demonstration County Project, New Energy Demonstration Town Project, Solar Photoelectric Building Application Demonstration Project, Pilot Project for a Low-Carbon Province, Region and Municipality, Pilot Project for Low-Carbon Industry Experimental Park, Pilot Project for Low-Carbon Community, Waste and Old Goods Recovery System Demonstration City Project, "Urban Mine" Demonstration Base Project, Remanufacturing Industry Cluster Zone Project, Demonstration Base Project for Comprehensive Utilization of Resources, Demonstration Project for Resource Utilization and Innocent Treatment of Urban Kitchen Wastes. China plans to build 80 waste and discarded goods recovery system demonstration cities, 100 new energy cities, 200 new energy demonstration counties, 1,000 new energy demonstration districts, 10,000 new energy demonstration towns, 5 remanufacturing industry cluster zones, 50 "urban mine" demonstration bases, 100 demonstration bases for comprehensive utilization of resources and 100 demonstration projects for resource utilization and non-polluting treatment of urban kitchen wastes by 2015. Fiscal, tax and financial policies are adopted to further support these pilot demonstration projects, which is the important means for the State to arouse the enthusiasm of local cities for realizing energy conservation, emission reduction and temperature control goals. For example, The Ministry of Finance increased the subsidy for photoelectric integration projects closely related to building materials and buildings from 6 yuan/W in 2011 to 9 yuan/W in 2012, and has provided a 20yuan/m^2 policy-oriented subsidy for green energy-saving buildings with energy-saving transformation and new green energy-saving buildings each year. 4.6 billion yuan of financial subsidies are expected to be granted by the State to build new energy demonstration counties, districts and towns during the 12th Five-Year Plan period.

7. Significant progress was made on pilot work on carbon finance and carbon emissions permit trading

In October 2011, the National Development and Reform Commission approved

Hubei Province, Guangdong Province, Beijing, Tianjin, Shanghai, Chongqing and Shenzhen to carry out pilot work on carbon emissions permit trading and required pilot regions to quickly develop plans for implementation of pilot work on carbon emissions permit trading and specify overall thinking, work goals, main tasks, safeguard measures and progress schedule, so as to explore and establish a carbon emissions permit trading market suited to China's national conditions within five years. According to the *Plan of Guangdong Province for Implementation of Pilot Low-Carbon Work*, Guangdong Province planned to try total quantity control of annual carbon dioxide emissions in the whole province and in part of industries, enterprises, and determined the plan for rationally allocating carbon emission permits and studied the carbon emission permit price mechanism and management measures, and strived to initiate carbon emissions permit trading in the second half of 2013. Carbon emissions permit trading requires carbon finance support. The Bank of Beijing set up the first low-carbon investment bank in the world in April 2011. It launched an "Energy Saving Loan" financial service program and offered loans totaling RMB 350 million to four enterprises on site. As of the end of 2011, the Bank of Beijing extended more than 70 loans amounting to more than one billion yuan through an "Energy Saving Loan" product to support enterprises in annually saving energy equivalent to 560,000t of standard coal in energy-saving projects relating to such fields as building and industry and annually reducing 1.4 million tons of carbon dioxide emissions. The Bank of Beijing offered an "Energy Saving Loan" one year earlier than the extension of loans for low-carbon energy projects by the first low-carbon investment bank in the UK. Construction of the carbon finance and carbon emissions permit trading market is an important move for China to shift the energy conservation and emissions reduction mode from a single executive order to the combination of administrative means and market guidance.

3.2 Several Key Points for Urban Low-Carbon Work in China during the 12th Five-Year Plan Period

In 2012, with breakthrough efforts starting on energy conservation, emission reduction and temperature control, dual control involving both total quantity and intensity was comprehensively exercised on fossil energy consumption and carbon emissions in China, which has become the important national policy for China to address climate change, promote economic transformation, further reform and opening-up. People in various cities shall recognize the State's determination to make intensified efforts in energy conservation, emissions reduction and temperature control, and grasp opportunities for developing efficient new industries with low consumption and low emissions, and focus on the following several tasks according to the *Overall Work Plan for Energy Conservation and Emissions Reduction during the 12th Five-Year Plan Period* and *Work Plan for Control of Greenhouse Gas Emissions during the 12th Five-Year Plan*

Period in the light of the actual situation in their regions and cities:

1. Proceed from a long-term and all-round development strategy to make overall planning for low-carbon development of cities

In recent years, many concepts, policies, laws and regulations, plans, actions and measures relating to low carbon, energy conservation, emission reduction and temperature control development have been put forward in order for socially, economically and environmentally sustainable development in China. Common concepts include environmental protection, energy conservation, emission reduction, temperature control, green economy, low-carbon economy, circular economy and ecological economy. What is the relationship between environmental protection, energy conservation, emission reduction, temperature control and low carbon? What is the relationship between the green economy, circular economy, ecological economy and low-carbon economy? What is the relationship between an *Overall Work Plan for Energy Conservation and Emissions Reduction during the 12th Five-Year Plan Period and a Work Plan for Control of Greenhouse Gas Emissions during the 12th Five-Year Plan Period*? Many party and government cadres at various places were not very clear about these, and as a result there was no planning for urban low-carbon development, and government work was not carried out in a coordinated, across-the-board way, and a package of official business was segmented, and officials were fruitlessly busy and inefficient, which did not improve for a long time. Proceeding from a long-term and all-round development strategy to develop overall planning for the low-carbon development of cities entails efforts to clarify relations among the above concepts, policies, laws and regulations, plans, actions and measures, and integrate urban resources, give play to comprehensive benefits of specific plans, actions and projects. In Xinjin County, Sichuan Province, low-carbon development planning was studied and developed, and energetic development of the green economy, low-carbon economy, circular economy and ecological economy served as a means for achieving important goals for environmental protection, energy conservation, emissions reduction and temperature control. in the whole country. Thus remarkable achievements were made with less effort during the construction of a felicific, scenic livable-in city.

2. Take total carbon emissions and intensity control as the highest goal to fully adjust the industrial development direction, boost upgrading of the industrial structure and transformation of energy utilization

Why is it necessary to take total carbon emissions and intensity control as the highest goal rather than considering environmental protection, energy conservation and ecology as the highest goal? The reason is that carbon emissions control shows a relatively high degree of comprehensiveness and coverage. Meanwhile, carbon emissions have become the core restrictive factor for social and economic development. Energy conservation and emission reduction are characterized by diminishing marginal utility and cannot be infinitely carried forward. Energy conservation and emissions reduction do not cover development or utilization of renewable energy, while low-carbon development requires carbon control but does not necessarily entail energy conservation but meanwhile calls for the development of new energy. In addition, the level of human development is positively related to the level of energy utilization; unconditional energy conservation will restrict further social development. Clean and renewable energies themselves are ecological, green and environmentally friendly, thus extensive development and utilization of them is the important direction for future social development. Therefore, compared with taking total energy and intensity control as the goal, taking total carbon emissions and intensity control as the highest goal to fully adjust the industrial development direction, boosting upgrading of the industrial structure and the transformation of energy utilization, is more scientific and beneficial for development of the local economy. In some sense, the release of the *Work Plan for Control of Greenhouse Gas Emissions during the 12th Five-Year Plan Period* is the important milestone which marks scientific energy conservation and emissions reduction in China.

As total carbon emissions and intensity control are taken as the highest goals to fully adjust the industrial development direction, boost upgrading of the industrial structure and transformation of energy utilization, specifically it is necessary to fulfill the following three tasks: firstly, vigorously develop new energy and new energy industries; secondly, restrict and phase out fossil energy intensive industries, energetically develop manufacturing with low energy consumption and high added value; thirdly, shift from "made in China" to "services from China". Xinjin County, Sichuan Province has ushered in economic transformation and an upgrading strategy focusing on large logistics, great tourism, extensive services and a highly livable environment, which is exactly a typical case of a shift from "made in China" to "services from China". Given that efforts were organized to execute the second batch of 15 million – 18 million kW wind power projects and the first batch of 3 million kW solar energy development plans covering the 12th Five-Year Plan period in China in 2012, coastal cities and cities with abundant solar energy resources can take opportunities for developing wind power and solar energy.

3. Make use of the carbon finance mechanism, carbon emissions permit trading market, new science and technology to drive the rapid development of temperature control, energy saving and new energy industries

For a long time China resorted to such energy conservation and emission reduction means as a closedown, suspension, merger, a switch to other production, a technical transformation and fiscal subsidy. Consequently, some negative problems occurred: (1) Failure to keep fair competition. Enterprises should be phased out through market competition. For some enterprises in China, their closedown, suspension, merger, switch to other production, were decided by government rather than by the market. Officials capitalized on this opportunity to engage in "crossing-red-line operations" with relevant enterprises, in which enterprises offered benefits and bonuses to officials for squeezing out competitors. There were low-cost mergers, closedown of enterprises' idle assets and their switch to other production, which as a matter of fact severely disrupted the law of fair competition. (2) Fiscal subsidies were inefficient. I) Fiscal subsidies were not put in place, which was a common problem that enterprises that had real intention of technical transformation confronted. II) Most of the projects for which fiscal subsidies could be obtained by application were those executed by people with close personal connections. Fiscal funding efficiency was much lower than that of the market funds due to relationship costs. (3) market prices, the market player structure and market supervision were fully distorted. Generally, the so-called closedown, suspension and switch to other production meant the closedown of small enterprises and the suspension of weak enterprises, while so-called fiscal subsidies generally flowed to large and strong enterprises. Thus small and medium-sized private enterprises were subject to a deteriorating operating environment. Selective fiscal subsidies and technical transformation distorted market prices. People with close personal connections and relationships disabled the market supervision.

The carbon emissions permit trading market and carbon finance mechanism involves the act of exchanging the energy conservation and emission reduction achievements of enterprises for funds, income and investments. It effectively overcomes the above three problems and is an effective mechanism for impelling temperature control, energy conservation and the rapid development of new energy industries in various places in the future. Meanwhile, it is likely that national fiscal subsidy funds will be reformed into a carbon fund through which incentives will be provided on the basis of funds obtained by exchanging the energy conservation and emission reduction achievements of enterprises. Under such circumstances it is essential for various cities to study the carbon finance mechanism, carbon emissions permit trading market rules, as early as possible. Certified emission reduction refers to the professionally appraised and measured emission reduction, which has been verified by the management department, recognized in the market and can be traded. Each certified emission reduction corresponds to certain fossil energy saving. Certified emission reduction is the important criterion for a country to evaluate emission reduction projects and provide subsidies and also the important subject for trading and exchanging for funds through the market. Undoubtedly, good use of certified emission reduction offers a source of funding and economic power for local energy conservation, emission reduction and an industrial upgrading transformation.

4. Comprehensively establish energy and carbon emissions statistical data and an internal evaluation system, and gain the initiative in emission reduction and evaluation

Work Plan for Control of Greenhouse Gas Emissions during the 12th Five-Year Plan Period clarifies efforts to speed up the establishment of a greenhouse gas emission statistical accounting system, and develops a basic statistical system and energy and carbon emissions accounting guidelines, as well as working out a temperature control evaluation indicator system for local governments at various levels and enterprises. In the era of pursuing rapid GDP growth and evaluating the performance of local government cadres by the economic growth rate, not only the greenhouse gas emissions statistical accounting system but also the national economic accounting system is distorted due to severe intervention by administrative evaluation goals. Currently, government officials are experiencing multi-goal evaluation of performance; especially after the growth of total greenhouse gas emissions is incorporated into performance evaluation goals, local governments at various levels no longer turn a blind eye to energy and carbon emissions. At present, local governments at municipal, county and lower levels except provinces, municipalities directly under the central government and autonomous regions are basically unclear about energy consumption and greenhouse gas emissions in their regions because there is no local energy consumption statistical system, and a greenhouse gas emissions statistical accounting system is seldom available. Therefore, it is very difficult to achieve low-carbon development planning at local level since key data are unavailable. Efforts to achieve overall evaluation and control of energy consumption and total greenhouse gas emissions will certainly be greatly compromised in such an unclear social environment. The Central Government has become aware of this and has decided to accelerate the construction of a greenhouse gas emissions statistical accounting system. Thus city governments at various levels cannot but attach importance to this. Otherwise, they will be in a passive position during emission reduction and evaluation. The National Development and Reform Commission has required relevant departments, in conjunction with developing a greenhouse gas emission list, to quicken the education of professionals and enhance their technical capability for developing local and industrial emission lists. Meanwhile, more statistical and survey work will be carried out in energy, industry, agriculture and forestry, relating to greenhouse gas emissions, and accurate information will be provided timely; also, a target-oriented responsibility system and evaluation indicator system for controlling greenhouse gas emissions will be gradually established. People in various cities shall also push ahead with this work according to arrangements made by the National Development and Reform Commission as soon as possible.

5. Properly and sufficiently utilize various policies released by the Central Government, including a pilot demonstration base construction policy and a fiscal subsidy policy to promote low-carbon development at local and municipal levels

Unlike developed capitalist countries, China practices a development mode in which the Party exercises leadership, and the government plays a leading role, and policy guidance is provided and enterprises follow such guidance while superior departments perform supervision, which is one of the important characteristics which define the development of a socialist economy with Chinese characteristics. In order to boost the development and utilization of renewable energy, energy conservation and emission reduction, the development of a circular economy and low-carbon economy, the Central Government has released a series of industrial policies, investment policies, price and charging policies, fiscal policies, tax policies and financial credit policies. For example, the *National 12th Five-Year Plan for the Development of a Circular Economy* stresses more support from governments at various levels, especially funds within the budget of the Central Government for a circular economy, and the establishment of a circular economy industry investment fund and venture capital fund as well as support for the "10-100-1000 Demonstration Project". If a financially underdeveloped city government fails to recognize such a development mode and its characteristics, it cannot make good and sufficient use of various policies released by the Central Government including a pilot demonstration base construction policy and fiscal subsidy policy to promote low-carbon development at local and municipal levels, and will certainly take a roundabout course.

Chapter 4

Technical Methods for Overall Energy Planning for Low-Carbon Cities and their Applications

Zhang Wang, Zhao Yan, Pan Xuehua

Abstract: The main technical route of integrated energy planning for a low carbon city includes three segments; Analysis and Forecast; Planning and Design; Implementation and Feedback. It is divided into five stages; the first stage is early analysis of integrated energy planning which includes city background and the emission environment balance, the strategy of the set target for energy-saving and emission-reduction, the determination of basic principles for energy planning and resources analysis based on available energy; the second one is analysis and forecast of dynamic energy load demand, covering scenario analysis and forecast of energy demand; the third one is integrated evaluation for city energy systems, which include loading capacity evaluation for key elements of an energy system, which including loading capacity evaluation for key elements of energy systems and infrastructure, and the evaluation of an urban integrated energy system; the fourth one is configuration and optimization of urban energy systems, this including urban configuration of integrated energy systems technology, the optimization and design of urban integrated energy systems; the last one is implementation and feedback for integrated energy planning, containing engineering problems, the running and management of energy systems, supervision and feedback for the implementation of integrated energy planning. Finally the study commented on three examples of application of urban integrated energy planning techniques.

Key Words: urban integrated energy planning; analysis and forecast; planning and designing; the feedback of implementation

Energy consumption in cities accounts for three quarters of the world's total energy consumption and generates three quarters of the world's pollution[1]; according to the estimation by the International Energy Agency (IEA), 76% of CO_2 emissions from energy consumption will come from cities by 2030. Scientific, rational overall planning for an urban energy system is of vital importance for mitigating the energy crisis, addressing climate change and fostering sustainable development of cities. Current urban energy planning studies in China mainly focus on two aspects, but have relatively obvious defects resulting in relatively large difficulties for actual operation of the planning: firstly, they emphasize macro planning of the overall balance between energy supply and demand, lack dynamic development, spatial distribution and specific technical support, become separated from urban planning to some extent, thus being difficult in developing a specific implementation plan and being operationally unfeasible; secondly, there is no coordination among four special plans concerning water supply, heat supply, gas and electricity in the urban infrastructure works system planning at the micro level, in which water, heat, electricity and gas are separately dealt with, and professional limitations exist. As a result, repeated construction of the infrastructure occurs and efficiency is not high. Studying a set of energy planning and optimization methods at the technical level and overall energy planning techniques, which reflect temporal and spatial changes of energy loads, has become a realistic and pressing issue for overcoming the above problems.

Based on reference to the definition of energy planning and connotations of a low-carbon economy and low-carbon city, we believe that the main technical route for overall energy planning for a low-carbon city is as follows (Fig. 4-1): (1) analysis and prediction–give a full survey and data analysis of the history and current situation of the city background and energy system, and then adopt a scenario analysis method to analyze and predict energy load demand; (2) planning and design–take energy conservation and carbon reduction as goals, be energy balance-oriented, consider renewable energy application as a measure to conduct overall evaluation of the urban energy system, and then use a suitable planning model to enable optimization to achieve overall goals for reducing greenhouse gas emission, saving energy, improving energy efficiency and promoting the application of renewable energy; (3) implementation and feedback–develop the corresponding safeguard measures, implement planning and collect feedback.

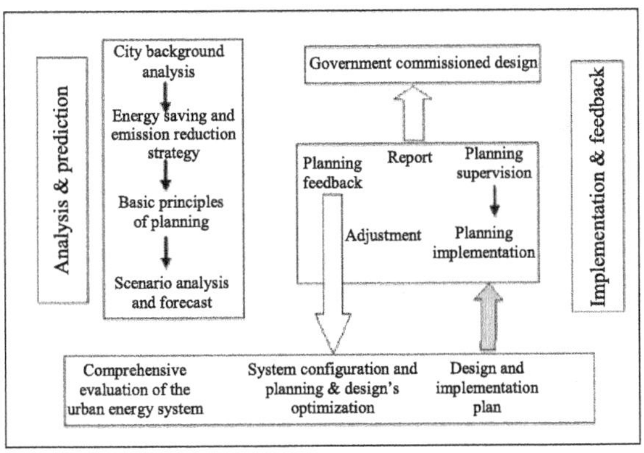

Fig. 4-1 Technical roadmap for overall urban energy planning

4.1 Preliminary Analysis of Overall Energy Planning

At the preliminary analysis stage, a survey of the basic situation of a city is conducted, and energy resources available for future city development and an energy environment balance system on which future city development will rely are analyzed, and its existing development mode is evaluated. Design personnel set feasible energy conservation and emission reduction goals for a city, and determine basic principles for energy planning according to the analysis report.

4.1.1 City background and energy environment balance

City background analysis is carried out by collecting and organizing main data and information about the current energy situation, the economic and environmental system of a city, and conducting overall evaluation of various energy resources for the economic and social development of a city and their distribution space. City background analysis mainly covers a city overview, acquisition of basic data and statistical data about a city, a visit to government departments in charge of energy, economy and the environment.

Energy environment balance analysis starts from analyzing a regional energy system from the perspective of rational utilization of energy, specifically as follows: (1) quantitative balance, namely whether its supply scale is determined according to real demand; (2) qualitative balance, namely whether full utilization of energy is achieved, energy is used according to the energy level and on a gradient basis; (3) energy storage balance, especially in the case of taking soil or underground water as a thermal and cold storage carrier; (4) value balance, investment return and use of a Programmatic Clean Development Mechanism (PCDM). A carbon footprint evaluation method is then applied to calculate the carbon emission load caused by energy production and consumption, waste discharge by specific activity and conduct in a specific area, namely emission of greenhouse gases including CO_2 generated by energy consumption in a specific area or specific activity. Energy and environment balance mainly involves evaluation of energy consumption in a city and its environmental impact. Different kinds of urban activities cause emission of different greenhouse gases, thus a city is divided into different energy areas according to the CO_2 equivalent emission indicator so that urban energy consumption is comprehensively, accurately evaluated[2].

4.1.2 Setting of strategic goals for energy conservation and emission reduction

Setting of strategic goals for energy conservation and carbon reduction is a process for a professional, democratic plan design and policy-making. Generally, a scenario analysis method is firstly adopted to set several scenarios of urban development and energy demand and carbon emissions from use of conventional energy in a scenario; energy consumption and emissions in various scenarios are then comparatively analyzed, and baselines for energy consumption and carbon emissions are set so as to drive the urban system to reach qualitative or quantitative indicators in terms of degree of satisfaction of energy demand, degree of urban environment improvement, degree of resource optimization utilization, employment impact, microeconomic impact; energy conservation and emission reduction target values are then determined, namely recommended target values for energy planning (including total urban energy consumption, fuel supply structure, total emissions or intensity of CO, NO_x, SO_2 and CO_2, etc.) are specified; finally, feasible plans put forward by city stakeholders are summarized and collected to develop an "action planning scheme" including series emission reduction goals, preferred measures and preferred action plan.

4.1.3 Determination of basic principles for energy planning

Urban energy planning shall follow the following eight basic principles: be hierarchical, people-oriented, achieve emission reductions, use marketization, achieve renewable energy maximization, achieve functional maximization, achieve energy-saving benefit maximization, achieve economic benefit maximization. Firstly, it must be be hierarchical: its basic hierarchy lies in taking energy conservation and demand reduction as the first and foremost carbon reduction principle for a low-carbon city, and the second hierarchy is utilization of afterheat and waste heat especially in the industrial park, while the third hierarchy is utilization of renewable and low grade heat sources, and the fourth hierarchy is utilization of renewable power, namely photovoltaic power generation, wind power generation and small-scale solar thermal power generation; secondly, it must be be people-oriented: energy planning is designed to meet the rational demand of users and residents; thirdly, it must achieve reductions: unit carbon emissions in a low-carbon city must be lower than the baseline; finally, marketization: different market mechanisms will result in different plans and different system configurations. The last four principles are successively described as follows: firstly, selection of energy shall follow maximization of low-carbon energy and renewable energy according to local conditions; secondly, energy planning shall coordinate with municipal facility planning so as to maximize infrastructure functions; thirdly, energy shall be utilized on a gradient basis and quality shall be geared to specific needs so that energy conservation benefits are maximized; fourthly, energy system planning, power supply, heat supply and the heating system shall be well-matched and optimized to maximize economic benefits[3].

4.1.4 Resource analysis methods for available energy

Resource analysis methods for energy available mainly include a resource analysis method for renewable energy and unused energy, demand-side energy resource potential analysis, emergy analysis of energy available.

1. Available resource analysis method for renewable energy

Characteristics of renewable energy: firstly, there is no upper limit for the total available quantity and it is regularly renewable. Secondly, the actual quantity of renewable energy used in energy supply depends on the process technology level for converting it into end-use energy. Table 4-1 lists several evaluation methods for renewable energy quantity[4].

Table 4-1 Several evaluation methods for the renewable energy resource quantity

Energy type	Resource quantity parameter	Unit	Resource quantity estimation method
Solar energy: optothermal, photoelectricity	Solar radiation intensity	MJ/(m^2·a)	Annually obtained electric output and heat output are estimated according to the technical performance of solar photovoltaic cells and solar water heaters etc.
	Full-year sunshine hours	h/a	
Wind energy	Wind energy density	W/m^2	Annual possible power generation is estimated according to the technical characteristics of wind power.
	Annual utilization hours	h/a	
Biomass energy: straw, firewood, energy crops, industrial, agricultural and urban wastes	Collected matter quantity	t	Annual possible biological development quantity or electric output is estimated according to different technical characteristics such as biomass energy forming, gasification, anaerobic digestion, power generation and liquidation etc.

2. Available resource analysis method for unused energy

Unused energies mainly include shallow soil heat storage energy, surface water heat energy and sewage source heat energy; their specific resource analysis methods are not repeated here due to space limitations; interested readers can see the details from references[4].

3. Demand side energy resource potential analysis methods

There are many kinds of demand side resources which are relatively complicated and include not only energy consumption reduced by improvement in power supply, distribution and lighting systems, energy saved by users' change of conduct, but also energy consumption reduced by peak and off-peak load regulation and consideration of a load staggering rate. For specific analysis, generally the main savings which may be implemented during the planning period are chosen. Several different calculation methods are available for demand side resources which rely on improving energy efficiency and energy use habits; common methods include a maximum technical resource potential analysis method, a cost benefit potential analysis method, a zero efficiency potential analysis method and conducting an energy saving analysis method.

4. Emergy analysis of available energy resources

Emergy refers to all energies which are created within a certain space during a certain time. Emergy analysis is conducted by taking emergy as a benchmark and converting different kinds of incomparable energies and all streams, including non-energy material flow and capital flow within a system, into emergy with the same standards to conduct data processing and give a system analysis. On the other hand, as an embodiment of emergy analysis results, evaluation indicators (Table 2-2) reflect the importance attached, in system analysis, to the environment and evaluation of sustainability. General steps for emergy analysis: (1) data collection, for example, regional spatial energy flow can be obtained by the GIS technique, or energy-related data can be acquired through the literature, including the Energy Statistics Yearbook etc; (2)plotting of an energy system diagram; (3) preparation of an emergy analysis table; (4) plotting of a comprehensive emergy system diagram; (5) establishment of an overall emergy indicator system; (6) dynamic simulation; (7) system development evaluation and strategy analysis.

Table 4-2 Part of emergy indicators

Indicators	Formula
Yield (*Y*)	$Y=R+N+F$
Percent Renewable (*PR*)	$R/(R+N+F)$
Energy Yield Ratio (*EYR*)	Y/F
Energy Investment Ratio (*EIR*)	$F/(R+N)$
Environment Load Ratio (*ELR*)	$(F+N)/R$
Energy Sustainable Index (*ESI*)	EYR/ELR

4.2 Dynamic Resource Load Demand Analysis and Forecast

Urban energy load is the foundation for energy system planning; influencing factors are classified and extracted on the basis of mass data survey accumulation and data analysis for planning of multiple cities; the quantitative relationship between loads and influencing factors is analyzed and summarized, and an energy load database and load forecasting model are developed[5], and scenario analysis is carried out; total urban energy demand, demands of various departments and their spatial and temporal distribution are forecast.

4.2.1 Scenario analysis method

Many parameters are uncertain or unknown at the beginning of overall urban energy planning for adapting to and mitigating climate change; for example, it is necessary to not only consider the most probable future energy development trends, but also study various possibilities for changing such trends and preconditions for realizing different possibilities, which entails comprehensive quantitative and qualitative analysis, and qualitative setting of several future possible scenarios, also an analysis of each scenario's background and variable intensity for providing a reference for policy-makers and final determination of the most probable scenario.

4.2.2 Energy demand forecast

After the current urban energy situation and available energy potential are specifically analyzed, and energy consumption characteristics, energy consumption structure and problems in utilization are identified, it is necessary to forecast a supply and demand balance of future urban energy in urban energy planning so as to provide a quantitative reference for developing a relevant energy development strategy, energy policies and measures. Therefore, urban energy planning necessitates efforts to set several scenarios such as population size, added value growth rates of primary industry, secondary industry and tertiary industry, the per-capita disposable income growth rate, price trends and energy efficiency, according to development trends in the population size of a city, its economic development, energy efficiency level, energy price, scientific and technological level, energy conservation policy, opening-up situation and living conditions of residents, and thus choose a proper energy forecast method and model to scientifically forecast the future energy supply and demand of a city.

A good energy forecast is of great significance for urban economic and social development, ecological environment improvement and reduction of greenhouse gas emissions; the energy forecast covers total demand, the subsector's demand, the supply and demand of various kinds of energies. Research on energy demand forecasting started in the 1970s; now many research achievements have been made, while relevant theories and methods are also becoming increasingly mature. In foreign countries, chaotic dynamics theory, times series analysis method, the DESON model, PILOT model and BESON model are mainly used to forecast and analyze the relationship between energy demand, energy consumption and economic growth[6,9], while domestic energy demand forecasting methods are broadly divided into two categories: firstly, a single model is adopted for forecasting; secondly, information about various single models is organized and a combined model is used for forecasting. More often used methods in the single model include the per-capita energy consumption method, elasticity coefficient method, regression analysis method, sector analysis method, econometric model method, input-output method, grey theory forecast[10] and artificial neural network forecast[11,12]. The energy consumption system is characterized by high complexity and nonlinearity, thus a single model cannot better forecast it. The combined forecast is conducted by rationally combining many single forecast model results obtained through certain arithmetical operations to build a new forecast model including various kinds of model forecast information. Empirical research shows that the accuracy of a combined model forecast is higher than that of a single model forecast. Currently, frequently used combined models include the AHP and GM(1,1) combined model[13], the combined forecast model based on GM(1,1) and ANN optimization[14], the combined model of grey GM and BP network[15,16], the combined model (GM-PLS)[17] based on optimal combination of partial least squares regression and GM(1,1), the forecast model based on optimization of nonlinear regression and grey forecast[18]. Given the current situation of research at home and abroad, medium and long-term urban energy supply and demand balance analysis, air pollutant discharge and greenhouse gas emissions from energy circulation and consumption, cost benefit analysis, a long range Energy Alternatives Planning System (LEAP) model can be adopted to forecast medium and long-term energy demand and greenhouse gas emissions from sub-industries and subsectors in a city[19], an input/output and energy system optimization model (I/O-INEJ model).

4.3 Overall Evaluation of Urban Energy System

For an urban energy system consisting of different energy supply and demand structures and various energy subsystems, it is necessary to use one overall evaluation technique, choosing suitable evaluation methods and evaluation indicators to further conduct vertical comparative evaluation of the current energy situation or planning at different development stages of a city, and horizontal comparative evaluation of different energy plans for a city. Specific evaluation indicators include thermodynamic properties, economic properties, environmental emission properties and other energy conversion properties, as well as reliability, bearing capacity and the carbon footprint of the energy system. Specific evaluation contents contain various elements of the energy system and their matching situation, the bearing capacity of the infrastructure including energy conversion equipment, energy system configuration based on new energy, an energy bus system and energy internet system based on unused low-grade energy.

4.3.1 Evaluation of energy system elements and bearing capacity of infrastructure

An urban energy system is composed of four large parts: hardware, software, organization, economy and financing. Hardware refers to the infrastructure for energy circulation and mainly includes energy production facilities (power stations, heat sources and gas sources), and energy transmission facilities (power transmission and transformation, power grid and pipe network). Software refers to energy utilization techniques, energy consumption concept, mode, relevant systems and management action. Organization refers to main energy consumers (enterprises/parks, families/communities), energy technology research institutions (universities or research institutions), energy-saving organizations and relevant bodies. Economy and financing means implementation of various plans, economic levers (subsidy and tax) prescribed by laws and regulations as well as financing plans. Energy system element evaluation is conducted by first establishing the energy supply and demand system adapting to the urban energy system and consistent with sustainable development principles and then proceeding from such establishment to research and develop an energy supply and demand element system evaluation technique. Such a technique should have description, diagnosis and treatment functions and can not only identify advantages and disadvantages of the urban energy supply and demand structure, supply and demand system involving different types and at different development stages, but also offer suggestions about the development of the urban energy supply and demand structure, the supply and demand system. Evaluation concerning the bearing capacity of the energy infrastructure starts from the establishment of a scientific and rational urban energy system infrastructure and enhancement of the bearing capacity of the infrastructure to develop an urban energy system infrastructure layout and bearing capacity evaluation techniques. This part should focus on researching the energy supply station site selection, evaluation of operating conditions of the main equipment, transmission and distribution network layout optimization technique.

4.3.2 Evaluation of integrated urban energy system

Evaluation of an integrated urban energy system can result in correctly selecting the energy system most suitable for a city and its functional zones so as to realize sustainable development of regional energy, reduce greenhouse gas emissions and pollutant discharge, maximize use of regional renewable energy resources such as solar energy, wind energy, biomass energy and geothermal energy, and optimize the regional energy structure and foster sustainable urban development through practice in building a low-carbon city, low-carbon park and low-carbon community. Methods for the evaluation of an integrated energy system include carbon footprint analysis, thermodynamic evaluation, environmental evaluation and a green community evaluation system, LEED-ND.

1. Carbon footprint analysis

Carbon footprint refers to total greenhouse gas emissions, expressed by CO_2 equivalent, generated by any tangible activities including individuals, organizations, events or products. The carbon footprint of an urban regional energy system refers to greenhouse gas emissions, expressed by CO_2 equivalent, generated by a regional energy system during operation. Analysis of the carbon footprint of a regional energy system helps determine the proportions of greenhouse gas emissions from elements in the whole system so as to effectively guide energy conservation and emission reduction renovation of the regional energy system.

Steps for analyzing the carbon footprint of an urban regional energy system: (1) set the energy system calculation boundary; (2) determine greenhouse gas emission sources in the system within the calculation boundary; (3) calculate greenhouse gas emissions.

2. Thermodynamic evaluation

Thermodynamic evaluation of a regional energy system covers energy efficiency evaluation based on the first law of thermodynamics, energy value evaluation based on the second law of thermodynamics. Energy efficiency is the ratio of energy output as income to energy input as an expense. Higher energy efficiency of a regional energy system shows more energy output with the same energy input. Energy value evaluation of a regional energy system is the ratio of the energy value of regional energy system output as income to energy value of a regional energy system input as expense. Generally the following steps are taken for calculating the energy value of a regional energy system: (1) plot the regional energy system diagram; (2) abstract the actual cycle into an ideal cycle; (3) determine the parameters at various state points; (4) provide an energy value balance equation; (5) calculate the energy value efficiency.

3. Environmental evaluation

At the regional energy planning stage, besides satisfaction of demands for regional cold energy heat and electricity, considerations must also be given to the degree of impact from the used energy system on the environment. Regional energy systems are available in various forms such as ice-storage systems, variable refrigerant flow systems, cogeneration systems and renewable energy systems. Environmental evaluation of different regional energy systems entails selection and consideration of different methods and indicators, including a lifecycle evaluation method, single indicator (global warming indicator, ozone depletion indicator and acid rain indicator) environmental evaluation.

4.4 Urban Energy System Configuration and Optimization

Energy structure and energy configuration should be rationally optimized on the basis of analyzing and forecasting an integrated urban energy system so as to achieve coordination and unification between energy, economic and environmental benefits and provide a basis and reference for choosing a strategic implementation scheme in energy planning.

4.4.1 Configuration technique for integrated urban energy system

1. Energy system configuration based on new energy

Urban energy systems based on new energy are regional energy systems based on new energy including renewable energy (solar energy, wind energy, biomass energy, hydro-energy, geothermal energy) and unused energy (shallow energy storage, surface water, subway exhaust, urban sewage and other temperature difference energy) rather than a conventional power supply, a cooling and heating supply system. Compared with a conventional urban power supply, cooling and heating supply systems, they are characterized by low energy supply intensity, low energy supply stability, low energy supply efficiency, low energy supply grade and uneven resource distribution. Therefore, urban energy system configuration based on new energy should follow the following four principles: (1) resource availability; (2) gradient utilization of energy; (3) circular recovery; (4) reduction.

Energy system modes based on new energy mainly include a distributed energy cogeneration system, integrated application of unused energy–energy bus mode, ground source heat pump. Renewable energies including solar energy, wind energy and biomass energy, are used according to specific energy composition modes of various cities to form an energy supply system with complementation by multiple energies including wind and photovoltaic hybrid power generation systems, domestic fuel cells, cold-heat-power combined supply and heat pump combination so as to enable a combined supply of power, heat and cold. This not only makes full use of resources, and improves energy efficiency, but also avoids the disadvantages of a single-energy power supply and relieves environmental pressure caused by conventional energy and secondary pollution caused by new energy. There are three key technologies for low-carbon urban energy systems: firstly, load calculation and load forecast at the regional level; secondly, optimization of the delivery pipe network and equipment; thirdly, optimization of operation strategy and operation control. The first and foremost basis for system selection lies in regional load characteristics and regional energy supply characteristics.

2. Energy bus system based on unused low-grade energy

Low-grade energy mainly refers to thermal energy close to ambient temperature and incapable of being directly used, which is widely available in soil, sun, water, air and industrial waste heat. The energy bus system is one kind of integrated low-carbon regional energy system with a large-scale application in regional renewable energy and unused energy, which is future heat pump/heat sink water which is based on renewable energy and unused energy and enables delivery to users through a pipe network as infrastructure. On the user side, water from the bus serves as a heat source/heat sink for a water source heat pump and returns to source after heat exchange, or is discharged (surface water), or circulates for heat exchange again (coupling with various "sources" and "sinks" through a heat exchanger), or is recharged (underground water). The energy bus system is suitable for a comprehensive community with peak and off-peak regulation of the air-conditioning load, or areas to which a large number of available natural cold sources are adjacent, which can lower water supply temperature, improve a unit's COP, increase the water supply and return temperature difference and reduce delivery energy consumption. Pipe network modes of the energy bus system include a single-source branch-shaped single-stage pump, single-source branch-shaped circulating pump, single-source branch-shaped multi-stage pump, multi-source branch-shaped circulating pump, multi-cycle circulating pump. Energy bus system equipment includes a source-side heat exchanger (ground heat exchanger and surface water heat exchangers), end equipment (water-source heat pump unit, water-source variable refrigerant flow unit) and cooling tower. Different energy bus system control strategies will affect the economic efficiency of the whole system design, operation effect and operating expenses, and thus its control shall follow the following two principles: firstly, fully utilize the outdoor climate to improve overall operating efficiency of the air-conditioning system; secondly, balance soil temperature, give consideration to not only the heat exchanger during operation in soil but also heat dissipated from the soil in the case of a stopping operation. Based on the above two principles, the system is dealt with by different control methods.

3. Energy Internet System

The energy Internet system means that some small energy stations or cogeneration equipment based on distributed energy technology and renewable energy within the region are networked into the energy Internet so as to make full use of renewable energy and unused energy within the region. Cold, heat and power within the region meet the needs and compensate each other. Energy stations in the energy Internet may be solar photoelectric, wind power generation, biomass energy cogeneration (including garbage power), domestic fuel cells, and various kinds of low-grade heat energy. Here are several key issues concerning the energy Internet mode: (1) genetic algorithms, graph theory and the multi-median principle are used for energy station site selection and arrangement; (2) energy in several energy supply bases is economically, effectively and timely allocated to energy users so as to give full play to production potential and the functions of each energy station, namely, the delivery issue involving optimum matching between energy demand and supply; (3) environmentally friendly, economical dispatching of the energy Internet is designed to rationally and effectively arrange capacity and output of DG equipment under the precondition of meeting load demand so as to minimize power generation heat supply costs, emission costs or total costs of the whole energy Internet; (4) optimal path design and layout optimization of the hot water pipe network between energy stations, which is a typical combinatorial optimization issue. The electricity production side of distributed energy is addressed by a looped network mode, namely an energy bus mode, while heat production of distributed energy is dealt with by a branched network, namely a honeycombed pipe system.

4.4.2 Optimization and design of integrated urban energy system

1. Optimization methods

Energy optimization methods include linear programming, nonlinear programming and dynamic programming, among which linear programming is widely used in practice and is mainly designed to give full play to limited resources and energy. The objective function can either involve the lowest cost, lowest energy consumption or lowest carbon emissions, or arrive at one kind of comprehensive optimization method by means of setting weights for various objectives. For a specific programming algorithm, see the relevant operational research literature.

2. Energy planning design software system

The above basic theories, analysis and forecasts are summarized to further develop an urban energy planning design software platform so as to provide designers and planners with a platform and tools; this software system adopts urban energy system energy load data and an energy conversion characteristics library as input condition for setting integrated urban energy system scenarios, and then integrates pipe network delivery and waterpower distribution calculation software, conducts analog calculation on the software platform, and uses an evaluation system to judge scheme feasibility, finally displaying the planning result through the geographic information system (GIS) and environmental pollutant discharge space distribution software.

4.5 Implementation of Overall Energy Planning and Feedback

Planning is subject to uncertainties of future development and multiple choices and thus is involved in many uncertainties during its implementation, namely development uncertainties. Therefore, it is very necessary to introduce a dynamic mechanism of diversified discussions and feedback at the implementation stage, which helps cope in a timely way with new situations and new problems, which gradually emerge along with temporal development. Project implementation, management supervision and feedback adjustment are three major steps at the implementation stage.

4.5.1 Engineering issues concerning energy system

Implementation of an urban regional energy system is a crucial link for carrying out a planning philosophy and design thinking, and it is characterized by large scale engineering, a long construction period, a close relation to planning, many adjustments and possibilities of change. Thus it is necessary to carefully arrange construction planning, construction organization design and various response plans, and conduct coordination and consultation involving planning design, administrative directors, equipment supply, electricity-related civil construction and user interfaces, and nip problems in the bud in a well-targeted way, so as to promote system construction as a whole in an orderly and coordinated manner.

Specific engineering issues: energy station site selection should be based on functional distribution and usage characteristics of regional energy consumers, and follow certain principles as far as possible; key points for civil construction of energy stations; requirements and principles for construction of a delivery and distribution pipe network; installation and debugging of main equipment within energy stations; process, methods and technical measures for construction of electrical works; engineering safety management which needs to be supported by a detailed engineering management system; engineering quality management which needs to be guaranteed by an engineering quality management system; project completion acceptance in which systems should be debugged, and data should be archived and transport management personnel should be put in place.

4.5.2 Operation management of energy system

People involved in construction of an urban energy system should recognize the objective and the great significance of operations management and take well-targeted measures from the beginning of project investment construction to lay down basic conditions for operations management in the process of system use. Operations management of a regional energy system should be performed using the following principles: (1) it is necessary to put safety and reliability first, and ensure safe, reliable operation of the system; (2) electricity, cold and heat generated by the system should meet users' demands; (3) when the load demand of the end user changes, the system should be capable of timely, quick adjustment and adaptation; (4) people should attach importance to system debugging, properly conduct debugging at the initial stage of operation and should be familiar with system performance characteristics; (5) people should proceed from details, continuously improve system energy efficiency to reduce costs; (6) the emergency response plan should be prepared to ensure that the system can stably, reliably provide services.

4.5.3 Supervision of overall energy planning

Supervision determines the quality of the specific action plan and repositioning of intended targets. Significant attention should be paid to the following contents during supervision: consistency of support policies of city government departments and strength of these policies; relevance among plans after a change in priority sequence; taking key plans as a benchmark for measuring the effects of planning implementation; adopting a plan implementation success rate and degree of overall plan implementation at each specific level, as indicators for evaluating energy supply and demand balance and achievements of CO_2 emission reduction planning.

With respect to the supervision of overall urban energy planning, the National Development and Reform Commission may take the lead in the improvement and innovation of a planning supervision system and mechanism by making reference to a planning inspector dispatching system of the Ministry of Housing and Urban-Rural Development. On the one hand, inspectors exercise overall supervision of implementation of overall urban energy planning and are not subject to management departments of local city governments, and can independently and fairly put forward supervision suggestions and opinions; on the other hand, with real-time supervision of planning implementation, inspectors can, in a timely way, discover actual circumstances and problems in planning implementation so that most of the laws and rules violation cases can be discovered early and nipped in the bud. Thus major laws and rules violation cases can be effectively avoided.

4.5.4 Feedback about overall energy planning

Planning feedback entails meticulous and in-depth analysis of various links in overall urban energy planning, and establishment of a relatively stable, institutionalized planning procedure with a feedback regulation function in line with its characteristics. Feedback regulation refers to one method for dynamic system control regulation in the case of a change of objectives in control theory. Its characteristic: objectives and the system movement locus are not predetermined but are repeatedly regulated according to feedback information. Though each regulation does not necessarily play a decisive role and has errors, higher adaptability, flexibility and accuracy of the whole system can be achieved[20].

Planning feedback regulation is a regulation mode covering the whole process for gradually approaching and reaching goals. In feedback regulation, an interest group consisting of city government, local development enterprises and community residents determines key indicators, collects relevant information and calculates implementation indices, and plays an important role in monitoring and evaluating planning implementation. Feedback opinions obtained during data monitoring and information summarization, adjustment of the scheme and revision of initial goals. exert an impact on series steps and procedures starting from "energy supply and demand balance and CO_2 emission reduction planning" at the planning design stage.

4.6 Technical Application of Overall Energy Planning: Case Evaluation and Analysis

4.6.1 Energy planning for Shanghai Expo Park

The Shanghai Expo Park covered an area of 5.28km^2, with the planned control area being 6.68 km^2, in which 1.4 km^2 of existing buildings were retained. Energy system planning for the Shanghai Expo Park followed the principle of maximizing system energy efficiency. Firstly, energy planning for this Park did not receive sufficient attention and thus overall planning did not concurrently fully reflect energy conservation and meet the special cooling and heating energy demands of the Shanghai Expo; however, from 2005, researchers firstly forecast the dynamic air-conditioning loads of building groups within the enclosed area of the Shanghai Expo Park, and mainly completed load forecasts relating to the following four aspects: (1) hourly dynamic air-conditioning load; (2) temporal distribution of air-conditioning load rate; (3) monthly cumulative air-conditioning load; (4) design of daily hourly load[21]. Air-conditioning load rate distribution (namely, the temporal frequency of the air-conditioning load) in the whole Expo Park was also analyzed in research. Secondly, with regard to the heating and cooling energy plan, neither the NGCC cogeneration plan with full reliance on natural gas nor the distributed energy plan with partial reliance on natural gas could be implemented in the Expo Park, in which the remaining plan could not be anything other than cooling technology for efficient utilization of power. Finally, a comprehensive plan based on a river water source heat pump was adopted. Professor Long Weiding from Tongji University and other experts drew the following main conclusions: firstly, regional energy planning should be one indispensable part of low-carbon city construction; secondly, integrated and comprehensive application of renewable energy and unused energy represents the inevitable choice and development trend in future regional energy development.

4.6.2 Regional energy planning for Guangzhou University Town

The planned area of Guangzhou University Town was about 18km^2; Guangzhou University Town is located on Xiaoguwei Island, Panyu District, Guangzhou. This University Town was planned to accommodate 0.2 million university students, while the total population on this island will be about 0.35 million. Energy demand in this University Town mainly includes electricity for teaching, scientific research, living and commercial activities, air conditioning refrigeration and domestic hot water without industrial loads. At the beginning of 2003, the Office of Guangzhou University Town Construction Headquarters entrusted South China University of Technology with setting up a research group for developing regional energy planning for Guangzhou University Town. Energy planning completed by this research group covered: (1) energy conservation optimization design of the main body of buildings; (2) distributed energy supply system planning; (3) renewable energy utilization planning; (4) the operating mechanism of the energy supply system. As constructor and operation manager of the energy system in Guangzhou University Town, we have developed the following first-hand experience and thinking about energy planning through practice in regional energy systems in recent years: firstly, regional energy planning is closely related to the functions and positioning of the overall planning for this region; secondly, insights into regional energy load characteristics and accurate load forecasts are the important part of regional energy planning; thirdly, the city background, namely, the regional geographical environment, ecological environmental conditions and weather conditions should be specifically analyzed, and the supply mode for energy products including cooling and heating within the region should be chosen according to local conditions; fourthly, implementation of regional energy planning cannot work without thc support of national energy policies, laws and regulations[22].

With respect to regional energy planning for Guangzhou University Town, the following issues need to be further studied: (1) according to operational data concerning the energy system over several years, there was a certain gap between the actually used air-conditioning load and the forecast load; (2) given that using regional heating and cooling, distributed energy can bring about significant environmental benefits and other social benefits, there is more hope that the State can make breakthroughs in relevant policies, laws and regulations for energy planning; the government should provide appropriate preferential treatment for users within the region in terms of the natural gas price and properly guided use of energy products supplied by the distributed energy system.

4.6.3 Low-Carbon energy planning for Wuhan International Expo City

Wuhan International Expo City is a multifunctional building complex, which mainly incorporates such functions as an exhibition hall, hotel, marine museum, conference center and integrates convention and exhibition, business, science and technology, tourism and cultural leisure activities. The total land area of this project is about 400 hectares; this project can accommodate 100,000 residents and a people flow equivalent to a daily average of 200,000. The total building area within the service scope of the Phase II energy station is about 500,000m^2, covering various buildings including hotel, stores, office buildings, marine park, and the energy station delivers heat, refrigeration and domestic hot water to the corresponding buildings. The Energy Conservation Center of the China Architecture Design & Research Group is responsible for low-carbon energy planning for the new area, while energy planning focuses on the overall demand for electricity, gas, cold and heat within this area, and the supply, conversion and matching. This organization conducted a comprehensive and systematic study of the overall energy plan for the Phase II project: based on a conditions analysis including the energy supply conditions, resource conditions analysis, building load forecast, key technology application case survey, put forward six combined energy system plans including water chilling unit+ gas fired boiler, water chilling unit+gas fired boiler+ice storage, cogeneration, river water source heat pump + gas fired boiler, river water source heat pump+cogeneration, combined cooling heating and power, and comprehensively weighed, compared and demonstrated energy supply reliability, energy-saving benefits, economic benefits and environmental benefits. Also analyzed energy price changes and policy environmental impact[23], finally drawing the following conclusions: (1) according to the time-of-use power price policy of Wuhan, hourly building load balance analysis of the Phase II subproject in this project, seasonal air-conditioner operation energy consumption and charge calculation, ice storage technology in Plan 2 was economically unfeasible in this project and was not recommended; (2) with comprehensive weighing of energy supply reliability, energy conservation emission reduction benefits, economic benefits, Plan 5 river water source heat pump+cogeneration was recommended; (3) Plan 5 entailed consideration of the investment bearing capacity, for which specialized departments could be entrusted to conduct a feasibility study of navigation, flood control and environmental impact, after approval was obtained from maritime affairs, water conservancy, land and environmental protection departments. It was necessary to coordinate with the electric power company with respect to matters concerning power grid connection to internal combustion engine power generation, and actively seek support from local preferential policies.

References

[1] Liu Chunxiang, Liu Hongyan: Energy Conservation Design——*Starting Point for Sustainable Development of Building and City, Journal of Liaoning Institute of Technology*, 2000(2):61-63.

[2] Yu Wei, Roberto Pagani: Urban Energy Planning Research——*European Urban Renewable Energy Strategy with Energy Conservation and Emission Reduction as Goals, Planner*, 2009(3):90-94.

[3] Wang Dengyun, Xu Wenfa: *Low-carbon City Construction and Regional Building Energy Planning, Heating Ventilation Air Conditioning*, 2011(4):17-20.

[4] Long Weiding, Bai Wei, Fan Rui, et al.: *Regional Building Energy Planning for Low-Carbon City*, China Architecture & Building Press, 2011

[5] Fu Lin, Zheng Zhonghai, Jiang Yi et al: *Urban Energy Planning Method Based on Dynamic and Spatial Distribution, Urban Studies*, 2008(SI):146-149.

[6] Weigend A.S.,"Time Series Analysis and Prediction Using Gated Experts with Application to Energy Demand Forecasts", Applied Articial Intelligence ,1996,10(6):583-624.

[7] Gevorgian V.,Kaiser M.,"Fuel Distribution and Consumption Simulation in the Republic of Armenia",Simulation,1998,71(3):154-167.

[8] John Asafu-Adjaye,"The Relationship between Energy Consumption,Energy Price Ad Economic Growth:Time Series Evidence from Asian Developing Countries",Energy Economics,2000,22(6):615-625.

[9] Soytas U,Sari R.,"Energy Consumption and GDP:Causality Relationship in G-7 Countries and Emerging Markets",Energy Economics,2003, 25(1):33-37.

[10] Deng Julong: *Grey Prediction and Decision-making*, Huazhong University of Technology, 1992.

[11] Fang Weihua: *Discussions of Issues Concerning Application of Artificial Neural Network Model in Hydroelectric Energy Sciences, Water Resources and Power*, 2004(3):71-73.

[12] Wu Jiansheng, Zhou Youjun, Jin Long: *Neural Network and Its Research Progress, Journal of Guangxi Teachers Education University: Natural Science Edition*, 2005(1):92-97.

[13] Lu Qi, Gu Peiliang, Qiu Shiming: *Building and Application of Combined Forecasting Model in China's Energy Consumption System, Systems Engineering-Theory & Practice*, 2003(3):24-30.

[14] Xing Mian: *Research on Nonlinear Optimization Combined Forecasting Model*

for Energy Development Trend, North China Electric Power University, 2002(3):64-67.

[15] Li Liang, Sun Tingrong, Huang Qiang et al: *Grey GM (1,1) and Neural Network Combined Energy Forecasting Model, Energy Research & Utilization*, 2005(1):110-113.

[16] Fu Jiafeng, Cai Guotian, Zhang Lei et al: *Combined Forecasting Model Based on GM and BP Network for China's Energy Consumption, Water Resources and Power*, 2006(2):1-4.

[17] Zhang Xiaomei, Zhang Shushen, Zhang Yun et al: *Forecast of Primary Energy Consumption by GM-PLS-based Combined Model, Journal of Liaoning Project Technology University*, 2006, (25):287-289.

[18] Zhang Ling: *Forecast of Energy Demand by Statistical Analysis Technique, Journal of Applied Statistics and Management*, 2001(6):27-30.

[19] Zhou Qingsheng: *Urban Planning Process and Feedback Regulation, Urban Problems*, 1989(5):10-14.

[20] Long Weiding, Ma Hongquan, Liang Hao et al: *Energy Planning for Shanghai Expo Park: Retrospect and Reflection, Journal of HV & AC*, 2010(8):61-69.

[21] Fu Jianping, Wu Shusheng: *Practice and Thinking of Energy Planning for Guangzhou University Town, Building Energy & Environment*, 2011(4):42-44.

[22] Xu Baoping, Liu Peng, Xu Wenlong: *Case Analysis of Low-Carbon Energy Planning for A New Area in Wuhan, Journal of HV & AC*, 2011(10):49-54.

Chapter 5
Integrated Technology for Efficient Coal Water Slurry Preparation and Combustion and Applications—Take Zhuzhou Lanyu Thermal Energy Technology Development Co., Ltd. as an example

Liu Jianwen, Chen Nan

Abstract: Since coal accounts for about 70% of the primary energy source in our country, clean utilization of coal will be a success or failure in deciding the low carbon development target of the economic transition. The Preparation & Combustion Technology or Coal Water Mixture (CWM) is at the application stage based on engineering approaches and scale. Also, integrated technology in the preparation & combustion of an efficient CWM, with intellectual property rights, has seen a more mature and reliable clean coal technology develop, based on an analysis of patents for CWM technology development over more than 20 years. The integrated technology used in the preparation and combustion of an efficient CWM will be an important prospect by promising new engineering applications in the development of energy-saving and emission-reduction by the application of low carbon emissions.

Key Words: CWM; Low Carbon (LC); Clean Coal technology (CCT); energy-saving and emission-reduction

5.1 Introduction

According to *China Human Development Report 2009/10—March Toward Sustainable Future of a Low-Carbon Economy and Society*, with the reform and opening up for more than 30 years, remarkable achievements have been made in China. However, China's extensive growth mode has been subject to an increasing decline in growth impetus, and high energy consumption with big pollution problems are looming large, and the environment and resources have increasingly become bottleneck factors which hamper social development. China should take advantage of the new wave–development of a low-carbon economy worldwide to speed up the transformation in the economic growth mode and shift China's economic development from a primary reliance on increasing the input of energy, resources and other elements to a primary dependence on improving efficiency[1]. China's 12th Five-Year Plan explicitly stated that in the next five years, marked effects will be produced in resource conservation and environmental protection, and energy consumption per unit of GDP and carbon dioxide emissions per unit of GDP would decrease by 16% and 17% respectively. Total discharge of main pollutants would significantly decline; both chemical oxygen demand and sulfur dioxide emissions would drop by 8%; both ammonia nitrogen and nitrogen oxide emissions would fall by 10%. Seven strategic emerging industries including energy conservation and the environmental protection industry would be vigorously developed, and emphasis would be placed on developing key technologies, equipment, products and services for efficient energy conservation, advanced environmental protection and the recycling use of resources in energy conservation and the environmental protection industry; major demonstration projects for energy conservation and environmental protection would be implemented; efficient energy conservation, advanced environmental protection and the recycling use of resources would be boosted.

Therefore, energy conservation and emission reduction tasks for China during the 12th Five-Year Plan period would be very arduous. Given that the energy resource situation of China is characterized by rich coal, lean oil and less gas, efficient and clean utilization of coal is of far-reaching significance for promoting energy conservation and emission reductions in China.

According to Professor Yao Qiang from Tsinghua University, coal combustion is the root cause of carbon dioxide, and the development of clean coal technology is an effective way of reducing carbon dioxide emissions before sustainable alternative energy sources are found.

Clean coal technology is the general term for new coal processing coal combustion, conversion and pollution control technologies, for reducing pollution and improving benefits in the whole process from coal development to utilization. Major development directions include coal gasification, liquefaction, efficient coal combustion and power generation technology, which is currently one of the leading technologies for countries in the world to solve environmental problems and also an important field for international high technology competition. In primary energy consumption of coal in China, 60% of coal is used for power generation, totaling about 1.5 billion tons, while 20% of coal is utilized in small and medium-sized industrial boilers, which is about 0.4-0.5 billion tons. Soot, carbon dioxide, sulfur dioxide, nitric oxide, from coal combustion greatly affect the climate and environment; how to improve coal energy efficiency and lower pollutant discharge is the pressing issue for China now and for a considerable time ahead.

Coal water slurry technology is one technology for the physical processing and reuse of coal[2]; integration of an efficient coal washing technology and coal water slurry technology is the foundation for clean coal technology, for efficient conversion including coal liquefaction and gasification. Coal water slurry technology has the advantages of high efficiency and energy conservation. Its combustion efficiency exceeds 98% and it saves 20% of coal. Based on the fuel price energy ratio in current market conditions, 2.42t and 2.02t of coal water slurry can replace 1t of diesel oil and 1,000m^3 of natural gas, while the coal water slurry fuel cost is only 31% that of diesel oil and 63% that of natural gas. The coal water slurry combustion temperature is 100-200℃ lower than that of fuel oil and gas; thus the NO_x emission is only about 50%, and after an efficient bag-type dust remover is provided, soot emissions from a coal water slurry fired boiler can reach the corresponding environmental standards for fuel oil and gas. Coal water slurry is a liquid-solid slurry fuel; using coal water slurry's property of containing water, integrated biomass coal water slurry preparation, efficient combustion technology, energy-oriented treatment of high-concentration organic sewage and sludge is currently the most realistic technical route for China to push forward energy conservation and emission reductions in small and medium-sized industrial boilers and develop energy conservation and the environmental protection industry.

5.2 Analysis of China's Coal Water Slurry Technology Development Patent Map

According to retrieval made with patent name "coal water slurry" as keywords on the patent literature retrieval homepage of the State Intellectual Property Office (as of November 20, 2011), since December 20, 1985, when Jiang Long et al. from the Photosensitive Chemical Research Institute of the Chinese Academy of Sciences filed an application for the first coal water slurry patent technology, applications for 436 patents, including 193 invention patents and 243 utility model patents, have been filed.

Patents are classified by technical field according to the composition of coal water slurry preparation and combustion industry chain, including 36 additive patents, 67 slurry preparation process patents, 194 combustion technology and equipment patents, 139 patents of other key technologies and equipment (see Fig. 5-1). As shown by this diagram, coal water slurry combustion technology and equipment and other key technologies and equipment are major fields for research and development in the coal water slurry industry. Currently, coal water slurry engineering practice also fully proves that China's coal water slurry preparation and additive technologies are very mature, while coal water slurry combustion technology and equipment technology progress have played an important role in promoting coal water slurry development.

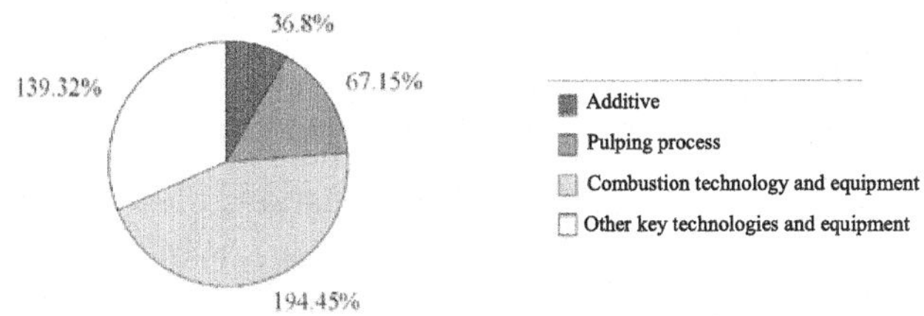

Fig. 5-1 Coal water slurry technology patent application distribution in China in 1985-2011

Patents are classified by organization engaging in research and development of coal water slurry technology and its equipment, including 97 coal water slurry patents applied for by universities and research institutions, 245 coal water slurry patents applied for by companies specializing in the development of coal water slurry technology and equipment, 94 coal water slurry technology and equipment patents applied for in an individual's name (see Fig. 5-2). According to Fig. 5-2, patents with development companies as patent applicants (patentees) account for 56%, while patents with universities and research institutions, individuals as patent applicants (patentees) make up roughly the same proportion. This shows that core forces for promoting coal water slurry combustion technology and equipment technology progress are organizations engaging in industry-university-research cooperation, namely, enterprises.

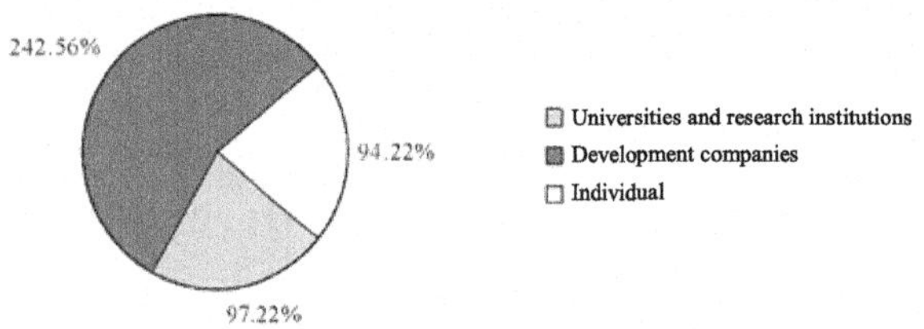

Fig. 5-2 Coal water slurry technology patent applicant (patentee) distribution in China in 1985-2011

Statistical analysis is conducted by patent publication time (see Figs. 5-3 and 5-4); according to Fig. 5-3, coal water slurry technology and equipment patent applications only accounted for 7.3% of total applications during the 15 years from 1985 to 2000, while such a figure reached 23% during the National 10th Five-Year Plan period; technical fields mainly included coal water slurry additives and slurry preparation processes. The State intensified efforts in energy conservation and emission reduction and thus boosted the rapid development of coal water slurry preparation and combustion technologies during the 11th Five-Year Plan period, thus patent applications accounted for 48% of total applications, technical fields concentrated on efficient coal water slurry combustion technology and equipment. Especially in 2011, as the State energetically developed and cultivated strategic emerging industries including energy conservation and the environmental protection industry, coal water slurry development enterprises firmly seized the strategic opportunity presented by the above move of the State to increase coal water slurry technology and equipment technology research and development and enhanced awareness of intellectual property protection. Patent applications accounted for 19.3% of total applications, while technical fields for which applications were filed extended to use of coal water slurry technology for biomass coal water slurry preparation by organic wastewater, pulverized coal and coal slurry co-combustion, coal slurry and domestic garbage co-combustion, coal slurry and sludge co-combustion, in order for efficient resource and energy-oriented utilization of sewage and sludge, which really reflect energy conservation and environmental protection characteristics of coal water slurry by making technical progress and proceeding from process centralization.

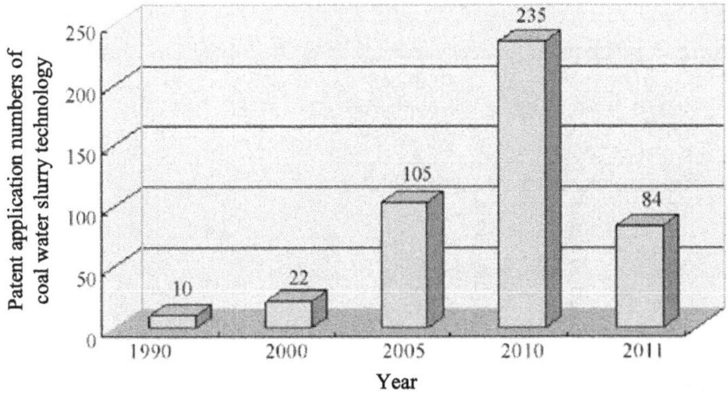

Fig. 5-3 Coal water slurry technology patent application distribution statistically analyzed by time in China in 1985-2011 (Accumulative on 10-year basis before 2000, accumulative on 5-year basis in 2005, 2010, accumulative in 2011)

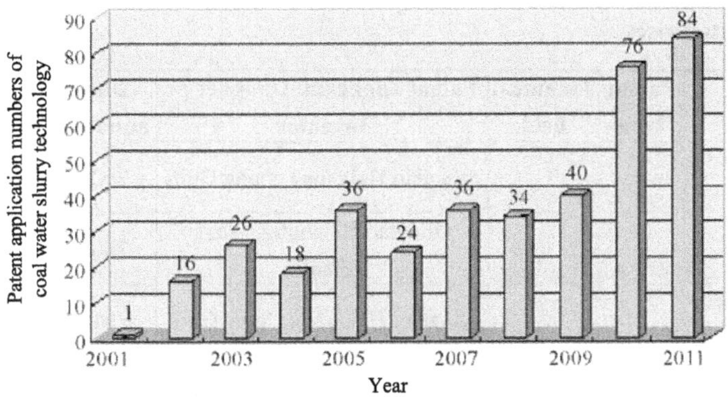

Fig. 5-4 Coal water slurry technology patent application distribution statistically analyzed by time in China in 1985-2011 (Statistically analyzed on the basis of actually published data in current year)

Engineering practice in coal water slurry technology popularization application proves that coal water slurry combustion technology and equipment technology progress has played a vital role, according to analysis based on patent applications filed by small and medium-sized coal water slurry fired boiler development enterprises (see Fig. 5-5, Table 5-1). Patent applications are dominated by those filed by Qingdao Haizhong, Zhuzhou Lanyu and Qingdao Vert which have filed 15, 9 and 7 patent applications respectively.

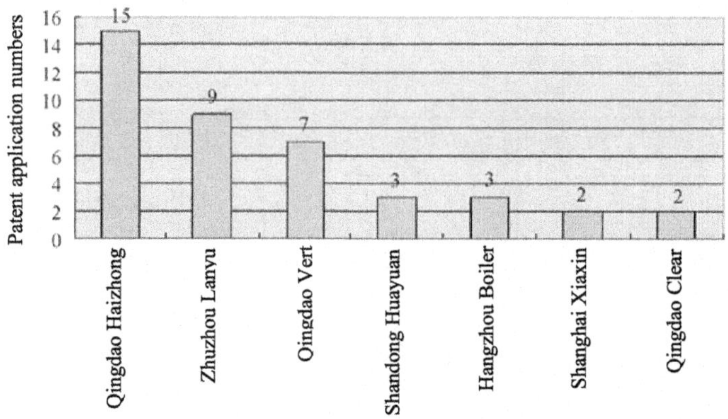

Fig. 5-5 Distribution of patent technology applications filed by small and medium-sized coal water slurry fired boiler development enterprises

Table 5-1 Analysis of patent technology applications filed by small and medium-sized coal water slurry fired boiler development enterprises

No.	Patent name	Patent type	Technical field	Patent applicant, Designer or Inventor	Date of the announcement
1	Coal-water slurry small industrial boiler	B	C	Qingdao Haizhong, Jiang Guili	2001-09-19
2	Coal-water slurry burner	B	C	Qingdao Haizhong, Wang Zhanlin	2001-09-19
3	Integrally movable coal water slurry fired boiler unit	A	C	Qingdao Haizhong, Jiang Guili	2010-05-19
4	Coal water slurry and fine pulverized coal fired burner	A	C	Qingdao Haizhong, Jiang Guili	2011-03-30
5	Coal water slurry and coal water slurry powder fired hybrid power boiler	A	C	Qingdao Haizhong, Jiang Guili	2011-04-27
6	Coal water slurry fired oxygen rich gasification boiler	A	C	Qingdao Haizhong, Jiang Guili	2011-08-17
7	Vertical coal water slurry fired industrial boiler	B	C	Qingdao Haizhong, Jiang Guili	2009-02-11
8	Integrally movable coal water slurry fired boiler unit	B	D	Qingdao Haizhong, Jiang Guili	2010-09-01
9	Coal water slurry and Chinese medicine residue fired hybrid power boiler	B	C	Qingdao Haizhong, Jiang Guili	2011-06-08
10	Special stirring filter for coal water slurry fired boiler	B	D	Qingdao Haizhong, Jiang Guili	2011-06-22
11	Coal water slurry and biomass fired hybrid power boiler	B	C	Qingdao Haizhong, Jiang Guili	2011-06-22

(Contd.)

No.	Patent name	Patent type	Technical field	Patent applicant, Designer or Inventor	Date of the announcement
12	Coal water slurry and industrial waste gas fired hybrid power boiler	B	C	Qingdao Haizhong, Jiang Guili	2011-06-22
13	Industrial liquid waste coal water slurry fired boiler	B	C	Qingdao Haizhong, Jiang Guili	2011-06-22
14	Special sulfur and dust remover for coal water slurry fired boiler	B	D	Qingdao Haizhong, Jiang Guili	2011-08-31
15	Coal water slurry oil-free ignition device	B	C	Qingdao Haizhong, Jiang Guili	2011-11-09
16	Stable combustion method and device for coal water slurry fired boiler	A	C	Zhuzhou Lanyu, Chen Nan and Zhu Henghui	2010-02-03
17	Biomass coal water slurry making, combustion method and integration system	A	B	Zhuzhou Lanyu, Liu Jianwen, Chen Nan, et al.	2010-02-03
18	Online soot blowing method and device for coal water slurry fired boiler	A	C	Zhuzhou Lanyu, Chen Nan and Xu Puzhou	2010-06-16
19	A suspension pulverized coal and coal water slurry fired compound combustion method and device	A	C	Zhuzhou Lanyu, Liu Jianwen and Chen Nan	2011-06-22
20	Oil-free ignition device for coal water slurry fired boiler	B	C	Zhuzhou Lanyu, Chen Nan and Huang Guohui	2010-05-26
21	Stable combustion device for coal water slurry fired boiler	B	C	Zhuzhou Lanyu, Chen Nan and Zhu Henghui	2010-05-26
22	Coal water slurry fired suspension and layered compound combustion boiler	B	C	Zhuzhou Lanyu, Chen Nan, Xu Puzhou et al.	2011-09-01
23	Online soot blowing device for coal water slurry fired boiler	B	D	Zhuzhou Lanyu, Chen Nan and Xu Puzhou	2011-09-01
24	A suspension pulverized coal and coal water slurry fired compound combustion device	B	C	Zhuzhou Lanyu, Chen Nan and Liu Jianwen	2011-09-01
25	Coal water slurry and medicine residue mixed combustion boiler	A	C	Qingdao Vert, Wang Zhanlin	2010-12-22
26	Fully automatic coal water slurry fired burner and control method for this burner	A	C	Qingdao Vert, Wang Zhanlin	2011-01-05
27	Coal water slurry ceramic slurry gun front end assembly	B	D	Qingdao Vert, Wang Zhanlin	2010-02-24
28	Coal water slurry and medicine residue mixed combustion boiler	B	C	Qingdao Vert, Wang Zhanlin	2010-02-24

(Contd.)

No.	Patent name	Patent type	Technical field	Patent applicant, Designer or Inventor	Date of the announcement
29	Dry bottom hopper type soot blower for coal water slurry fired boiler	B	D	Qingdao Vert, Wang Zhanlin	2010-02-24
30	Fully automatic coal water slurry fired burner	B	C	Qingdao Vert, Wang Zhanlin	2010-04-28
31	Mixed coal water and slurry medicine residue burner	B	C	Qingdao Vert, Wang Zhanlin	2010-04-28

Note: patent type A——invention patent, B——utility model; technical field A——additive, B——slurry preparation process, C——combustion technology and equipment, D——key process technology and equipment.

Patent technologies for which applications have been filed by three enterprises basically concentrate on such fields as coal water slurry combustion technology and equipment, and other key technologies and equipment. Only applications filed by Zhuzhou Lanyu have extended to coal water slurry preparation, combustion technology integration and other key technologies and equipment; with respect to patent application publication (announcement) dates, it is found that small and medium-sized coal water slurry fired boiler patent applications started in 2001 and were predominantly filed in 2010 and 2011, suggesting that coal water slurry combustion technologies and equipment have entered the mature period. With regard to patent technology inventors (designers), both Qingdao Haizhong and Qingdao Vert only have one inventor (designer), while Zhuzhou Lanyu has displayed scientific research team spirit. Both engineering practice and market operations prove that though all are private enterprises, they have provided cooperation from teams with a professional background involving coal water slurry preparation, efficient combustion technology, environmental engineering and low-carbon technology, and Zhuzhou Lanyu has been widely recognized by users thanks to its scientific and technological innovation competitiveness and advantages in technical performance of coal water slurry fired boilers.

It should be noted that users should keep in mind the following fundamental philosophy during initiation of coal water slurry fired boiler projects, bidding and construction: “patent technology is not equivalent to mature technology”, on-the-spot investigation of tenderers with respect to the engineering operation of technologies and products should be conducted, which is a very important part of the new initiation of a project. For example, a Japanese wholly-owned company and Thailand Chia Tai tracked and surveyed domestic coal water slurry technology development for more than three years before adoption of coal water slurry technology and boiler and finally chose Zhuzhou Lanyu patent technology and complete equipment.

5.3 Process Principle and Application of Integrated Technology for Efficient Coal Water Slurry Preparation and Combustion

5.3.1 New developments in coal water slurry technology in China

1. Slurry preparation process innovationsc[3]

With continuous progress in slurry preparation technology, in order to expand slurry-making coal varieties, technical treatment measures which enhance advantages and avoid disadvantages are adopted according to characteristics of different raw material coals to research and develop new slurry preparation processes including "more crushing and less grinding, classified grinding", "enhancing ultra fine grinding, strengthening stirring and shearing" and " parallel connection of multiple grinders, optimized gradation", and create multiple international patents including "a method for using low-rank coal to prepare high concentration coal water slurry", "biomass coal slurry preparation process technology", "biomass coal water slurry preparation, combustion method and integration system", "secondary wet grinding of multiplex slurry for slurry preparation", which meanwhile result in derivatives such as Shenhua environmentally friendly coal water slurry, low volatile component (petroleum coke) coal slurry, brown coal slurry, biomass coal slurry, blended coal slurry, coal slurry for gasification, multiplex gasified slurry and instant pulverized coal. Currently, these new coal water slurry varieties have been applied on the market.

Increasingly mature "optimized-gradation slurry preparation process technology" is adopted to prepare high concentration coal water slurry with low-rank coal which, as internationally recognized, can't be used to prepare concentration coal water slurry, which fully demonstrates progress and innovations in coal water slurry preparation technology in China in recent years and is of positive significance for further expanding the scope of slurry-making coal and promoting the development of the coal water slurry industry. Mixing such wastes as papermaking black liquid, industrial wastewater (sewage) and municipal sludge with pulverized coal at a certain proportion to prepare coal slurry for the furnace kiln through a specific slurry preparation process not only enables utilization of waste resources but also saves raw material coal and water for slurry, reduces production costs, which leads to huge market development potential.

2. Development of new type coal water slurry additivesc[3]

As coal water slurry additives play a crucial role in the performance of coal water slurry, many domestic organizations have attached greater importance to research and development of coal water slurry additive products, and have made a series of new marketable products, which satisfy various coal water slurry performance requirements and show higher cost performance. Series coal water slurry additives researched and developed in China in recent years have unique characteristics in raw material selection and the production process. For example, NDF series additives researched and developed by Nanjing University can meet the requirements of multiple slurry preparation processes and highly adapt to coal varieties; an additive synthesized by copolymer dominated by organic acid carboxylic acid as chosen by Kunshan Dikun Fine Chemical Co., Ltd. is characterized by high dispersing performance and relatively low costs; ZDFS series additives developed and made by Beijing Zidong Environmental Water Treatment Agents Factory are a compound of multi-component surfactants and are highly compatible with other domestic additives in terms of performance; an HNF additive produced by Huainan Synthetic Materials Factory can shows dual functions for addressing dispersibility and stability of coal water slurry; polyacrylic acid series additives recently developed by China University of Mining & Technology (Beijing) are EAO, EA3 and ST2 series and 13 additives synthesized by structural design and modification of polycarboxylate surfactants and are obviously superior to naphthalene additives in terms of increasing slurry concentration and decreasing additive use level. The National Coal Water Slurry Center considered the characteristics of low-rank coal varieties represented by Shenhua, optimized the main structure of additive molecules and adjusted composition ratios of hydrophobic and hydrophilic groups in molecular weight to continuously improve matching between molecular structure and surface properties of Shenhua coal and prepare SHPF series additives with high cost performance; an SAF high load dispersant successfully tested by South China University of Technology has broad prospects for development. In addition, great progress has also been made in new type multifunctional non-ionic additives researched and developed in China.

3. Innovations and applications of combustion technology

With increasing coal water slurry varieties in recent years, in order to meet combustion requirements of coal slurry of different types, new developments have been made in the coal water slurry combustion mode, combustion equipment, furnace construction, environmental control of soot.

(1) Diversification of combustion technology

As coal water slurry varieties increase, in response to combustion characteristics of various kinds of coal slurry such as ignitability of coal slurry, its combustion stability, clinkering property, ash's stickiness and abradability, the coal water slurry combustion mode has also developed from single mist spray—suspension combustion to fluidized—suspension combustion, suspension—layered combination combustion, multi-air distribution and cyclone combustion, catalytic combustion and low-temperature, low oxygen combustion of coal water slurry. With respect to combustion equipment, coal slurry atomizing nozzles and combustors have become diverse, are available in various types and widely applied, and have been greatly improved in structural design, material selection, atomization quality and rationality of air distribution quantity, and basically meet the requirements of various kinds of coal slurry for temperature and intensity at the time of ignition, combustion and complete combustion[3]. Furthermore, with respect to suspension,,a layered combination combustion mode has also opened up a new path for combustion of biomass (sludge) coal slurry, low volatile component difficult-to-burn slurry and coal slurry with poor slurrying property. Layered combination combustion technology can achieve co-combustion of combustible solid wastes (such as Chinese medicine residue, biomass solid wastes), high concentration organic wastewater (high concentration COD), organic waste gases and biomass coal water slurry. An energy and environment integration system is built in large and medium-sized enterprises and industry parks to promote ecological construction of large and medium-sized enterprises and industry parks.

(2) Reliability improvement innovations of combustion equipment

Coal water slurry is a liquid-solid slurry fuel in which ash inevitably contains alkali metal oxides including potassium oxide and sodium oxide. Combustion equipment inevitably causes slagging during combustion, while slagging is the key factor which affects the reliability and long-term operation of a coal water slurry fired boiler. In order to cope with slagging problems in combustion, a great variety of key technologies and equipment for practical engineering applications, such as anti-coking structure at the bottom of a coal water slurry fired boiler furnace, online soot blowing method and equipment for a coal water slurry fired boiler have been researched and developed in China in recent years.

Fuel characteristics of coal water slurry in terms of containing water determine difficulty in ignition and stable combustion of coal water slurry. In view of this, Chinese coal water slurry combustion equipment research and development personnel put forward primary and secondary furnace layers—suspension coal water slurry combustion technology and boiler (namely, double furnace coal water slurry boiler[4]) for the first time, while this combustion technology can enable stable combustion within a coal water slurry fired boiler within 10%–110% load range. Currently, there are national invention patents including stable combustion method and equipment for a coal water slurry fired boiler, coal water slurry suspension—layered combination combustion boiler, oil-free ignition device for a coal water slurry fired boiler.

5.3.2 Integrated technology for efficient coal water slurry preparation and combustion

1. Process principle of integrated technology for efficient coal water slurry preparation and combustion

New type integrated technology and equipment for efficient biomass coal water slurry preparation and combustion system is a coal water slurry system industry chain technology, independently developed by Zhuzhou Lanyu Thermal Energy Technology Co., Ltd., and was funded by the Innovation Fund for Technology-Based Firms under the Ministry of Science and Technology in 2010 (10C26214304669), in which core independently innovated intellectual properties of the Company— green low-carbon technologies such as primary and secondary furnace layers—coal water slurry suspension combustion boiler, integrated membrane separation, power ultrasonic treatment and efficient mechanically centrifugal separation, are relied on to carry out energy and resource-oriented utilization of sludge and organic wastewater. The integrated technology process can be divided into five modules: efficient organic wastewater pretreatment module, efficient sludge pretreatment module, biomass coal water slurry preparation module, biomass coal water slurry combustion module and purified water biochemical treatment module.

Efficient organic wastewater pretreatment module: According to the physicochemical characteristics of organic wastewater, firstly adjust its pH value, and then filter fibrous matters and/or suspended solid particles, use a membrane separation device to conduct molecular level concentration. Concentrated liquor serves as raw material for preparing coal water slurry, while penetrating fluid undergoes biochemical treatment before use or discharge; separating membrane model selection depends on the minimum molecular weight of organic matter in wastewater; steam-water pulse or ultrasonic is adopted for online cleaning in order to guarantee efficient operating performance of the membrane separation device. Organic wastewater undergoes membrane concentration to separate out organic matter, and organic matter concentration in penetrating fluid is very low, thus this greatly improves biodegradability in subsequent biological treatment, not only recovers biomass energy but also significantly lowers organic wastewater treatment costs.

Efficient sludge pretreatment module: According to the physicochemical characteristics of sludge, firstly adjust its pH value, and then provide power ultrasonic treatment with throw-in type ultrasonic vibration plate, use efficient centrifugal separation technology to dehydrate sludge which has undergone the above ultrasonic treatment to below 50%, preferably below 30% (final dehydration rate of sludge is determined by comprehensive economic comparison; in a laboratory test, three frequencies 22, 28, 40kHz and three different power densities, different ultrasound irradiation times are chosen, and optimal process parameters are determined, and then system energy consumption analysis is given by considering the subsequent sludge incineration process); dehydrated sludge undergoes filtration to remove fibrous matters and/or coarse suspended solid particles and is used as raw material for preparing coal water slurry or is directly put into a coal water slurry fired boiler for co-combustion with biomass coal water slurry prepared in Step 1.

Biomass coal water slurry preparation module (Fig. 5-6): Combustible slurry-like solid-liquid suspended matter which is prepared on the basis of the conventional coal water slurry preparation process and by partially and/or fully replacing pulverized coal and water with wastewater containing organic matter and dehydrated sludge is hereby defined as biomass coal water slurry. Select, in certain proportions, concentrated organic wastewater generated in Step 1 and sludge dehydrated through ultrasonic treatment in Step 2 and superfine pulverized coal and water, add a small quantity of dispersant and/or stabilizer, conduct matching and stirring to prepare biomass coal water slurry and the performance of the biomass coal water slurry product such as heating value, viscosity, volatile matters and grain granularity distribution. As this project is designed to maximize comprehensive utilization of organic wastewater and/or sludge, it is suitable for preparing furnace-front slurry preparation, and the laboratory test is provided with a stabilizer for testing its slurry ability, while industrial application may dispense with, or be provided with, a small quantity of stabilizer to reduce production costs.

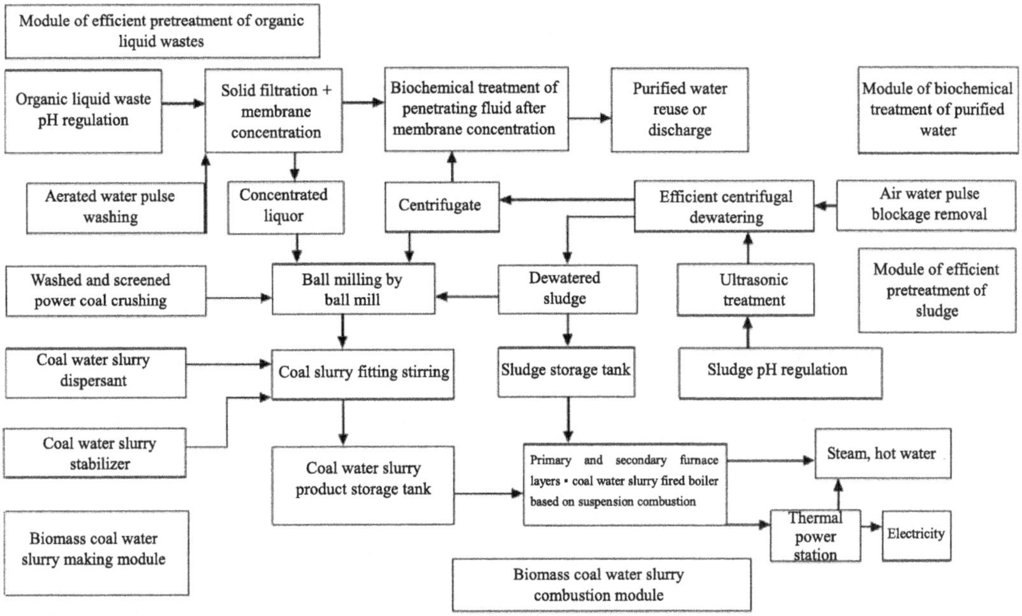

Fig. 5-6 New type efficient biomass coal water slurry preparation and combustion process module

Biomass coal water slurry combustion module (Fig. 5-6): This module mainly relies on primary and secondary furnace layers—suspension combustion technology (Fig. 5-7); it is a new type coal water slurry combustion technology which resulted from more than 20 years of coal water slurry combustion engineering experience of Zhuzhou Lanyu Thermal Energy Technology Co., Ltd., for which patent application was filed by the Company in 2009 and a national invention patent has been granted; more than 30 boilers have been applied in Changsha, Zhuzhou and Xiangtan in Hunan Province, Nanning in Guangxi. In particular, one Japanese wholly-owned company in Guangxi and listed company Zhuzhou Times New Material carried out a market survey on coal water slurry fired boilers nationwide and finally adopted this new type of coal water slurry fired boiler, and Zhuzhou Times New Material installed 4t, 6t and two 10t coal water slurry fired boilers. The former two boilers have been operated for several years and 10t coal water slurry fired boilers have been put into operation with good operating conditions.

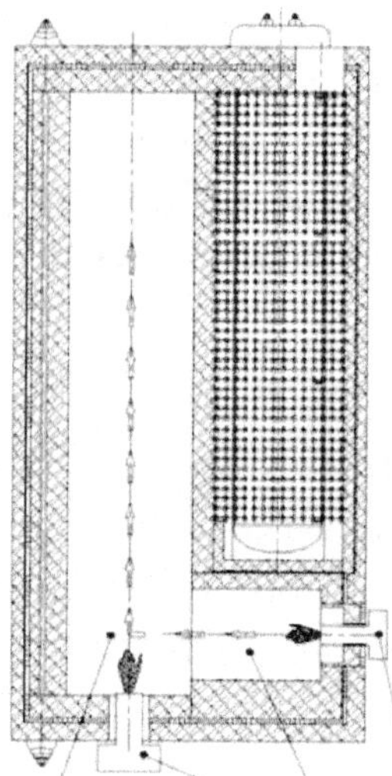

Fig. 5-7 Structure of primary and secondary furnace layers—suspension combustion coal water slurry fired boiler furnace

Note: 1 is the secondary combustor, 2 is the secondary combustion chamber, 3 is the primary combustor, 4 is the primary combustion chamber.

In Fig. 5-7, 1 is the secondary combustor, 2 is the secondary combustion chamber, 3 is the primary combustor, 4 is the primary combustion chamber. The arrow in this diagram refers to flue gas flow direction of the secondary combustor. As shown in Fig. 5-7, the boiler features primary and secondary furnace construction including a primary combustion chamber 4 and secondary combustion chamber 2 which show a 7-shaped arrangement and are provided with the corresponding combustors 3 and 1.

The secondary combustor 1 is installed above the secondary combustion chamber 2; the secondary combustion chamber 2 is characterized by a small furnace, fast temperature rise and easy ignition so that it can enable stable combustion in the case of small load combustion in the boiler, and the resulting high temperature heat source enters the front part of the primary combustion chamber 4, as shown by the arrow direction in this diagram, to increase the temperature of the primary combustion chamber furnace so as to ensure that coal water slurry can be fully burnt when the boiler is in low load operation. With interaction between the primary combustor 3 and secondary combustion chamber 2, stepless regulation and stable combustion is achieved while the combustion load in the coal water slurry fired boiler is in the broad range of 10%-110%.

Technical principle of primary and secondary furnace layers—suspension combustion: the boiler features primary and secondary furnace design in which the secondary furnace is the secondary combustor, and the primary furnace is the primary combustor; the secondary furnace can perform ignition and load regulation; the furnace has a special air distribution design and dispenses with a blower; there is a moving grate at the bottom of the primary combustion chamber, and air distribution occurs at the furnace bottom so as to lower the furnace bottom temperature and make coking difficult. The furnace is not provided with a fire wall so as to effectively prevent direction injection of coal water slurry cause coking. An online soot blowing device is provided for online soot blowing and deslagging, and it is not necessary to stop the furnace. Production is not affected so that the boiler can be continuously, stably and reliably operated for a long time. Suspension-layered combination combustion is conducted; coal water slurry is injected into the furnace by the slurry gun; most of the atomized relatively fine slurry particles are suspended in the air for combustion, while a small part of atomized relatively coarse unburnt slurry particle blocks scatter on the fire grate at the furnace bottom for secondary combustion in the form of layered combustion and are fully burnt by the air supply at the bottom of the fire grate.

Primary and secondary furnace layers-suspension combustion technology has the following characteristics: (1) oil-free ignition, economical operation; (2) broad load regulation range; (3) high thermal efficiency of boiler, wide scope of fuel application; (4) long-time continuous stable operation; (5) good environmental protection effects on boiler operation (Table 5-2).

2. Theoretical innovation in integrated technology for efficient coal water slurry preparation and combustion

(1) Breakthroughs in coal water slurry preparation technology: In coal water slurry preparation technology initial oil and pulverized coal are used to prepare oil coal slurry → water replaces oil to prepare coal water slurry → wastewater and sludge containing organic matter replace clear water or partially pulverized coal to prepare biomass coal water slurry, so that coal water slurry technology really produces dual effects – energy and environmental protection.

Table 5-2 Performance indicator comparisons of layered-suspension combustion, fluidized-suspension combustion and mist spray-suspension combustion coal water slurry fired boiler

Performance index boiler	Layer-coal water slurry fired boiler based on suspension combustion	Fluidization-coal water slurry fired boiler based on suspension combustion	Atomization-coal water slurry fired boiler based on suspension combustion
Ignition method	Oil-free ignition	Oil, gas ignition	Oil, gas ignition
Combustion temperature	Low-temperature combustion	Low-temperature combustion	High-temperature combustion
Combustion efficiency	98.5%-99%	98%	96%
Desulfurization performance	Desulfurizing agent can be added on fire grate and into slurry so as to enable desulfurization within boiler.	Limestone can be used as fluidizing medium for desulfurization within boiler.	Desulfurizing agent can be added to slurry so as to enable desulfurization within boiler.
Load regulation	10%-110%	30%-110%	50%-100%

(Contd.)

Performance index boiler	Layer-coal water slurry fired boiler based on suspension combustion	Fluidization-coal water slurry fired boiler based on suspension combustion	Atomization-coal water slurry fired boiler based on suspension combustion
Fuel adaptability	Economical coal water slurry, national standard III coal water slurry, biomass coal water slurry and biomass combustible solid, liquid waste.	High volatile matter high-quality slurry, biomass (sludge) slurry, low volatile matter difficult-to-burn coal slurry, coal slurry with poor slurry forming property	High volatile matter high-quality slurry
Starting performance	Suitable for frequent starting	Starting can be frequent	Not suitable for frequent starting
Boiler output	Boiler output reaches 100%	Boiler output reaches 95%	Boiler output is 70%
Exhaust temperature	High load ≤ 180℃ ; medium load ≤ 120℃ ; low exhaust loss.	High load ≥ 230℃ ; medium load ≥ 200℃ ; high exhaust loss.	High load ≥ 230℃ ; medium load ≥ 200℃ ; high exhaust loss.
Dust removal and desulfurization method, environmental protection effect	bag-type dust remover + wet desulfuration is adopted and there is no smoking from ignition from furnace shutdown; dust content <10mg/nm3; bag service life 1-15 years; dry dust removal, without secondary pollution; no power consumption, low operating costs, small footprint	Electric dust remover: high price: wet dust removal; easy to generate secondary pollution, with black smoke caused at the time of ignition, high water consumption (6t/h boiler, daily water consumption 290t), highly frequent maintenance of induced draft fan, chimney, large area of sedimentation basin.	Electric dust remover: high price; high power consumption; wet dust removal: easy to generate secondary pollution, with black smoke caused at the time of ignition, high water consumption (6t/h boiler, daily water consumption 290t), highly frequent maintenance of induced draft fan, chimney, large area of sedimentation basin.
Long-term operation performance	High online soot blowing, slagging prevention property; operation can be continued for more than one year.	Slagging prevention; continuous operation is longer than atomization-suspension combustion.	Regular boiler shutdown for coke, slag removal is needed; continuous operation for 15-30 days.

(Contd.)

Performance index boiler	Layer-coal water slurry fired boiler based on suspension combustion	Fluidization-coal water slurry fired boiler based on suspension combustion	Atomization-coal water slurry fired boiler based on suspension combustion
Project	Since 2000 when the first boiler was installed at Dongyuan Hotel in Zhuzhou City, Hunan Province, about 40 first-fourth-generation products have been efficiently operated in Changsha-Zhuzhou-Xiangtan Region, Nanjing, Ningbo, Huai'an etc.	Project demonstration stage	Since 2001, when small and medium-sized coal water slurry fired boilers were installed, there have been applications across the country, and part of them have been replaced by double-hearth layer-coal water slurry fired boilers based on suspension combustion.
Operation mode	Provision of slurry making, boiler design, heat supply system design, key equipment for dust removal and desulfurization, energy outsourcing and energy contract management technical services (BOT)	Supply of boilers	Supply of boilers

With intensified efforts in pollution control in China, treatment of sludge generated during municipal sewage purification, wastewater which contains heavy metals and pathogens, and other difficult-in-biodegradation wastewater have become pressing and difficult problems. Coal water slurry technology can process, utilize and burn them to greatly simplify complicated multi-factor environmental difficulties and completely demonstrate the broad environmental protection significance of coal water slurry technology and fully embody the objectives, nature and functions of the resource/energy/environment integration system.

(2) Breakthroughs in coal water slurry combustion technology: Conventional coal water slurry combustion is performed in mist spray–suspension combustion mode; this suspension combustion mode is disadvantageously characterized by easy coking, combustion of only highly volatile coal water slurry, use of heavy oil or natural gas for ignition, the necessity for regulating the load above 50% and failure to maintain long-time operation. Integrated technology is used in the primary and secondary furnace layers–suspension combustion technology for addressing matching between coal water

slurry combustion and physicochemical properties specific to its product and providing equipment support for large-scale industrial application of coal water slurry (Table 5-2).

(3) Improvement of coal water slurry industry chain operation mode: The coal water slurry industry chain covers steam coal washing and dressing, coal water slurry preparation, design and manufacturing of the coal water slurry fired boiler, design of the coal water slurry fired boiler room, installation, debugging and operation of the coal water slurry fired boiler; many domestic companies only engage in one or two links in the industry chain such as steam coal washing and dressing + coal water slurry preparation or design, manufacturing, installation and debugging of the coal water slurry fired boiler. With respect to integrated technology, other links except steam coal washing and dressing (direct purchase of washed and dressed steam coal which meets quality requirements for slurry preparation) are in a unified complete industry chain operation by the Company so that process management is optimized.

(4) System integration of the coal water slurry industry technology chain: Based on the integration innovation principle, combine suitable modern high technologies with coal water slurry preparation and combustion technology to build a complete coal water slurry industry technology chain, provide technical support for large-scale industrial application of coal water slurry.

(5) Modularization of coal water slurry preparation and combustion system integration: Various links in the biomass coal water slurry industry chain are modularized and standardized to simplify and optimize the system process and offer diverse choices for coal water slurry users.

3. Engineering applications of integrated technology for efficient coal water slurry preparation and combustion

(1) Centralized steam supply system for Zhuzhou Longquan Washing Industry Park

This project was solely built by Lanyu Technology Company using independently innovative intellectual properties; it includes a 200,000t/year coal water slurry production line and 20t/h coal water slurry fired steam boiler which supplies steam to washing enterprises in this Washing Industry Park, while untreated high concentration organic wastewater from a centralized sewage treatment plant is partially extracted. Part of the coal water slurry products made by the coal water slurry production line are used in this project, while the remaining parts are supplied to Changsha, Zhuzhou and Xiangtan, Hengyang, Yiyang. The Company and washing enterprises make a consumption settlement through a steam flow meter; as garment processing enterprises only earn small profits, some of the washing enterprises use a biomass boiler to produce steam, thus a centralized heat supply based on coal water slurry fails to be applied in the whole industry park. However, it plays a very important role in turning Zhuzhou City into a National Sanitary City and National Model City for Environmental Protection.

(2) Popularization and application of coal water slurry in Nanning, Guangxi (Table 5-3)

Nanning Municipal Government, Guangxi, has set up a specialized coal water slurry popularization office for extending the application of coal water slurry and has released *Measures of Nanning City for Prevention and Control of Sulfur Dioxide Pollution from Coal-Fired Boilers, Implementation Opinions of Nanning Municipal Government Concerning Popularization and Application of Coal Water Slurry and Administrative Measures of Nanning City for Special Funds for Popularization and Application of Coal Water Slurry.* Lanyu Technology Company has made investments in coal water slurry preparation technology to become a shareholder and has built a coal water slurry factory with local strategic investors, to which the Company offers system technology services concerning coal water slurry preparation by means of independently innovative intellectual properties. This mode will be adopted in Ningbo market, to be launched soon.

(3) Energy outsourcing operation service of coal water slurry fired boiler room in Zhuzhou Municipal Finance Bureau

A sub-bureau of Zhuzhou Municipal Finance Bureau is located in Zhuzhou Hexi National Hi-Tech Development Zone and has high environmental requirements; a 2×2 t/h coal water slurry fired boiler was completed at the beginning of the century, which supplies steam and hot water to office buildings and living quarters. Since 2007, the Company has been responsible for system operations, maintenance and management (including supply of coal water slurry fuel), for which settlement with residents and this sub-bureau is made according to the steam and hot water used.

(4) Popularization of coal water slurry in Changsha, Zhuzhou, Xiangtan and surrounding areas

Lanyu Technology Company undertakes more than 90% of business concerning coal water slurry fired boilers and furnace kiln renovations and construction projects in this region, and provides coal water slurry products which match boilers and furnace kilns and other premium supporting services. After users shifted to coal water slurry technologies and products, significant energy conservation and environmental protection benefits have been delivered in industrial production and household energy consumption.

(5) 6t/h coal water slurry-based centralized heat supply in Zhuzhou Fortune Chemical Industrial Co., Ltd.

Products made by Zhuzhou Fortune Chemical Industrial Co., Ltd. mainly include sodium dimethyl dithiocarbamate, potassium dimethyldithiocarbamate, metham sodium, rubber accelerators TMTD, NOBS, TMTM, a new type flotation agent, beneficiation reagent, terpenic oil, among which sodium dimethyl dithiocarbamate is this Company's star product; its monthly output reaches about 1,200t, and production capacity is more than 2,000t each month. One 2t/h coal water slurry fired boiler was built in 2007, which makes a centralized heating supply for the chemical production process. In 2010, a supporting 6t/h coal water slurry fired boiler was built due to expansion, for which Lanyu Technology Company was responsible for its investment. Construction and operation management and commercial settlement is made on the basis of an annual 50,000t of steam as the quantity.

(6) Centralized heat supply for aerated concrete block production process in Zhuzhou Xiawan Building Materials Co., Ltd.

Zhuzhou Xiawan Building Materials Co., Ltd. has an aerated concrete block production line with annual production capacity being one million cubic meters; the original steam pressure process received steam from a coal-fired boiler, the lower heating value of coal was 5,500 kcal/kg, and coal consumption by aerated concrete blocks was 30kg/m^3. Given the steam use level for one million cubic meters annually produced by this Company, two 15t/h coal water slurry fired boilers replaced original coal-fired boilers. The body is made of SZS double drums, with a longitudinal D-shaped layout of all water pipes, in which there are primary and secondary combustion chambers, and a moving grate is arranged at the furnace bottom, which is characterized by fast ignition, high thermal efficiency, wide scope of applications for coal water slurry, a broad load regulation range (20%-100%). After this project was operated, coal water slurry as fuel was compared with coal, and the operating cost was reduced by about 10%. It delivered significant energy conservation and emission reduction benefits; it resulted in more significant economic benefits compared with natural gas and oil. This project is a typical case in which the replacement of coal by coal water slurry technology brings about significant economic benefits.

(7) Centralized heat supply based on coal water slurry in Maotai Town, Guizhou

The change in the 10t coal-fired chain-grate boiler into a coal water slurry fired boiler in China Tobacco Guiding Cigarettes Factory marked the popularization and application of coal water slurry in Guizhou; Lanyu Technology Company provided design and combustion technology; renovation was completed and the operation was started in May, 2010; currently this boiler is under normal operation with excellent performance from various indicators: thermal efficiency improved by 27.8%—29.8%, the burnout rate increased by 23.6%, sulfur dioxide declined by 99%, and the annual total fuel consumption fell from 7,357t standard coal to 4,782t standard coal, down 35%. Projects currently under operation and construction include a 2×10t/h coal water slurry centralized heat supply project in Qingzhen Pharmaceutical Industry Park and a 3×15t/h coal water slurry centralized heat supply project in Maotai Town.

Table 5-3 Comparisons of popularization and application of coal water slurry between Nanning City, Guangxi and Dongguan City, Guangzhou

Project	Nanning City, Guangxi	Dongguan City, Guangdong
Government policy support	Measures of Nanning City for Prevention and Control of Sulfur Dioxide Pollution from Coal Fired Boilers, Implementation Opinions of Nanning Municipal People's Government Concerning Popularization and Application of Coal Water Slurry, Administrative Measures of Nanning City for Special Fund for Popularization and Application of Coal Water Slurry.	Implementation Opinions of Dongguan City for Popularization and Application of Coal Water Slurry, Pilot Program of Dongguan City for Popularization and Application of Coal Water Slurry.

(Contd.)

Project	Nanning City, Guangxi	Dongguan City, Guangdong
Technical support mode	Integrated: coal water slurry making technology and efficient combustion technology.	Distributed: coal water slurry making technology and coal water slurry boiler technology.
Technical support organization	Zhuzhou Lanyu Thermal Energy Development Co., Ltd.	China Coal Research Institute, other coal water slurry boiler manufacturers.
Popularization operation mode	Government guidance, market operation; practice of access system for coal water slurry production enterprises and coal water slurry boiler equipment. All coal water slurry boiler equipment entering Nanning City must be produced by enterprises with coal water slurry boiler production qualifications as prescribed by the State and enjoy good reputation among demonstration enterprises. Centralized heating by coal water slurry: Nanning —ASEAN Economic Development Zone.	Operation is monopolized by large enterprises; coal water slurry combustion technology is not mature; reconstruction costs are high; operation effect is not good.
Coal water slurry production capacity and market consumption	Coal water slurry production capacity is 400,000t/year; actual market consumption is about 150,000t/year.	Phase I coal water slurry production capacity is one million tons/year; actual market consumption is about 200,000t/year.
Implementation effect	Coal-fired boilers below 4t/h in enterprises with heavy pollution and failure to reach emission standard within urban ring road, and used not for the purpose of industrial production within urban high-speed ring road, and used in industrial enterprises on both sides of urban roads were replaced in a centralized way from 2007 to 2010. Special fund worth RMB 6 million is allocated from municipal finance every year and is specially used for popularization and application of coal water slurry.	Currently, only six enterprises have applied for pilot application, four among them satisfy acceptance conditions. According to incomplete statistics, ten enterprises have used coal water slurry in Dongguan City in three years. At present, it seems that popularization of coal water slurry is one-sidedly driven by the government, while many enterprises have not yet formed a scale. There are two coal water slurry producers in Dongguan City, among which one produces coal water slurry for its own use and the other shifts to produce natural gas.

Source: *Nanfang Daily*, September 7, 2011.

(8) Construction plan for Zhuzhou Ecological Washing Industry Park

This project is a key project in the Zhuzhou 11th Five-Year Plan for Environmental Protection; it is based on engineering experience from Longquan Washing Industry Park and incorporates integrated technology for biomass coal water slurry preparation and combustion system to provide standard workshops and steam for washing enterprises in this Park; it was planned to take three years to build an ecological washing industry park for sewage-based preparation of slurry and centralized heat supply where 25 enterprises are located (80 enterprises in the long term). 500,000t/year coal water slurry preparation, a 3×25t/h coal water slurry fired boiler and 52.5mW steam supply are available.

(9) Construction plan for new energy distribution center for high-concentration wastewater-based slurry preparation, sludge co-combustion and cogeneration in Qingshuitang, Zhuzhou City

Based on the guiding concept of wastewater slurry, sludge co-combustion, cogeneration, energy conservation and environmental protection, circular demonstration, adopt an integrated technology for biomass coal water slurry preparation and combustion system, and carry out unified planning, system integration, unified construction, unified management and market-oriented operation mode, and unify high-concentration organic wastewater and ammonia-nitrogen wastewater treatment, sludge treatment and energy distribution by means of a biomass coal water slurry system integration system so as to make resource and energy-oriented utilization of wastewater (sludge); build a new energy distribution center for high-concentration wastewater-based slurry preparation, sludge co-combustion and cogeneration.

5.4 Prospect for Development of Integrated Technology for Efficient Coal Water Slurry Preparation and Combustion

Integrated technology for efficient coal water slurry preparation and combustion originated from Huamei Nanfang Coal Water Slurry Co., Ltd. (former Zhuzhou Coal Washery), one of the four largest coal water slurry centers at the initial stage of coal water slurry technology research and development in China, and developed at the Lanyu Thermal Energy Technology Development Co., Ltd. established by scientific and technical personnel, and has developed internationally advanced, leading domestic integrated technology for coal water slurry preparation and a combustion system with independently innovative intellectual property, which has been in operation for nearly 20 years and has been highly recognized by the market and energy conservation emission reduction promotion planners and executors in terms of operation reliability, business operation mode innovation, energy conservation benefits and environmental protection benefits. In the current general international environment for promoting energy conservation, emission reductions and actively addressing climate change, integrated technology for efficient coal water slurry preparation and combustion will enjoy broad development prospects based on the following several aspects:

1. Energy contract management services based on integrated technology for efficient coal water slurry preparation and combustion

Fully utilize technical reliability, energy conservation and the environmental protection performance of integrated technology for efficient coal water slurry preparation and combustion to perform energy contract management services.

2. Distributed system with multi-energy complementation for combined supply of heat, electricity, cold energy

Coal water slurry serves as a leading clean energy, while renewable energies including solar energy, wind energy and terrestrial heat act as supplementary energy. With respect to the operation mode, use of renewable energies is maximized to maximize energy conservation, carbon emission reduction benefits.

3. Thermoelectric system for energy and resource-oriented utilization of sewage and sludge

Characteristics of water for coal water slurry preparation are utilized, and sewage-based slurry preparation, co-combustion of sludge and biomass coal water slurry are achieved to enable energy and resource-oriented utilization of sewage and sludge and realize unification and coordination of energy environment goals.

4. Thermoelectric system for disposal of combustible solid wastes such as domestic garbage, etc.

Coal water slurry serves as auxiliary energy for the ecological disposal of garbage and is conducive to energy and resource-oriented utilization of domestic garbage.

5. Ignition system for inorganic mineral roasting

Coal water slurry replaces heavy oil and natural gas for sintering ignition, which is designed to reduce production costs with the condition of guaranteeing environmental protection benefits.

6. Ecological construction of energy environment integration system for an industrial park

With respect to industry parks relating to food processing and biological medicine, integrated technology for efficient coal water slurry preparation and combustion is adopted to integrate high concentration organic wastewater and biomass solid wastes generated by manufacturing enterprises in the industrial park and coal water slurry preparation and combustion process, and enable integrated energy environment treatment within the industrial park, circularly utilizing resources and building an ecologically circular industry chain.

References

[1] Representative Office of United Nations Development Programme in China, *China Human Development Report 2009/10——March Toward Sustainable Future of Low-Carbon Economy and Society*, China Publishing Group Corp., China Translation & Publishing Corporation, 2010.

[2] Jia Chuankai: *Application of Coal Water Slurry Fired Boiler and Energy Conservation, Emission Reduction Analysis, Clean Coal Technology*, 2011(3).

[3] Jia Chuankai, Wang Yanfang, Wang Xiuyue et al. : *Application and Development Trends of Coal Water Slurry Technology, Coal Processing & Comprehensive Utilization*, 2011(4).

[4] He Xinwen: *Brief Discussion of Application of Double-Furnace Coal Water Slurry Fired Boiler, Friend of Science Amateurs*, 2011(9).

Chapter 6

Applied Research on Urban Low-Carbon Transportation Technologies in China

Zhang Taoxin Xie Shixiong Yangying

Abstract: This article analyzes the application status and trend in low-carbon technology for urban road traffic. The results showed the following: 1) Not all new energy vehicles have an effect on carbon emissions reduction in China's current coal-based power structure. The ideal new energy vehicle is the compressed or liquefied natural gas vehicle over the mid-term; the alcohol ether vehicle is not suitable for big development, the electric vehicle will increase carbon emissions in China's current energy structure, but with the transformation of China's energy structure and the development of carbon capture as well as storage technology, the vehicle with the greatest potential for reducing carbon emissions is the electric vehicle, and should be regarded as the direction for the development of new energy vehicles. 2) In order to maximize the carbon emissions reduction of urban road traffic, the focus is on the integrated application of energy saving technology in motor vehicles at the current time, paying attention to urban road technology, and achieving green driving as well as smart management of urban transport. Based on the research, three strategic choices and five main measures for the low-carbon technology development of China's urban road traffic are proposed.

Key word: Urban; low-carbon transport; low-carbon technology; strategies; measures

How to curb increasing carbon emissions in a more effective way has become the focus of attention from countries in the world. In 2009, vehicle fuel consumption in China accounted for nearly 1/3 of total apparent oil consumption; carbon emissions in the automotive fuel lifecycle reached 0.48 billion tons carbon dioxide equivalent[1] and were 25% of total carbon dioxide emissions of 1,918 million tons generated by fossil fuel consumption in China in 2008[2]; road transport has become one of the main sources of carbon emissions in China. With the acceleration of urbanization, carbon emissions resulting from energy consumption in urban road transport will grow more quickly. Carbon emissions from urban road transport have a direct bearing on the smooth realization of China's energy conservation and emission reduction goals and the development of China's low-carbon economy.

Science and technology are the primary productive forces; technological progress factors are leading factors for reducing carbon emissions[3]; in 2006, Energy Technology Perspectives, the International Energy Agency (IEA) stated through scenario analysis that with the impact from key energy technologies, global carbon emissions can return to the current level by 2050[4]. Therefore, in order to promote urban low-carbon transport construction and reduce carbon emissions from urban transport in China, it is necessary to give top priority to low-carbon technologies, analyze and research low-carbon technologies in the urban transport field to identify the direction for developing low-carbon urban transport technologies.

6.1 Current Situation of Carbon Emissions from Urban Transport

6.1.1 Civil automobile

The number of private cars in China increased from 16,089,100 in 2000 to 78.02 million (excluding three-wheeled automobiles and low-speed trucks) in 2010, up 3.85 times, with a compound average growth rate of 21.82%. According to Fig. 6-1, China's automobile output increased from 2.07 million to 18.27 million during 2000-2010, up nearly 7.83 times, with a compound average growth rate of 31.28%. In 2010, China's automobile output accounted for 23.5% of the world's output and ranked first in the world.

In 2008, every 1,000 persons had 593 and 809 cars in Japan and USA respectively, while such a figure was only 37 in China; thus the opportunity for increasing the number of cars in China is very huge.

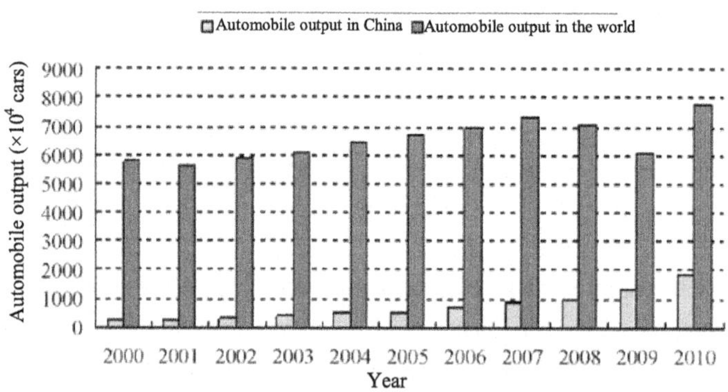

Fig. 6-1 Automobile output in China and the world during 2000-2010

Source: Data from China Association of Automobile Manufacturers (http://www.caam.org.cn/newslist/a40-1.html).

6.1.2 Energy consumption and carbon emissions from urban road transport

According to Fig. 6-2, fossil energy consumption in China's road transport increased annually by 11.99% during 2000-2008, which was 3.11 percentage points higher than the annual average growth rate of China's fossil energy consumption in the same period; carbon emissions from urban transport grew annually by 12.62%, which was 3.13 percentage points higher than the annual average growth rate of China's total carbon emissions in the same period. Urban road transport has become one of the fields with the fastest growth in fossil energy consumption and carbon emissions in China.

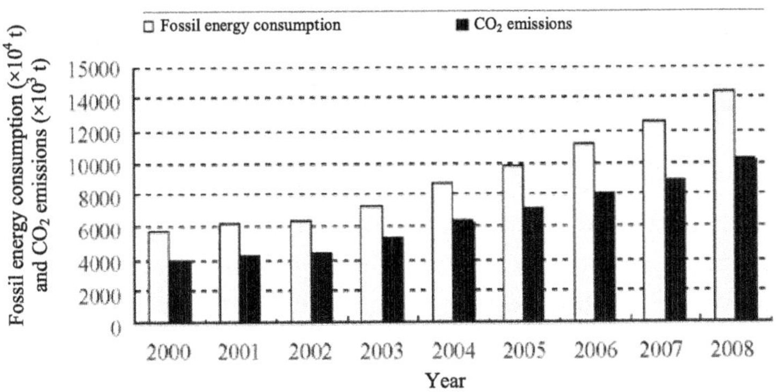

Fig. 6-2 Fossil energy consumption and CO_2 emissions from urban roads during 2000-2008
Source: Data from the World Bank and Carbon Dioxide Information Analysis Center (CDIAC).

Energy consumption and carbon emissions from roads dominate energy consumption and carbon emissions from the transport industry; energy consumption from road transport accounted for 81.2% of energy consumption for all transport in the USA in 2002, while carbon emissions from the transport sector made up 32.3% of total carbon emissions in the USA, almost all of which came from various fossil fuels for vehicles; energy consumption for various types of road transport also accounted for 87% of energy consumption for the whole transport sector in Japan in 2005, while carbon emissions from transport made up 22.4% of total carbon emissions in Japan, 96% of which came from various passenger vehicles[5]. Energy consumption for urban road transport accounted for 5.34% of total energy consumption and 62.69% of energy consumption for transport, warehousing and postal industries in China in 2008; though it was lower than that in the USA and Japan, urban road transport constituted the main consumer in the transport industry. We should adopt various means including low-carbon technologies to prevent, as far as possible, the situation in which road transport in developed countries can not only promote economic and social development, bring convenience to people's lives, but also exert a negative impact on climate change.

6.2 Research on Low-Carbon Technology for Urban Transport

Low-carbon technology refers to the technology for effectively controlling carbon emissions and making people reduce carbon emissions during production and daily living. Generally, low-carbon technology is divided into the following three categories:

Carbon reduction technology: it refers to the technology for saving energy and improving energy efficiency in high energy consumption and high emission fields, including various advanced energy use technologies and complementary energy recycling technology, in various sectors including industry, transport and building, such as an integrated gasification combined cycle and smart power grid, etc.

Carbon-free technology: it refers to the technology for renewable energy such as solar energy, wind energy, hydroenergy, biomass energy.

Carbon removal technology: such as carbon dioxide capturing and sequestration technology

Low-carbon technology for urban road transport means application of low-carbon technology in the urban transport field. Urban transport includes road transport, rail transport, waterway transport and air transport; the following discussions focus on new energy and new energy automobile technologies, motor vehicle energy conservation technologies, road technologies, intelligent management techniques relating to road transport.

6.2.1 New energy automobile technology

1) Analysis of carbon emissions during the full lifecycle of new energy automobiles

(1) Pure electric vehicle

A pure electric vehicle refers to a vehicle powered by a vehicle-mounted power supply and whose wheels are driven by the motor. A pure electric vehicle is the result of integration based on the rapid development of new advanced technologies including modern automobile technology, electrochemistry, new materials, new energy, microelectronics, electric drive technology, electronic computer intelligent control.

The motor generator and vehicle-mounted battery are key components, among which the battery is the most critical one; a common battery in an electric vehicle mainly includes a lead acid battery, nickel-cadmium battery, hydrogen-nickel battery and lithium battery. Currently, energy stored per unit weight of a storage battery is very little, the post-charging mileage is not ideal; the service life of a high-capacity battery is relatively short and its usage cost is high, thus the commercial operation of a high-capacity battery is restricted and it can only be used for some limited distances. In China, pure electric vehicle technology based on a lead-acid storage battery has become relatively mature, while most of the other technologies are at the initial stage.

A pure electric vehicle dispenses with the internal combustion engine as power equipment and features a simple basic structure; it basically does not cause CO_2

emissions during operation and is characterized by non-pollution, low noise, high energy efficiency and easy maintenance. As a vehicle-mounted power supply for a pure electric vehicle, electric energy is derived from extensive sources and is inexhaustible, coming from either fossil energy such as coal or renewable energy such as hydroenergy, wind energy, solar energy and thermal energy.

Compared with conventional vehicles, an electric vehicle relies more on the electrical energy production structure in a country, even it is also essentially different between different regions in the same country. In China, extraction of 1t of coal from a coalmine causes the emission of approximately $6m^3$ CO_2 on average.[6] From the perspective of a full lifecycle, if CO_2 from coal extraction is not considered, take 2008 for example, CO_2 emission/1,000m from an electric vehicle would be 5%-57% less than that from a conventional gasoline car in China, except that the 320EV-Lifan electric vehicle which used electricity from North China Power Grid and Northeast China Power Grid would cause more carbon emissions than a conventional gasoline car. However, if CO_2 from coal extraction is considered, take 2007 for example, CO_2 emission/1,000m from an electric vehicle would be 1.8-6.4 times more than that from a conventional gasoline car in China (for specific calculation, see Appendix). Therefore, given the current coal power structure and technical level in China, it is difficult for the current popularization and use of electric vehicles to effectively reduce carbon emissions.

As CO_2 emissions from electric vehicles mainly come from extraction, processing and transportation of vehicle fuels and vehicle manufacturing, CO_2 capturing and sequestration technology may reduce CO_2 emissions from extraction, processing and the transportation of vehicle fuels by nearly 80%[7]. China abounds in renewable energy resources and enjoys resource conditions and the technical potential for large-scale development. Therefore, the development of a pure electric vehicle can not only help China's automobile industry break away from excessive dependence on oil to a large extent, but also brings about a very big potential for reducing carbon emissions.

Though a pure electric vehicle is completely free from fuel expenses, the full lifecycle cost of a pure electric vehicle is higher than that of a petroleum fuelled car; factors which affect the full lifecycle cost of a pure electric vehicle mainly include purchase cost and battery cost; in addition, underdevelopment of infrastructure (such as public charging equipment) and lack of supporting facilities cause inconvenience. Therefore, the key to improving the economical efficiency of a pure electric vehicle lies in developing more low-cost and durable batteries to reduce the full vehicle production cost of a pure electric vehicle so as to reduce its market price and battery cost within its service life and promote the development of the pure electric vehicle market.

(2) Hybrid electric vehicle

Hybrid electric vehicle (HEV) refers to the vehicle in which more than two power sources are installed and can work in a coordinated way. Its key technology is a hybrid power system. According to different power system structures, it can be divided into a series hybrid electric vehicle (SHEV), parallel hybrid electric vehicle (PHEV), parallel-series hybrid electric vehicle (PSHEV).

A hybrid electric vehicle entails less modification of existing automobile manufacturing technology and social infrastructure and enjoys better fuel economy; generally, a hybrid electric vehicle can save fuel oil by 30%-50% during operation compared with a conventionally fuelled car. The result of a hybrid bus operation test in Jinan City in 2010 showed that a hybrid electric vehicle can save about 13.5L oil/100km compared with an ordinary bus and the fuel saving rate reached 30%[8]. A hybrid electric vehicle can reduce total CO_2 emissions by 30% during operation compared with a conventionally fuelled car[9]. In China, hybrid electric vehicle technologies have rapidly developed; some models have entered the technologically mature phase; major technical difficulties for the hybrid electric vehicle at the current stage include battery technology, motor technology, internal combustion engine technology and full vehicle energy management techniques. Currently, the hybrid electric vehicle is still relatively inefficient and also uses more gasoline or diesel oil. With continuous breakthroughs in car battery technology, especially lithium battery technology, the hybrid electric vehicle will develop in the direction of a pure electric vehicle.

As one transitional technical proposal, a plug-in hybrid electric vehicle (PHEV) enjoys the advantages of both a conventional hybrid electric vehicle and pure electric vehicle and a technical advantage for fully utilizing electric energy to reduce conventional fossil fuel consumption; thus it is one kind of feasible clean, energy-saving, easy-to-use vehicle at the current stage and has become a current hotspot for research, development and popularization in the electric vehicle field[10].

(3) Fuel cell vehicle

A fuel cell vehicle is a new type of vehicle in which a fuel cell serves as power source and chemical energy in the fuel is directly converted into electric energy for the power drive; a fuel cell vehicle mainly includes a hydrogen fuel cell vehicle, biofuel (grain/non-grain) fuel vehicle. Its biggest difference from a hybrid electric vehicle and pure electric vehicle lies in completely dispensing with the combustion process and instead combining hydrogen with oxygen through an electrochemical method to directly generate electricity and heat. Oxygen is extensively available in water, mineral fuels and various carbohydrates; fuels used in a fuel cell vehicle come from diverse sources including natural gas, methanol, propane, gasoline, diesel oil, coal, coal bed gas and solar energy, wind energy, biomass energy, ocean energy, terrestrial heat and other renewable energy.

Among current various technical proposals for hydrogen production, the technical proposal for conducting water electrolysis of energies dominated by fossil fuels such as coal to obtain hydrogen causes more carbon emissions than gasoline vehicles and is also significantly worse than the technical proposal for obtaining hydrogen through methanol reforming and gasoline reforming in terms of carbon emissions, while the technical proposal for methanol reforming and gasoline reforming is worse than the proposal for obtaining hydrogen from coal, and the proposal for obtaining hydrogen from coal is worse than the proposal for obtaining hydrogen from natural gas. For the same technical proposal for hydrogen production, the liquid hydrogen proposal causes more carbon emissions than the gaseous hydrogen proposal[11]. As a primary energy for developing a hydrogen source for a fuel cell vehicle, natural gas enjoys many advantages in diverse technical routes for hydrogen production, high economic competiveness, high energy utilization efficiency and good environmental benefits, and it should be currently the preferred energy for hydrogen production for fuel cell vehicles[12]; if solar energy, wind energy, biomass energy, ocean energy, terrestrial heat and nuclear energy can also be used as energy for hydrogen production, carbon emissions will be greatly reduced.

In addition, a fuel cell vehicle is also characterized by non-pollution, high energy efficiency, low noise, good power and control system. Thus it is also one kind of ideal vehicle from the perspective of energy utilization and environmental protection. However, the fuel cell cost is high, and fuel preparation, transport and storage consume lots of energy, and preparation technology has not yet become mature, and the manufacturing cost is high. Hydrogen fuel injection equipment also needs to be specially manufactured. Currently, the fuel cell vehicle is still at the stage of research and trial use, and its technical development has just started in China.

(4) Hydrogen engine vehicle

A hydrogen engine results from some adaptative renovations of the ordinary internal combustion engine, while a hydrogen engine vehicle is the vehicle which obtains power by co-combustion of hydrogen (or other auxiliary fuels) and air for generating energy; it is one relatively good technical solution for better using the huge stock of assets currently available in the automobile industry and gradually shifting from conventional vehicles to new energy vehicles; there is no essential difference between a hydrogen engine and ordinary internal combustion engine. Both the hydrogen engine vehicle and fuel cell vehicle use hydrogen, the difference between them lies in the hydrogen fuel use mode: the former involves direct combustion of hydrogen for generating kinetic energy to make the engine run and drive the vehicle, while the latter involves the reaction of hydrogen with oxygen within the fuel cell for generating electric energy which drives the vehicle.

Besides non-pollution and low emissions, the hydrogen engine vehicle also enjoys some unique advantages such as relatively low requirements for hydrogen, high combustion performance and mature internal combustion engine technology. Compared with a conventional energy vehicle, a vehicle which uses hydrogen as engine fuel can achieve an energy conversion rate higher than 40%, while noise is low and the driving range is equivalent to that of a gasoline vehicle. Meanwhile, for economic considerations, hydrogen comes from extensive sources since electric energy, wind energy, solar energy, hydro energy, geothermal energy and nuclear energy can be converted into hydrogen. Merely a slight renovation of a gasoline engine can enable hydrogen combustion. Hydrogen engine vehicle technology generally remains at the initial stage, and the gap between China and developed countries in the world in hydrogen engine vehicle technology is much smaller than that in the conventional automobile industry. Like a fuel cell vehicle, construction of public facilities for hydrogen production, storage and injection is also a technical bottleneck, which restricts wide application of hydrogen in automobiles. Future prospects for hydrogen application in automobiles hinges on a breakthrough in hydrogen production and transportation technology.

(5) Alcohol ether vehicle

An alcohol ether vehicle refers to a vehicle in which methanol gasoline, ethanol gasoline, methanol and ethanol serve as fuels. Alcohol ether vehicle technology is relatively mature; after modification, a conventional internal combustion engine can adapt to different ethanol gasoline fuels. An ethanol vehicle is rapidly developing in countries with abundant ethanol resources such as the USA and Brazil; vehicle fuels with different ethanol proportions of 10%, 22% and 85% were popularized and applied in the USA and Brazil. According to relevant tests in the USA, Denmark, Japan and Austria, when dimethyl ether serves as vehicle fuel, waste gas pollution is much lower than that from currently used diesel oil, and a compression ignition engine can reach European III emission standards without any after treatment measures. Alcohol ether vehicle technology is at the initial stage in China.

Compared with conventional diesel oil production with crude oil as raw material, coal-based dimethyl ether production causes relatively high power consumption, which is 2.5 times that in conventional diesel oil production; however, coal-based dimethyl ether significantly lowers standard emissions and carbon emissions during use in vehicles. From the perspective of a full lifecycle, carbon emissions from an alcohol ether vehicle with coal-based dimethyl ether as fuel are higher than in the case of conventional diesel oil, which is mainly attributable to the fact that CO conversion during production and relatively high coal consumption cause more carbon emissions. As energy conversion efficiency in coal-based dimethyl ether production is enhanced to lower energy consumption, and solar energy, wind energy, biomass energy, ocean energy and terrestrial heat are used for coal-based dimethyl ether production, and low-carbon technologies such as coal bed gas power generation and carbon capture are comprehensively utilized, carbon emissions from an alcohol ether vehicle with coal-based dimethyl ether as fuel will be greatly reduced.

From the perspective of a full lifecycle, CO_2 emissions from an alcohol ether vehicle fueled by dimethyl ether made from natural gas are lower than that from conventional vehicles, while an alcohol ether vehicle fueled by dimethyl ether made from biomass can significantly lower CO_2 emissions[13]. China has become the third largest biofuel ethanol producer and user in the world, only second to Brazil and the USA. The direction for the future development of China's fuel ethanol industry is large-scale utilization of non-grain ethanol. Therefore, the key to determining the prospect for future development of fuel ethanol lies in costs and technologies. Currently, alcohol ether vehicle fuel preparation costs are relatively high, and it is difficult to effectively reduce CO_2 emissions. It is only used as a supplementary technical solution for replacing diesel vehicles in regions with abundant alcohol ether.

(6) Natural gas vehicle

A natural gas vehicle refers to a vehicle with natural gas as fuel and can be divided into compressed natural gas vehicle (CNG), liquefied natural gas vehicle (LNG) and liquefied petroleum gas vehicle (LPG). A natural gas vehicle only entails some necessary modifications of a convention internal combustion engine for adapting to natural gas fuel. Compared with a conventional fuel vehicle, a liquefied petroleum gas vehicle can reduce CO_2 emissions by 20%, while a compressed natural gas vehicle and liquefied natural gas vehicle can lower total CO_2 emissions by 25%[9]. In addition, the total waste gas emission from a natural gas vehicle is less than 10% that from a conventional vehicle, and fuel gas contains no lead, benzene or sulphur and is characterized by low pollution and a high degree of safety.

Natural gas vehicle technologies have become increasingly mature, widely developed and utilized. A natural gas vehicle was predominately used as bus and taxicab in the past and has gradually been extended to medium and heavy-duty trucks, goods vans, municipal garbage transport vehicles, and engine emissions performance has also reached relatively high requirements. Currently, overall technology for natural gas vehicles in China lags behind the world's advanced level by a relatively large margin, but the natural gas vehicle is rapidly developing in China. China enjoys a larger proportion of natural gas resources than that of petroleum, which is a very favorable condition for developing natural gas vehicles. Power performance of natural gas vehicles is relatively low, and the driving distance is short. It is not easy to carry the gas, and large-scale use requires the establishment of the corresponding gas stations and pipelines for delivering the natural gas and involves many factors including urban construction planning, funds, environment and safety. Infrastructure investments are massive and costs are relatively high, which have become bottlenecks for the development of natural gas vehicles in China to some extent.

(7) Solar-powered vehicle

A solar-powered vehicle refers to a vehicle in which the vehicle body is directly used to convert solar energy into electric energy as a power source. Though solar energy is pollution-free inexhaustible renewable energy, solar energy must be stored in a storage battery; it is difficult to make technical breakthroughs in solar batteries for vehicles, thus its price is exorbitantly high. No great progress will be made in solar-powered vehicles and it is difficult for its popularization and application in the next 10-20 years. Solar-powered vehicles may become an important vehicle type in the long term.

2) New energy vehicle technology development route

With the growth of motor vehicles and increasing shortage of oil resources, new energy vehicle technologies will be gradually developed and utilized. Different new energy vehicle technologies are at different development stages and each new energy vehicle technology has different carbon emission characteristics; Table 6-1 shows carbon emissions, economic efficiency and technical levels of several new energy vehicles based on a full lifecycle (compared with fossil energy). Therefore, a new energy vehicle development route should be determined according to local energy resource conditions.

Table 6-1 Carbon emission, economic efficiency and technical levels of new energy vehicles based on full lifecycle

Vehicle type	Using energy types	Carbon emissions	Economy	Technical level
Pure electric vehicle	Coal-Fired Power	high	Poor	Development period
	Electricity generation by renewable energy	extremely low		
Hybrid electric vehicle	Mixture of coal power and fossil energy	relatively high	Very poor	Mature period
	Production of mixture of electricity and fossil energy by renewable energy	low		
Fuel cell vehicle	Hydrogen production by energy dominated by fossil fuels such as coal	high	Very poor	Initial period
	Hydrogen production by electricity generated by renewable energy	Very low		
Hydrogen engine vehicle	Hydrogen production by energy dominated by fossil fuels such as coal	high	Poor	Initial period
	Hydrogen production by electricity generated by renewable energy	Very low		
Alcohol ether vehicle	Alcohol production by biomass on the basis of energy dominated by coal power	relatively high	Relatively poor	Mature period
	Alcohol production by biomass on the basis of renewable energy	relatively low		
	Ether or alcohol production from coal as raw material	high	Relatively poor	Initial period
	Ether production based on other raw materials	low		
Natural gas vehicle	Compressed natural gas	low	Poor	Mature period
	Liquefied natural gas	low	Poor	Development period
	Liquefied petroleum gas	relatively low	Poor	Development period

According to Table 6-1, if the power supply for an electric vehicle (pure electric vehicle and hybrid electric vehicle) comes from coal power, from the perspective of a full lifecycle, carbon emissions from both a pure electric vehicle and hydrogen engine vehicle are not less than from a conventional vehicle. It is difficult to change, in the short term, the situation in which China's power mainly comes from coal; thus application of electric vehicles cannot significantly reduce carbon emissions in the short term. Once electric energy for electric vehicles mainly comes from hydroenergy, wind energy and solar energy and other renewable energy, and power generated by renewable energy is used to produce hydrogen, for the full lifecycle, except carbon emissions at the vehicle manufacturing and scrapping stages, very little carbon emissions are caused by a hydrogen engine vehicle and pure electric vehicle at other stages. Especially when battery technology becomes mature, a pure electric vehicle will hardly cause carbon emissions. Therefore, a pure electric vehicle should become the direction for developing a new energy vehicle; however, it is inappropriate to make it the focus of medium and long-term development from both a technical perspective and from a carbon emission perspective.

As shown in Table 6-1, from the perspective of a full lifecycle, a natural gas vehicle produces better effects on carbon emission reduction, and its technologies are also relatively mature. China has more natural gas reserves than oil reserves, but per-capita reserves are relatively low and they are non-renewable energy resources. Thus a compressed natural gas or liquefied natural gas vehicle can serve as an ideal new energy vehicle in the near and medium terms.

Fuel for a liquefied petroleum gas vehicle is a by-product of oil exploitation or processing, and cannot serve as a stable fuel source. Its economical efficiency is relatively poor, and its carbon emission reduction cannot also be comparable with other natural gas vehicles; therefore, liquefied petroleum gas is only a makeshift solution in the short term.

According to Table 6-1, if alcohol ether used in an alcohol ether vehicle comes from energy dominated by coal power or is extracted from natural gas or coal, from the perspective of the full lifecycle the alcohol ether vehicle cannot effectively reduce carbon emissions. The development of bioethanol liquid fuel is subject to issues concerning the raw material source; the preparation of ethanol from plants consumes a good deal of energy and water and increases water pollution; given shortages of per-capita water resources and cultivated land resources in China, it is inappropriate to vigorously develop an alcohol ether vehicle.

Though currently a hybrid electric vehicle cannot significantly reduce carbon emissions from the perspective of a full lifecycle, as a transition to a pure electric vehicle, it is a better technical choice in the short and medium term.

Therefore, the new energy vehicle development route focus on concurrent development of multiple new energy vehicle technologies and development towards a pure electric vehicle.

6.2.2 Motor vehicle energy conservation technology

1) Green driving technology

Driving energy conservation means that a good driving habit is developed to reduce fuel oil consumption during use so as to make the most economical, most efficient trip with minimum fuel oil consumption and carbon emission. The driving skill of a driver exerts a great impact on fuel consumption. In the case of driving the same car under the same conditions, oil consumption differences caused by different operations of the driver can reach 20%-40% and even more; thus the potential for driving energy conservation is huge[14].

The Ministry of Transport drew upon well-proven experience in energy-saving driving at home and abroad to prepare an *Energy-Saving Driving Manual* which specifies eight links in operation techniques, ten rational use habits and correct maintenance, to provide easy-to-operate feasible green driving methods for drivers. New energy vehicle technologies can tremendously reduce carbon emissions; however, mature application of new energy vehicle technologies requires a long arduous process and cannot be achieved overnight, while drivers should foster a green driving philosophy since saving even one drop of fuel helps make a little reduction in carbon emissions. Green driving technology is feasible without input; assuming that existing automotive technologies and equipment are unchanged, even a 30% fuel saving can also produce a very good carbon emission reduction effect.

2) Driving efficiency improvement technology

(1) Driving resistance reduction technology

Driving resistance on the road mainly comes from air and rolling friction. Technology for local optimization design of the vehicle body, technology for overall optimization of appearance and technology for improving the body surface quality can effectively mitigate air resistance. Each 10% reduction in air resistance can lower fuel consumption by 0.15L/100km[15] so that carbon emissions from the vehicle can be effectively decreased.

Low rolling resistance tire technology can reduce rolling frictional resistance to driving. Overall, a low rolling resistance tire can decrease vehicle fuel consumption by 0.2L/100km and CO_2 emissions by 4g/1km[16]. A radial tire is currently preferable to a low rolling resistance tire since its abrasion performance is 30%-70% higher than that of an ordinary bias tire and rolling resistance decreases by 20%-30%, while vehicle fuel consumption can fall by 5%-8%[17]. If the pressure of a new type vehicle tire is increased, fuel consumption can be further reduced and carbon emissions from the vehicle can be further lowered.

(2) Vehicle lightweight technology

Vehicle lightweight technology is designed to integrate lightweight design technology, lightweight materials and lightweight manufacturing technology under the condition of meeting vehicle use requirements and controlling costs to make the vehicle light. Fuel consumption in a vehicle is directly related to vehicle weight and volume, and vehicle weight has the greatest impact on fuel consumption; 37% of fuel saving in a vehicle relies on reduction of vehicle weight[18]; if the total vehicle weight is reduced by 10%, fuel consumption can be reduced by about 8%[17], leading to decreasing carbon emissions from the vehicle. Vehicle components and relevant parts are optimally designed, and high-strength light materials such as high-strength steel, aluminum-magnesium-titanium alloy, plastics, high-ductility aluminum alloy plate, various fiber reinforced materials are chosen, and new manufacturing technologies including laser welding of tailored blanks, internal high-pressure forming, thermal forming of high-strength steel and high-strength steel rolling and pressing are adopted to manufacture vehicle parts so as to effectively enable a full vehicle to become lightweight.

3) Energy-saving technology for vehicle engine operation

Modification of a conventional vehicle engine for improving engine performance can increase the fuel utilization rate, reduce energy consumption and carbon emissions.

(1) Variable cylinder management technology for oil saving: This technology is used to make part of the cylinders always work and another part of the cylinders work under high load and stop under low load so that fuel can be saved by 10%-20%[15].

(2) Lean combustion technology: This technology is employed to deliver excessive air to stably, fully burn fuel in a gasoline engine, which can enhance effective thermal efficiency by about 30% [19]so as to save energy and reduce emissions.

(3) Gasoline direct injection technology: A gasoline direct injection engine is operated according to engine load conditions to enable stratified lean combustion in the case of low load and homogeneous combustion in the case of high load. A gasoline direct injection engine breaks away from conventional gasoline engine injection mode and is simple to operate; it electronically and accurately controls fuel injection within the cylinder so that every drop of fuel can be completely burnt, which not only reduces fuel consumption and carbon emissions, but also outputs higher torque and power to provide the vehicle with swift and strong power. Overall, gasoline direct injection technology can save energy by 2%-3%; though currently the extent of its application in China is very low, it is a very promising development direction for a conventional gasoline engine[16].

(4) Turbocharging technology: Vehicle engine air inflow is increased to improve engine power, mechanical efficiency and thermal efficiency so that the fuel consumption rate after engine turbocharging decreases by 5%-10%[20]; for a small passenger vehicle, the energy conservation effect can reach 4.2%-4.8%[16]. Turbocharging technology is an important energy-saving technology for vehicle engine operation and has just been preliminarily applied in China. It should be energetically popularized and applied in vehicles.

(5) Variable valve technology: Variable valve technology can be realized by various means, while all of these means can improve gasoline engine fuel economy and power and reduce carbon emissions to varying degrees. There are a series of relatively practical variable valve mechanisms in foreign countries[21]; currently the most widely used one is the vane type variable cam phasing mechanism. Variable valve technology can save energy by 2%-3%; compared with foreign countries, variable valve technology is seldom applied in China.

(6) Start-stop technology: This technology means that when a vehicle stops advancing and the engine is in idling condition, if safety is guaranteed the engine is stopped to reduce fuel consumption at the time of vehicle idling so as to save fuel and lower emissions. Vehicle start and stop frequently occurs amidst urban traffic congestion, so start-stop technology produces a very obvious fuel saving effect. When the vehicle with a start-stop function runs on the road, fuel saving reaches 7.6% and the overall fuel saving effect is 3% in the case of a warmed up engine; the fuel saving reaches 7% and

overall fuel saving effect is 4% [22]in the case of a cold engine. Start-stop technology is more applied in the European auto market and is expected to become a standard part of Chinese passenger vehicles.

(7) Crankshaft integrated starter generator technology: Vehicle starter and generator are integrated, and a generator with relatively high transient power directly replaces a conventional starting dynamo. The energy saving effect can reach 8.6%-8.9%[16]; crankshaft integrated starter generator technology is one low-cost energy-saving technology between hybrid electric vehicles and conventional vehicles and is worthy of popularization and application.

(8) Automatic transmission technology: This technology can free the driver from frequently using the clutch pedal as is the case when driving a manual transmission vehicle, which saves fuel and is easy to operate, convenient, stable and comfortable. Currently, internationally, manual transmission mainly includes a hydraulic mechanical automatic transmission (AT), automated mechanical transmission (AMT), mechanical continuously variable transmission (CVT), infinitely variable transmission (IVT), dual clutch transmission (DCT) [23]while its key technologies are electronic technology, electric hydraulic control technology and sensing technology. Overall, a 6-speed AT can produce a 1.4%-3.4% energy saving effect, while CVT and DCT can save energy by 4%[16]. Currently, most of the passenger vehicles in the USA and Japan are equipped with automatic transmissions, while in China very few automatic transmissions are available in passenger vehicles, so there is a very big space for development.

(9) Electric power steering technology (EPS): EPS is one power steering technology for directly relying on the motor to provide auxiliary torque and it can save energy by 1%-2%[16]. 19% of cars were provided with EPS in the Chinese auto market in 2008[24]; replacement of the hydraulic steering system by EPS will become a trend.

4) Vehicle energy recovery technology

Relatively mature vehicle energy recovery technology is the use of a flywheel for recovering energy, which is expected to reduce fuel consumption by 20%[25]. A flywheel is lightweight and is available at a low price, and it can efficiently save energy and is most effective for running with frequent stops and starts; therefore, it can produce maximum energy conservation and emission reduction effects in the congested urban transport environment.

Overall, each energy-saving technology can bring about certain energy conservation and emission reduction effects; however, in order to maximize vehicle energy conservation and emission reduction effects, it is necessary for integrated application of individual technologies. For example, General Motors Corporation carries out an integrated application of in-cylinder direct injection technology, direction injection turbocharging, homogeneous charge compression ignition and an intelligent fuel management system into the corresponding products under various brands, which improves fuel economy and reduces carbon emissions on the basis of guaranteeing power. Taking another example, Volkswagen integrates series technologies capable of significantly lowering fuel consumption and carbon emissions such as start-stop technology and regenerative braking energy recovery technology. into the "BlueMotion Technology Series" and has started the application in mass-produced vehicles.

6.2.3 Road Technologies

1) Slow road transport design

Slow traffic is one kind of green low-carbon traffic for sustainable development, which not only eases traffic jams, reduces environmental pollution but also promotes rational utilization of resources.

(1) Bicycle lanes should be arranged on all roads (except low-speed local roads), and should be at least 6m wide, and safe bicycle parking space should be built near buildings, roads and stations.

(2) Special slow network roads (only walking, bicycles and public transport are permitted) should be built in urban areas, and the interval between two special slow roads should not exceed 800m.

2) Microcirculation design of roads

Change conventional superblock road network design, increase urban branch road network density, create a street intensive network, improve the travel environment for walking, bicycles and motor vehicles, form road microcirculation. Road microcirculation design techniques mainly include:

(1) At least 50 intersections per $1km^2$ are planned in urban areas, and diverse street block dimensions and road surfaces are designed according to road type and main service functions, and diverse traffic mode choices including motor vehicles, bicycles and walking are provided.

(2) Passable roads are integrated; neighborhood areas can be connected at least every 300m; efficient unidirectional double roads replace main trunk road with the width more than 45m.

(3) Special bus roads and rapid bus transit network are built; at least one special bus road is guaranteed every 800-1,000m; the distance from all residential and office places to a local bus station and to regional public transport stations does not exceed 400m and 800m respectively. Transfer frequency for most of passengers is reduced as far as possible. One integrated diversified transport system is built to ensure seamless transfer between all exiting transport modes.

3) Roads with new materials and new technologies

(1) Green road gelled materials. Road gelled materials made by industrial waste residues as main raw materials serve as substitute products for road cement, which not only results in low production costs and good pavement performance, reduces environmental pollution caused by industrial waste residues, liquid wastes and waste gases, but also makes energy consumption be only 1/3-1/2 of ordinary Portland cement and does not cause carbon emissions resulting from cement production[26]; thus it is one

kind of low-carbon road construction material.

(2) Foamed asphalt. Production of 1t of bituminous mixture consumes 6.5-7.2kg 0}sp}#}/sp} diesel oil and 2.5-3.0 kwh electricity, and causes 19.4kg CO_2 emissions; furthermore, harmful bituminous substances increase with the rise in the hot-mix asphalt mixture production temperature[27]. Foamed asphalt technology means that a certain quantity of normal temperature water is added to hot asphalt to improve bituminous viscosity and reduce mixing temperature during bituminous mixture production so as to save energy and lower carbon emissions.

4) Intelligent management technology

An intelligent transportation system (ITS) (Fig. 6-3) is one large system which integrates many high technologies including electronic vision technology, electronic sensing technology, measuring technology, judgment processing technology, database technology, information technology, data communication transmission technology, control and servo mechanism, computer technology, man-machine interaction technology, human body mechanism engineering, traffic engineering and road guidance technology, which takes vehicles as nodes and is based on a network to perform real-time, accurate and efficient comprehensive transportation management so as to foster coordinated development among people, vehicles, roads and networks. An urban intelligent transportation management system based on intelligent transportation technologies not only guarantees safe urban transport and maximization of the transport infrastructure supply capacity, but also reduces transport fuel consumption and carbon emissions and improves the quality of the transport environment. ITS has become an internationally recognized best way for addressing road transport development issues.

In March 1995, the USA in the National ITS Program Plan explicitly specified seven large fields concerning intelligent transportation systems including a travel and transportation management system, travel demand management, public transportation operations system, commercial vehicle operations system, electronic payment system, emergency management system, advanced vehicle control and safety systems. Currently, the application of intelligent transportation in the USA has exceeded 80%.

Japan's transportation system consists of one integrated center control and ten subsystems including bus priority system, transportation information system, intelligent image system, safe driving system, pedestrian information system, emergency traffic priority system, emergency warning system, environmental protection system, dynamic guidance system and vehicle travelling management system.

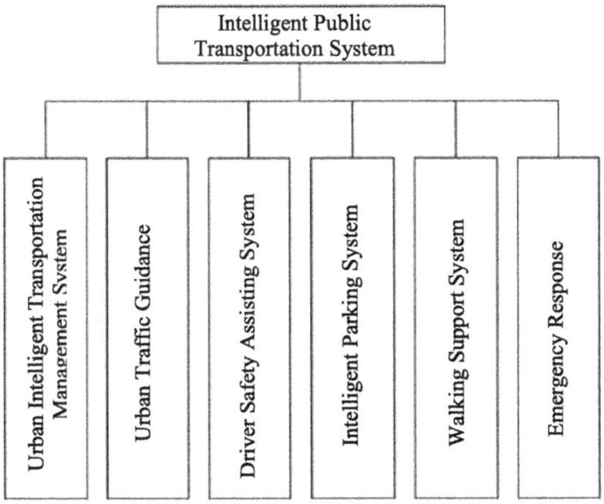

Fig. 6-3 General structure of urban intelligent transportation management system

ITS not only serves as an effective way for addressing urban transportation issues but also promotes revolutionary change in the transport industry. In Europe, adoption of ITS technology not only has reduced economic losses caused by traffic accidents, traffic congestion and environmental pollution by 555-560 billion EUR, but also has increased market scale by 100 billion EUR[28]. According to Japan's estimation, with ITS adoption, by 2025 traffic accidents will decline by 50% and average vehicle speed will increase by 10km/h, and traffic congestion will be completely eliminated. This will reduce time loss by 0.56 billion yen/year and economic losses by 1.23 million yen/year, fuel consumption by 25% and CO_2 emissions by 15%[28].

Though China is at the initial stage in this aspect, data shows that the road intelligent transportation market scale exceeded 22 billion in 2008 and will annually grow by an astonishing 25% in the next five years. Use of existing intelligent transportation technologies can reduce vehicle energy consumption by at least 15%, nitric oxide emissions by 15%, congestion by 15% and traffic accidents by 15%[29].

6.3 Urban Transportation Carbon Reduction Technology Development Strategy and Measures

6.3.1 Strategy choice

Given China's natural resources and social economic development conditions, with regard to current urban low-carbon transportation construction in China, apart from the focus on energy conservation and emission reduction in the urban transportation field, the latest developments in transportation carbon reduction technologies in the world should also be closely followed and urban transportation carbon reduction technologies in future decades should also be studied so as to "overtake other cars on a curved road". In the near and medium term (2011-2030), China's development strategy for urban transportation carbon reduction technologies should follow the most cutting-edge technologies in the world and focus on utilizing existing theoretical and technical research achievements in developed countries and fully tapping its own potential to change the situation of the lag in technologies, with the emphasis on the following three directions:

1) Accelerate transformation of the energy system, build a renewable energy network

Fossil energy is a main source for carbon; mere dependence on fossil energy during the development of new energy and new energy vehicles cannot fundamentally solve urban transportation carbon emission problems, and does not contribute to addressing shortages of fossil energy. Thus it is essential to speed up transformation of the energy system and vigorously develop and utilize renewable energies such as solar energy, wind energy, hydroenergy, thermal energy and biomass energy, and gradually get rid of dependence on fossil energy, building a renewable energy network. With support from a renewable energy network, new energies are used to drive urban low-carbon transportation development.

2) Energetically popularize urban transportation energy saving technologies and develop electric vehicles

Urban transportation energy saving technologies such as motor vehicle energy saving technology and urban road design construction based on new materials and new technologies change rapidly and can effectively reduce carbon emissions, and deliver relatively high economic efficiency. Thus they should be vigorously popularized. An electric vehicle is the best vehicle which meets zero emission standard requirements during use, and also enjoys other advantages such as low noise; however, currently electric vehicle technologies are still at the initial stage, while it is difficult to make substantive breakthroughs in key technology for electric vehicles, thus battery technology in the near and medium term; in addition, the proportion of electric energy from renewable energy in China's power grid needs to be further increased in the near and medium term. From the perspective of a full lifecycle, popularization of the electric vehicle in China neither delivers good economy nor significantly reduces carbon emissions in the near and medium term. Therefore, in the near and medium term, China should take electric vehicle technology as the development direction and devote preliminary scientific research, funds and personnel to key technologies for electric vehicles so as to make technical and market preparations for the popularization and application of the electric vehicle in the medium and long term (after 2020).

3) Realize intelligent management of urban transportation

With the efforts in recent years, one batch of professional management information systems with high technology content for road transportation information construction, such as road tolls, monitoring, communication, road surface, road administration management and emergency management have been built and the construction of an intelligent transportation management system has started or is underway in many cities in China. Therefore, Chinese cities should build on the existing foundation to expedite the development of intelligent transportation and strive to basically materialize intelligent management of urban transportation in the near and medium term.

6.3.2 Implementation measures

1) Promote distributed energy system and a smart power grid

Renewable energy power generation is one of the various renewable energy development utilization modes; as shown in the above analysis, if electric energy from renewable energy is used for electric vehicles or as preparation of new energy, the best effect on transportation carbon reduction can be produced. A certain foundation relating to renewable energy power generation and grid connection techniques has been built; for example, currently, a distributed energy system and smart power grid can optimize peak loads and is capable of fully utilizing less used and relatively scattered hydroenergy, wind power, solar energy, geothermal energy and biomass energy to generate electricity which is connected to a large power grid so as to decrease the proportion of fossil energy power generation in the power grid. China should intensify popularization of a distributed energy system and smart power grid in various cities and create a market mechanism to make early breakthroughs in key technologies for large-scale power generation based on renewable energy and power grid connection.

2) Develop strict fuel economy standards, laws and regulations for carbon emission restrictions

Actions will be taken to improve internal combustion engine efficiency, enable power hybridization and use hybrid electric vehicles, even pure electric vehicles and hydrogen fuel cell vehicles, which can reduce fossil fuel use and carbon emissions. Along with the development of electric vehicles, China should impose stricter restrictions on fuel consumption and carbon emissions from conventional fuel vehicles, and make vehicle fuel consumption and the emission control level meet the latest international standards as soon as possible. Also supplementarily adopt a vehicle-mounted diagnosis system for real-time monitoring. Laws and regulations will be developed, and manufacturing enterprises with relative insufficiency in research on vehicle energy conservation and emission reduction technologies, or that lack attention to such research, will be phased out to continuously promote corporate progress in vehicle energy conservation and emission reduction technologies.

3) Establish an international standard system

Standardization is an important guarantee for promoting the healthy development of new energy vehicles. China's standards for new energy vehicles will be developed in line with international standards to ensure a technically advanced level of standards and give full play to the technology orientation role of standards.

4) Boost applications of intelligent transportation technologies

With the rapid increase in the number of motor vehicles, especially the skyrocketing number of private cars, contradictions between traffic supply capacity of the infrastructure and huge traffic demand become increasingly acute, and carbon emissions increase. Therefore, besides construction of a relatively perfect low-carbon public transportation system, it is necessary to allow full play to the potential of the infrastructure to improve transport efficiency and further accelerate intelligent transportation system construction focusing on information technology as well as expanding the application scope of intelligent transportation technologies.

5) Apply policy instruments to promote low-carbon technology progress

Measures will be taken to reinforce fiscal and tax support, and impel research, development, marketing, popularization and application of new energy vehicles. Consumers will be encouraged to adopt low-carbon technologies and a low-carbon lifestyle, and support the development of new energy vehicles. Motor vehicle drivers will be provided with energy saving technology driver training, and will be guided to develop good energy-saving driving habits to realize the most economical travel with minimum fuel consumption and carbon emissions.

6) Attach importance to the cultivation of low-carbon technology talents

Progress, popularization and application of low-carbon technologies for urban transportation entail support from sufficient intelligent resources; the State should intensify the training of low-carbon technology personnel to build a large contingent of high-caliber research and development forces. Institutions of higher learning, where conditions permit, should actively offer courses concerning low-carbon technologies for urban transportation or set up relevant specialties to develop low-carbon technology talents for various industries and various sectors. Enterprises should take advantage of a competition mechanism to absorb and train a group of their own low-carbon technology research and development researchers as a technical backbone, and fund them to carry out subject research, learning and communication relating to new energy vehicles, and stimulate corporate research and development activities.

References

[1] Energy and Transportation Innovation Center. Low-Carbon Fuel Standard and Policy, [2011-10-27],http: //www.icet.org.cn/dtjt.asp? Cataid=A0014&id=216.

[2] CDIAC, World's Countries Ranked by 2008 Total Fossil-fuel CO_2 Emissions [2011-07-17].

[3] Sun Jianwei, Zhao Rongqin, Huang Xianjin et al: China's Carbon Emission Accounting and Its Factors Breakdown Research During 1995-2005, Journal of Natural Resources, 2010(8).

[4] IEA: Energy Technology Perspectives 2006 [DB/OL], [2015-11-20], http://wenku.baidu.com/link?url=UQoesURrEsUM4NvE5ZacHn8kAk5HgZTj5uMfmZEgJFQs6UVEHQ2s8zH7IiTT7DInLglOo7tRc83cYCwsTuxIJQtrbsZ7g6BBAtZSE_7Mudq.

[5] Wu Wenhua: Energy Consumption and Emission in China's Transport Industry and Comparisons of Typical Countries, Energy of China, 2007(10):19-22.

[6] Zhang A'ling, Shen Wei, Han Weijian et al: Analysis of Lifecycle of Vehicle Alternative Fuel, Tsinghua University Press, 2008.

[7] He Xiaoliang: Hybrid Electric Vehicle: the Only Way for New Energy Vehicles. Science and Technology Daily, March 12, 2011.

[8] Min Haitao, Cheng Meng: Analysis of Environmental Impact and Environmental Efficiency of New Energy Vehicle, Tractor & Farm Transporter, 2007(8):105-106.

[9] Qin Kongjian, Chen Haifeng, Fang Maodong et al: Research on Plug-in Hybrid Electric Vehicle Emission and Energy Consumption Evaluation Method, Automobile Technology, 2010(7):11-15.

[10] Li Qiang, Yang Jianhui, Li Qing et al: Analysis of Lifecycle of Hydrogen Source for Fuel Cell Vehicle, Research of Environmental Sciences, 2003(3):60-61.

[11] Qiu Tong, Sun Baiming, Hong Xuelun et al: Development of Fuel Cell Vehicle with Natural Gas as Raw Material, Natural Gas Industry, 2003(9):1-4.

[12] Zhang Liang: Research on Energy Consumption, Environmental Emission and Economic Efficiency in Lifecycle of Vehicle Fuel Coal-based Dimethyl Ether, Doctoral Dissertation, Shanghai Jiao Tong University, March, 2007.

[13] Zou Xiaobo: Discussions about Vehicle Oil Saving Technology, Highways & Automotive Applications, 2008(6):36-39.

[14] Yang Yinghui: The principles and measures of automobile energy saving. Hunan Agricultural Machinery, 2011(5):28-29.

[15] Industrial Economy Department of Development Research Center of the State Council, Society of Automotive Engineers of China, Volkswagen Group (China), Development Report on China's Automobile Industry, Social Sciences Academic Press, 2011.

[16] Cai Fengtian, Xie Yuanmang: Influencing Factors for Fuel Consumption in Vehicle Operation and Vehicle Energy-Saving Technologies, Energy Conservation & Environmental Protection in Transportation, 2006(1): 28-33.

[17] Geng Xuejian, Fan Enzhuo: Discussions about Measures for Vehicle Energy Conservation and Emission Reduction, Science & Technology Information, 2011(7):369-370.

[18] Mei Juan, Fan Qinhua, Zhao Youcai et al: Carbon Emission and Control Technologies for Transportation Field, Chemical Industry Press, 2009.

[19] Zhang Junhong, Li Zhigang, Wang Tiening: Development Retrospect, Current Situation and Outlook for Turbocharging Technology for Vehicles, Small Internal Combustion Engine and Motorcycle, 2007(1):66-69.

[20] Wang Libiao, He Bangquan, Xie Hui et al: Research Progress in Engine Variable Valve Technology, Automobile Technology, 2005(12):4-8.

[21] Wei Guangjie, Wu Qiong, Tu Anquan: Research and Application Development of Vehicle Engine Start-stop Technology, Journal of Xihua University (Natural Science Edition), 2011(5):14-17.

[22] Wu Guangqiang, Sun Xian'an: Overview of Development of Automatic

Transmission for Vehicle, Journal of Tongji University (Natural Science Edition), 2010(10):1478-1483.

[23] Zuo Jianling: Classification and Application Characteristics of Electric Power Steering System of Vehicle, SAIC Motor, 2009(12):27.

[24] Cao Bin, Zhang Haiqiang: Volvo Cars Will Test Flywheel Kinetic Energy Recovery Technology, Economic Information Daily, June 9, 2011.

[25] Yin Zhifeng, Cheng Lin: Current Research and Development Trends of Green Road Gelled Materials, Materials Review, 2007(5):94-97.

[26] Liu Shijie: Low-carbon, Environmentally Friendly Formed Asphalt Technology and Its Applications, Road Machinery & Construction Mechanization, 2011(7):7-10.

[27] Xie Fei: Future Intelligent Vehicle and Intelligent Vehicle Transportation System, Automobile Technology, 1997(7):53-57.

[28] Yang Xuecong, Internet of Vehicles: New Entry Point for Intelligent Transportation, Economic Daily, August 16, 2011.

[29]Robert Earley, Liping Kang, et al: Report on Background of Sustainable Development of Electric Vehicle in China, [2015-11-20]. http://doc.mbalib.com/view/b4cb49ab9a7748d1dc2971b0384b3f5e.html.

[30] Climate Department of National Development and Reform Commission: Announcement Concerning Baseline Emission Factors for China's Regional Power Grid in 2009. 2015-11-20, http://www.doc88.com/p-6921849094693.html.

Appendix Calculation of Carbon Emissions in Lifecycle of Electric Vehicle

Data concerning consumption of fossil energy such as raw coal for power generation, the carbon emission coefficient and power supply quantity from various regional power grids in China in 2007 come from the literature[30]; mining of 1t of coal consumed 34.4kWH power and 26.7kg raw coal respectively, and 1t of coal emitted 6m^3 CO_2 [6], and power line loss was 6.64%; the full lifecycle carbon emission coefficient for use of power from regional power grids by electric vehicles in 2007 was calculated without regard to carbon emissions from other links including coal washing and transport and carbon emissions generated during exploitation and transport of other fossil energies consumed in power generation. Calculation formula: carbon emission (t CO_2/MWh) = power consumption/100km by electric vehicle x [raw coal consumed in power generation (t) x (6×1.965+0.0344×carbon emission coefficient of power grids in 2007+0.0266×1.8252) ÷ power supply quantity from power grids in 2007 (MWh)]÷ 0.9336. Please refer to Tables 6-2 to 6-6 and Fig. 6-4 for more details.

Table 6-2 Full lifecycle carbon emissions from use of power from regional power grids by electric vehicles in 2007 (kg/100km)

EV type	North China Power Grid	Northeast Power Grid	East China Power Grid	Central China Power Grid	Northwest Power Grid	Southern Power Grid	Hainan power grid
E6-BYD	159.4	175.7	130.3	182.9	167.1	143.5	115.8
M1 EV-Cherry	106.3	117.2	86.6	122.0	111.4	95.7	77.2
Benni EV-Chana	75.9	83.7	61.9	87.1	79.6	68.3	55.1
QQ3 EV-Cherry	91.1	100.4	74.3	104.5	95.5	82.0	66.2
5008 EV-Zotye	91.1	100.4	74.3	104.5	95.5	82.0	66.2
GA6380 EV-Gonow	121.5	133.9	99.0	139.4	127.3	109.4	88.2
Freema EV-Haima	121.5	133.9	99.0	139.4	127.3	109.4	88.2
620 EV-Lifan	113.9	125.5	92.8	130.7	119.4	102.5	82.7
320 EV-Lifan	121.5	133.9	99.0	139.4	127.3	109.4	88.2
ULLA-GWM	75.9	83.7	61.9	87.1	79.6	68.3	55.1

Table 6-3 Percentage of more carbon emissions from gasoline vehicles than the corresponding electric vehicles with respect to regional power grids in China in 2007 (%)

EV type	North China Power Grid	Northeast Power Grid	East China Power Grid	Central China Power Grid	Northwest Power Grid	Southern Power Grid	Hainan power grid
E6-BYD	457.0	514.1	354.2	539.2	484.0	401.5	304.7
M1 EV-Cherry	459.9	517.2	356.5	542.5	487.0	404.1	306.8
Benni EV-Chana	348.2	394.1	265.4	414.3	369.9	303.5	225.6
QQ3 EV-Cherry	419.9	473.2	323.9	496.6	445.0	368.1	277.7
5008 EV-Zotye	339.4	384.4	258.2	404.2	360.6	295.6	219.2
GA6380 EV-Gonow	477.7	536.8	371.0	562.9	505.6	420.1	319.7
Freema EV-Haima	395.2	445.9	303.7	468.2	419.1	345.8	259.8
620 EV-Lifan	413.1	465.6	318.3	488.8	437.9	362.0	272.8
320 EV-Lifan	549.9	616.4	429.9	645.8	581.3	485.1	372.2
ULLA-GWM	288.0	327.7	216.4	345.2	306.8	249.3	181.9

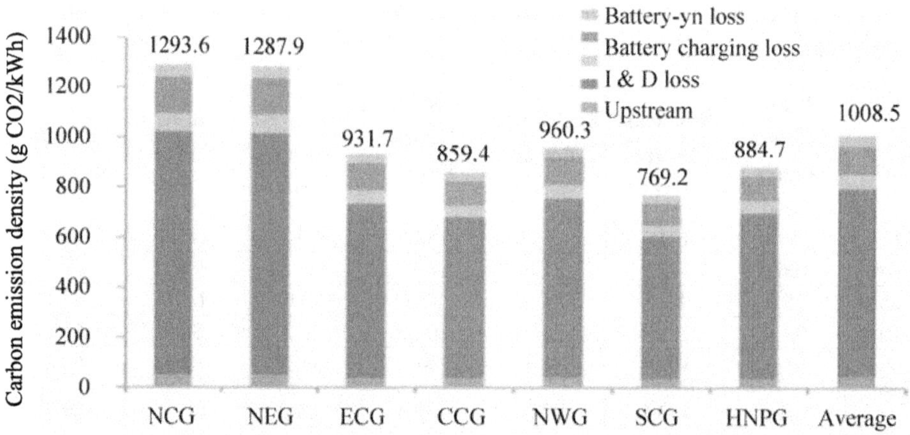

Fig. 6-4 Carbon emission density from mine shaft to oil tank with respect to power grids in China in 2008 (CO_2 kge/kWh)[29]

Note: NCG–North China Power Grid, NEG–Northeast China Power Grid, ECG–East China Power Grid, CCG–Central China Power Grid, NWG–Northwest China Power Grid, SCG–South China Power Grid, HNPG–Hainan Power Grid, Average–National average.

Table 6-4 Carbon emissions from gasoline vehicles in 2008 (kg/100km)

EV type	E-BYD	M1 -Cherry	Benni -Chana	QQ3 -Cherry	5008 -Zotye	GA6380-Gonow	Freema -Haima	620 -Lifan	320 -Lifan	LLA-GWM
Value	28.6	19.0	16.9	17.5	20.7	21.0	24.5	22.2	18.7	19.6

Table 6-5 Carbon emissions from electric vehicles in regional power grids in China in 2008

EV type	North China Power Grid	Northeast Power Grid	East China Power Grid	Central China Power Grid	Northwest Power Grid	Southern Power Grid	Hainan power grid
E6-BYD	27.2	27.0	17.5	18.0	20.2	16.2	18.6
M1 EV-Cherry	18.1	18.0	11.6	12.0	13.4	10.8	12.4
Benni EV-Chana	12.9	12.9	8.3	8.6	9.6	7.7	8.8
QQ3 EV-Cherry	15.5	15.5	10.0	10.3	11.5	9.2	10.6
5008 EV-Zotye	15.5	15.5	10.0	10.3	11.5	9.2	10.6
GA6380 EV-Gonow	20.7	20.6	13.3	13.8	15.4	12.3	14.2
Freema EV-Haima	20.7	20.6	13.3	13.8	15.4	12.3	14.2
620 EV-Lifan	19.4	19.3	12.5	12.9	14.4	11.5	13.3
320 EV-Lifan	20.7	20.6	13.3	13.8	15.4	12.3	14.2
ULLA-GWM	12.9	12.9	8.3	8.6	9.6	7.7	8.8

Table 6-6 Percentage of more carbon emissions from gasoline vehicles than the corresponding electric vehicles with respect to regional power grids in China in 2008 (%)

EV type	North China Power Grid	Northeast Power Grid	East China Power Grid	Central China Power Grid	Northwest Power Grid	Southern Power Grid	Hainan power grid
E6-BYD	5	5	39	37	30	44	35
M1 EV-Cherry	5	5	39	37	29	43	35
Benni EV-Chana	24	24	51	49	43	55	48
QQ3 EV-Cherry	11	12	43	41	34	47	39
5008 EV-Zotye	25	25	52	50	44	55	49
GA6380 EV-Gonow	2	2	37	35	27	41	33
Freema EV-Haima	16	16	46	44	37	50	42
620 EV-Lifan	13	13	44	42	35	48	40
320 EV-Lifan	-11	-10	29	26	18	34	24
ULLA-GWM	5	5	39	37	30	44	35

Chapter 7

Research on Urban Waste Recycling Technology Applications

Li Xiaoyong, Liu Jianwen, Lu Peng, Zhao Xianchao

Abstract: Wasting urban refuse can cause serious harm to the ecological environment and a great waste of resources. The recycling and utilization of municipal waste can produce huge economical and social benefits.The low-carbon technology in urban waste treatment reduces the amount of refuse, keeps the level of pollution low, allows for recycling, and so on. We should promote the recycling of urban waste, the use of low-carbon technology, the development of a low-carbon economy and sustainable social development.

Key Words: Urban waste; recovery utilization; resource utilization; low carbon technology

With economic development, rapid rise of industry and increasing income levels, more and more consumer goods processed and made in an industrialized production mode enter people's lives; they not only enrich and facilitate life but also cause piles of urban waste whose composition becomes increasingly complicated, while old systems and methods for disposal of garbage fail to satisfy the requirements for development of China's economy and environmental protection. Thus environmental pollution caused by waste becomes increasingly severe and a large quantity of resources are wasted, resulting in vicious uncontrollability over time. Meanwhile, waste collection and disposal can create social value and is an important force for promoting economic development. Currently, collection, disposal and recycling of urban waste have become a pressing issue for environmental protection in various countries and also the paramount measure for controlling global environmental pollution.

7.1 Category of Urban Waste and Current Pollution

7.1.1 Classification and sources of urban waste

Currently there is no consensus on the classification of urban waste, and classification methods and types are not the same; overall, several common classification methods are shown below: (1) Waste can be divided by recycling nature into recoverable garbage, unrecoverable garbage and harmful garbage, among which recoverable garbage includes waste paper, plastics, metals and glass; unrecoverable garbage refers to unrecyclable garbage such as fruit peel and residue; harmful garbage includes waste which needs to be separately collected and disposed of such as batteries, fluorescent tubes, electronic elements, paint buckets; (2) Waste can be divided by waste characteristics into wet garbage, dry garbage, bulky garbage and harmful garbage. Wet garbage mainly refers to biological garbage with a relatively high moisture content such as fruit peel. Dry garbage mainly refers to garbage with a relatively low moisture content which can be directly recycled such as waste paper, scrap metal and waste plastics. Bulky garbage refers to bulky solid garbage whose removal and transportation are difficult such as old furniture, worn out electrical appliances. Harmful garbage refers to garbage which does actual or potential harm to human health or the environment and also includes heavy metal which does harm to human health or waste from toxic matter. (3) Waste can be divided by waste source into industrial garbage and domestic garbage. Industrial garbage is waste matter which is generated along with various for-profit industrial activities such as various kinds of industrial garbage and building garbage discharged along with industrial production and commercial garbage such as packaging paper scraps and plastics from commercial activities. Domestic garbage refers to various waste matter from residents consumption including kitchen garbage dominated by vegetables, fruit peel and coal ash and household domestic garbage dominated by old plastics, newspapers, batteries, ironwork, old household appliances and furniture. (4) Waste can be divided by waste composition into organic garbage and inorganic garbage. Organic garbage is also called combustible garbage including such organic combustible components as waste paper pulp, waste plastic rubber, leather, oil, coal cinders. Inorganic garbage is also called incombustible garbage including such inorganic components as ceramics, glass, bricks and tiles, worn out electrical appliances, furnace slag.

7.1.2 Pollution situation of urban waste

Currently in China, accumulated waste has reached 6.5 billion tons, and urban waste annually grows by 8.98% on average; urban waste stood at more than 30 million tons in 2006 and annually increases by more than 1.6 million tons. At present only about 50% of urban waste is disposed of in China. The overall recycling rate of waste in China is not high and much lower than the world's average level; a considerable part of recyclable waste is not recycled. Harmful components and chemical substances in waste can be directly or indirectly delivered into the human body through environmental media–air, soil, surface or underground water, thus threatening human health and spreading diseases, causing potential near-term and long-term harm to human beings: (1) Polluted water environment: Solid waste dumped into rivers or oceans will pollute the water and severely jeopardize survival conditions for aquatic organisms and the utilization of water resources. With impregnation by rainwater and decomposition of solid waste, piled or buried waste enters rivers, lakes and seas along with natural rainfall and surface runoff, or leachate enters nearby soil and infiltrates into underground water, causing pollution to surface water and underground water and endangers human health, social and economic development. (2) Polluted atmospheric environment: Some organic solid waste is decomposed by microorganisms at suitable temperature and humidity and releases toxic gases; granular waste residue and garbage are spread by gales to very far places, causing atmospheric dust pollution; solid waste gives rise to harmful gases and dust during transport and disposal. Incineration is currently the relatively popular way for disposing of waste; however, incineration causes lots of harmful gases and dust and jeopardizes human health. (3) Polluted soil environment: Soil serves as the foundation for the terrestrial ecosystem on the earth and also the material base for human survival and development. If waste is heaped up or no proper anti-seepage measures are taken in landfill, harmful components undergo weathering, rain and erosion from overland runoff and infiltrate into the soil. The soil is deprived of decomposition capability, and the ecological balance in the soil is broken, the soil quality decreases and even no grass and trees can survive. (4) Encroachment on land resources: According to statistics, urban waste piled up over years are comes to about five billion tons; total solid waste reached 1.2 billion tons only in 2004, while wastes rapidly increase every year. Massive waste is exposed in city outskirts; nearly 2/3 of cities nationwide have been besieged by urban waste; statistics shows that waste accumulated over the years has occupied cultivated land of more than 0.55 billion square meters, and precious land resources are wasted and China's economic and social development has been restricted.

7.1.3 Significance of urban waste treatment and recycling

China is a county where there is resource shortage and many natural resources are unrenewable. A considerable part of urban waste including paper, plastics, glass and metals in Chinese cities are completely recyclable and can save enormous resources and energy while mitigating environmental pollution and delivering huge economic benefits and social benefits. According to statistics, when scrap iron, scrap aluminum pots and waste paper are treated and remade into steel products, aluminum products and paper, the proportion of saved energy is astonishing and air pollution is substantially reduced. According to expert estimation, if iron and steel scrap replaces ore to make 1t of steel, 74% of materials can be saved; replacement of papermaking raw material by 1t of waste paper can save 40kW power, 400kg coal, 300t water and 50kg caustic soda.

Urban waste collection and disposal can result in environmental protection and rational utilization of natural resources. Waste reduction, harmless disposal and resources recycling have become the important topics for the government, society and industries in the 12th Five-Year Plan. Scientific and efficient recycling of urban waste is an effective way for reducing resource consumption, environmental costs and maximizing social and economic benefits. Establishment of a new urban waste collection system and urban waste recycling companies will provide more employment opportunities for cities and promote China's green environmental protection undertaking. Therefore, we should recognize the importance and significance of urban waste recycling and attach great importance to urban waste recycling by carrying out it as an important project.

7.2 Current Situation and Problems Concerning Recycling of Urban Waste

7.2.1 Current situation of recycling of urban waste

In China, waste is disposed of mainly in the following three ways: landfill, incineration and composting. As of the end of 2008, there were 509 harmless disposal plants in China, including 407 sanitary landfill plants, 14 composting plants and 74 incineration plants. (1) Landfill is currently the first and foremost way for dealing with urban waste in most Chinese cities and is characterized by less investment and high capacity, and more than 90% of urban waste is disposed of in this way. Most of the landfills are only simple ones and only a small number of sanitary landfills are real ones in China due to construction and operation costs, management measures and environmental protection technologies. (2) Incineration is a waste disposal method adopted by many developed countries. 60% of waste is incinerated in the USA and Japan. Generally, combustibles make up 30% of urban waste, some of which can be burnt directly or by merely adding some auxiliary fuels. Waste dwindles to about 5% of the original volume after incineration, and incineration leads to thorough sterilization and disinfection. Furthermore, heat from incineration can be used to generate electricity and supply heating. Domestic waste incineration equipment can be broadly divided into simple incinerators, a localized incineration facility and integrated incineration facility. Common incinerators include bed combustion incinerator, grate incinerator, dispensing incinerator, packing layer incinerator, single-function and multifunctional incinerator; incineration entails massive investment, high energy consumption and requires an enormous quantity of domestic waste, and is relatively suitable for large cities. (3) Compost treatment of urban waste is basically at the stage of research, testing and gradual development; the compost treatment rate is not high in China.

Currently, methods for recycling of urban waste are available to a limited extent and mainly include composting and waste collection and recycling in China, while sanitary landfill is used for recovering methane gas in a few cities. Urban waste in a small quantity is recycled in China and only accounts for 5% of total waste, 70% of which is disposed of by simple landfill without any income and only 2% is delivered for simple composting, while less urban waste is recycled by other methods. Moreover, in China few types of urban waste are recycled, and urban waste contains many useful substances but a small quantity of components are utilized, while composting mainly focuses on some perishable matter and organic matter, while other matter has not yet been utilized. In addition, recycling of urban waste in China delivers poor economic benefits; the costs for recycling of waste are relatively high and the matter recycled is of low grade.

7.2.2 Problems in recycling of urban waste

Collection and disposal of urban waste mainly presents the following problems: (1) Failure to reduce waste at source. In recent years, various kinds of domestic garbage, especially packaging garbage, have soared, and massive disposable garbage has existed, accounting for more than 20% of total garbage in China. (2) Hybrid collection of garbage increases difficulties in recycling of waste. Currently, urban garbage is still predominately collected in a hybrid way and is gathered by containers at fixed points, transported to transfer stations after collection and then delivered to landfill. Many usable materials and toxic, harmful matter are mixed in the garbage, making it more difficult in recovering waste in the garbage. Thus garbage is characterized by high organic matter content, high moisture, low heating value and complicated components and heating value in incineration disposal is low. Also the quality of products in compost treatment is poor, and landfill pollution is high. (3) Garbage disposal technology level is low, and there is hidden pollution. Currently, in China garbage is disposed of mainly by landfill supplemented by composting, incineration and other disposal methods. As the technical level is relatively low, and management is backward, and funds are not duly made available and there is no relevant policy and legal support, the disposal effect is not ideal and even severer secondary pollution is triggered. (4) Garbage pollution accidents frequently occur and the general public has strong views. As technologies lag behind, and the management system is less developed, and garbage plant pollution accidents frequently occur, the general public becomes jittery at the construction of garbage plants, often lodges complaints and appeals to the higher authorities. (5) Funds for recycling of urban garbage are insufficient. In China, funds for the disposal of urban garbage mainly come from the government and are available in limited amounts, while construction of a large sanitary landfill plant or incineration power plant requires a lot of funds, thus the infrastructure for recycling of urban garbage is poor. (6) The management system, laws and regulations are underdeveloped. China's existing management system can't cater to the needs for recycling of urban garbage, and garbage management is always exclusively performed by the government, and the environmental sanitation department concurrently acts as supervisor, manager and executor. No effective supervision and competition mechanism is built in, thus restricting the development of garbage management; city environment and sanitation management forces are multifarious and disorderly, involve many levels and are not coordinated, with failure to dispatch manpower to key places, which is unfavorable for the recycling of urban garbage. Though laws and regulations for preventing and controlling urban environmental pollution and sanitary management regulations have been promulgated in China, the corresponding sub-laws and detailed rules for implementation are unavailable, causing certain difficulties in law-based management. (7) Recycling of garbage does not attract the attention of the whole population, and residents have faint awareness of environmental protection and recycling. Garbage has not yet been collected on a sorted basis in most Chinese cities, increasing difficulties in recycling of urban garbage. There is less awareness about environmental sanitation and recycling in society, which hampers smooth implementation of various management systems, prevention and control measures[8-12].

7.3 Applications of Technologies for Recycling of Urban Waste

7.3.1 Overview of technologies for recycling of urban waste

As urban waste comes from different sources and has different properties, it is recovered and disposed of in different ways. Based on different waste sources and disposal modes, waste can be divided into domestic waste, industrial waste, building waste and electronic waste[1,10,11].

1) Domestic waste

Domestic garbage makes up a large proportion of urban garbage and is widely distributed, and its sorted collection and disposal are relatively difficult. Generally, urban domestic garbage is divided into organic garbage and inorganic garbage. According to reports, 0.18-0.2 billion tons of urban garbage is annually generated and urban garbage annually increases by 8.8% on average in China. Collection of domestic waste should start from the source, namely, waste is separately placed at residents' homes; Germany's practice may be used for reference: vegetable leaves, fruit peel, used batteries, glass bottles, waste paper are separately collected and then classified into garbage cans in different colors. Besides domestic garbage, "barrier garbage", namely, obsolete old sofas, wardrobes and electric appliances are collected four times a year. Currently, common methods for disposing of domestic waste include landfill, incineration, composting and high temperature decomposition in China.

(1) Landfill

It is one way of disposing of waste by selecting a reasonable site and taking waterproofing, anti-seepage and covering measures. Its advantages are small investment, low disposal costs, high handling capacity, simple operation, while its disadvantages are as follows: the occupied area is large and also increases with the growth in domestic garbage, the site is remote and the haulage distance is relatively long. It is difficult to select a site and leakage of harmful matter from solid waste may cause secondary pollution to underground water. Landfill is currently the first and foremost method for disposing of domestic garbage in most Chinese cities; there were 407 domestic garbage landfills nationwide at the end of 2008. Nearly 85% of urban domestic waste is disposed of by landfill.

(2) Garbage incineration

It is one way for thoroughly oxidizing and decomposing organic matter in garbage at high temperature and converting carbon and hydrogen into carbon dioxide and water during combustion, killing viruses and bacteria under high temperature so as to effectively reduce both quantity and weight; post-combustion residue only accounts for 5%-20% of the original garbage; it is suitable for domestic garbage with relatively high combustible content. Incineration is the common garbage disposal technology in developed countries in the world. Currently, plasma incineration and two-stage gasification incineration have also been gradually popularized in China. Garbage incineration covers a small area and causes less pollution, while the resulting thermal energy can be utilized; however, the investment is large and operating costs are high; moreover, gases (such as Dioxin) and ash from incineration may cause secondary pollution. Currently, urban garbage disposed of by incineration only accounts for 5% in China. Research and application of domestic garbage incineration technology in China started in the mid-1980s; there were 74 domestic garbage incineration plants nationwide at the end of 2008; with economic development in southeast coastal regions and some large and medium-sized cities in China and the increasingly low heating value of domestic garbage, construction of domestic garage incineration plants has been put on the agenda in many cities and is being actively carried out at the stage of rapid development.

(3) Composting

Composting refers to a disposal method through which solid wastes are placed under specific conditions and undergo a natural or microbial strain process. Fermentation and temperature rise to degrade organic matter and achieve harmless disposal while generating organic fertilizer after screening and treatment or deeply processing it into organic compound fertilizer. Composting can be divided into simple composting, aerobic high temperature composting and anaerobic composting. Simple composting is characterized by small project scale, a low degree of mechanization, dominance of a static fermentation process, and incomplete environmental protection measures, plus relatively low investment and operation costs. Generally, simple high temperature composting technology is mainly applied in small and medium-sized cities. Aerobic high temperature composting is characterized by a relatively large project scale, relatively high degree of mechanization, general adoption of a dynamic or semi-dynamic aerobic fermentation process, relatively complete environmental protection measures, investment and operation costs higher than those for simple composting technology. After entry into the 21st century, with the development of composting technology, aerobic high temperature composting is applied again in parts of Chinese cities. Anaerobic composting has the following characteristics: generally the project scale is relatively large; the degree of mechanization is relatively high; a wet or dry anaerobic fermentation process is generally adopted; the fermentation period can be shortened to 15-20 days; after collection, methane gas can be used for power generation; the domestic garbage recycling rate is relatively high; investment and operation costs are higher than those for aerobic high temperature composting; the covered area is smaller than that of aerobic high temperature composting. There are many application examples of the anaerobic digestion technique in Europe; currently, some Chinese cities are preparing for developing a project for anaerobic digestion disposal of domestic garbage. Currently, composting mainly focuses on urban domestic garbage, sludge from sewage treatment plants, human and animal excreta, agricultural waste and waste from the food processing industry. However, it is difficult to completely decompose organic matter; the composting duration is long; the covered area is large and the fertility of organic fertilizer is relatively poor. Composting accounts for 10%-20% of total garbage disposal in China, and its proportion has significantly decreased in recent years. Composting of urban domestic garbage has existed for a long time in China; however, currently the composting-based disposal rate is not high for various reasons; by the end of 2003 there were 70 urban domestic garbage composting plants nationwide and the composting-based disposal rate was nearly 10%.

(4) High temperature decomposition

It refers to the chemical decomposition process in which combustible solid wastes are decomposed at high temperature in an oxygen free or oxygen deficient condition and are finally turned into combustible gases, oils and solid carbon. Thermal decomposition is suitable for urban solid wastes, sludge and industrial waste such as plastics, rubber. Combustible gases, oils from thermal decomposition can be recycled; their recycling property is good and environmental pollution is low; it reduces secondary pollution from incineration and the quantity of waste which needs to be disposed of by landfill.

(5) Microbiological disposal

The former four disposal methods are currently common ones for solid domestic waste from urban residents; however, each method has advantages and disadvantages. Currently the available microbiological treatment method enjoys broad development prospects. The microbiological treatment method works by using the metabolism of microorganisms to decompose solid wastes and make them harmless. Earthworm cultivation is one approach for the microbiological disposal of garbage. One earthworm can swallow garbage equivalent to 2-3 times its weight in a day. Excreta resulting from swallowing and treatment by earthworms are high-quality, odorless, harmless, efficient multifunctional biological fertilizers and can be used for floriculture and production of fruits and vegetables. The microbiological fermentation technique is designed to produce high protein foods and feeds through sustained and rapid propagation of microorganisms. The microbiological fermentation technique involves less investment and is simple and convenient, and results in thorough disposal while not causing secondary pollution; actually, reestablishment of one material recycling process not only eliminates environmental pollution but also turns waste into valuables.

2) Industrial waste

Common industrial waste includes iron and steel scrap, nonferrous metals waste, waste rubber, waste plastics, wastepaper, waste glass and chemical slag. Methods for recycling various wastes are shown in Table 7-1.

Table 7-1 Common methods for recycling industrial waste[①]

Type of industrial waste	Methods for recycling various wastes
iron and steel scrap	Remelting, as needed.
waste plastics	To produce a variety of building materials; to produce a variety of building materials; to recover fuel or monomer.
nonferrous metals waste	Remelting, as needed.
wastepaper	Making recycled paper; making synthetic wood.
waste glass	The use of a part of the glass in place of the use of glass;
waste rubber	Desulfurization, manufacture of reclaimed rubber; crushing, as filler rubber industry or building materials industry.
Metallurgical slag	Manufacture of glass ceramics and artificial granite; instead of sand and gravel as building materials; backfill.
Tailings	Making wall materials and artificial granite; replacing sand and gravel as building material; filling.
Chemical slag	Manufacture of various chemical by-products; incineration, recycling of heat; landfill.

Industrial waste is recycled through the following main players: firstly enterprises which generate waste; secondly enterprises which specialize in recycling waste. Costs for recycling industrial waste include manpower, material and financial costs for recycling waste. Income from recycling of industrial waste includes the value of recycled waste and income from resource saving. As resources are scarce, as long as there is the corresponding technical support, income from recycling of industrial waste will be much higher than costs in the long-term perspective.

3) Building waste

Waste from the construction field mainly includes the following three categories: soil residue, mud and building waste. Building waste materials include metals such as iron and steel, copper and aluminum, and nonmetals such as concrete, bricks, bamboo and timber. Building waste is recovered in the following two ways: firstly, waste resulting from demolition of buildings including such building materials as concrete, timber and glass are collected on site on a sorted basis and recycled. Secondly, waste from new construction of buildings is collected on the construction site on a sorted basis or recycled, or delivered to a relevant place for recycling treatment. The mode for recycling metals, rubber, plastics and glass in building waste is the same as that for recycling similar waste among industrial waste; a mechanical or chemical disposal method can turn waste bamboo and wooden decoration materials into wooden artificial boards and sawdust boards, while the remaining combustibles serve as fuel for generating heat and slag can be used as organic fertilizer. Modes for recycling waste specific to the construction industry are shown in Table 7-2.

① Zuo, Hongying & Yang, Zhongzhi, "The Classification, Collection and Reuse of Urban Waste", *Productivity Research*, 2006 (8): 115-116.

Table 7-2 Building waste recycling methods[①]

Waste category	Waste recycling method
Waste common concrete	Recycled aggregates, recycled coarse aggregates can completely or partially replace stones, while recycled fine aggregates can partially replace sands for recycling.
	Recycled concrete additives can partially replace concrete and sands.
	Similar structural lightweight aggregate concrete and its members can make structural lightweight aggregate concrete members with load bearing and heat preservation functions.
Waste bricks and tiles	Baking-free masonry cement.
	Cement admixture.
	Recycled baked bricks and tiles.
Waste aluminous cement concrete	It can be made into recycled fireproof aggregates, recycled aluminous alumina, recycled cement expansive material etc.

4) Electronic waste

Electronic waste is called electronic trash including various waste batteries, waste computers, communication equipment, TV sets, washing machines, refrigerators and precision electronic instruments phased out from some enterprises and public institutions. Electronic waste contains a large quantity of heavy metals and radioactive elements, which will do severe harm to the environment and directly jeopardize human health if they are not specially disposed of. Electronic waste should be professionally recovered and disposed of by a specialized department to maximize utilization of resources and reduce pollution. An electronic waste recycling process is defined at the following levels: (1) Reuse, namely repair or upgrading of waste electronic products for extending their service life; (2) Cyclic regeneration, including collection and reuse of dismantled components and parts and recycling of materials. Electronic waste recycling and its methods mainly include pyrogenic collection, wet collection, mechanical separation and biological collection. Recycling can extract precious metals from electronic waste and deliver considerable economic benefits.

7.3.2 Applications of new technologies for urban waste recycling

Waste exerts a huge impact on the urban environment and development as well as residents' lives. Conventional disposal methods either endanger human health and pollute air and water sources or destroy farmland and waste recyclable resources. Thus they are not the effective ways for disposing of urban domestic waste. Therefore, people start exploring new methods and give consideration to the current situation and existing methods of urban domestic waste disposal; the effective method for disposing of and recycling urban domestic waste should be based on classification at source and separate utilization.

① Zuo, Hongying & Yang, Zhongzhi, "The Classification, Collection and Reuse of Urban Waste", *Productivity Research*, 2006(8): 115-116.

Currently, many countries are dedicated to researching comprehensive utilization of waste; change and utilization of waste has become one promising industry[5-6,13].

(1) Kernel combustion: heat from garbage is used to incinerate garbage, and a large charging volume is adopted. After garbage is ignited, several heat kernels are formed at the bottom of the material layer, and heat kernels gradually expand to ignite and completely burn the whole material layer. Other auxiliary fuels may not be used to dispose of domestic waste with low heating value and high moisture; this is a new method for garbage incineration.

(2) Hi-tech garbage sorting and disposal: Physical properties of different kinds of garbage are utilized and methods including manual roughing, gravity separation, magnetic separation and airflow classification are employed for classification before recycling. This method makes the best use of everything to minimize pollution, and it is one of currently the most advanced garbage disposal methods. As its one-off investments are excessively huge, it is difficult to promote and apply it in most countries except in a few developed countries.

(3) Waste-based power generation: a power station in New York annually disposes of 600.000t of waste and generates 0.25 billion KWH power. Heat energy from combustion of 4t of waste in foreign countries is equivalent to 1t of kerosene. 70 waste power stations have been built in the USA; waste power generation accounts for 5% of total power in the UK. 50% of industrial waste is used for power generation and heating in Switzerland.

(4) Plastics regeneration: a mixture of various plastics is made into "timber", and composite plastics which can be nailed and sawn like timber are turned into benches, picnic tables and waterside timber piles, by means of mould pressing. Plastic waste undergoes oxidation at high temperature and high pressure to reduce plastics into oil; 1t of plastic waste can be reduced to 800kg of oil. Waste plastics gasification disposal: there are large enterprises for producing liquid ammonia through waste plastics gasification disposal in Japan, in which the core technology lies in gasifying waste plastics to make liquid ammonia and then turning ammonia into various derivates which serve as raw materials including ash for various products so as to recycle all waste and enable zero emission of waste.

(5) Urban garbage composting technology: with organic matter in an aerobic condition, part of organic matter is oxidized into simple inorganic matter through metabolic activity of aerobic microorganisms so as to provide energy necessary for biological life activity, while another part of organic matter is converted into nutrient substances necessary for the living body; microorganisms play a very important role in composting. Microorganisms for composting come from nature or manually screened special microbial strains are used for inoculation to increase the composting reaction speed. Waste-based preparation of waste means that wastes are rapidly fermented under aerobic conditions and 50 ℃-60℃ below which general bacteria, worms and ova can be killed; fermented waste makes an excellent organic fertilizer. Generally, waste is converted into organic fertilizers in China.

(6) Technology for extraction of ethanol and gas from domestic garbage: it mainly

uses hydrochloric acid as catalyst to hydrolyze light components (containing plenty of cellulose) resulting from garbage flotation under high temperature while extracting sugar and making sugar water fermentation and them enabling reduction to ethanol which generates electric energy during hydrolysis. Concentration of ethanol extracted at the laboratory is 1.6% and its concentration after distillation can reach as high as 95%.

(7) Steam-oxygen garbage brick technique: a steam-oxygen garbage brick is one kind of hydrocolloidal silicate building material and is hydrated calcium silicate and hydrated calcium aluminate with cementing power which are generated by effective reaction between active SiO_2 and Al_2O_3 (mainly available in coal ash) from inorganic garbage and CaO from lime in certain temperature and humidity conditions; its hydration product has properties similar to that of ordinary silicate hydrate, thus a garbage brick has strength and mechanical properties necessary for the walls of buildings.

(8) Collection and dismantling of waste household appliances: Japanese TERM Corporation is mainly engaged in collection and dismantling of waste electronic products such as waste household appliances, environmental testing, and delivers significant benefits through disposal of waste electronic materials. The dismantling workshop is fully based on line production, in which each process is completed by only 2-3 persons and the recycling rate reaches 90%. Its kinescope cutting technology is internationally advanced; its efficient and perfect operating process highly deserves to serve as a reference for counterparts in China.

(9) Vehicle battery regeneration: a Japanese corporation has developed one kind of activating agent which uses an impulse wave to remove attachments through a special liquid additive to increase the surface area of the plate electrode and electric power storage. A battery treated by such an additive has a service life more than 80% that of a new battery and can be repeatedly recycled 3-4 times.

(10) Waste-based manufacturing of composite materials: more than 95% of raw materials for waste-based composite materials are various solid wastes, while their costs are much lower than that for primary resources obtained through extraction and separation. Thus this not only solves problems concerning environmental pollution from waste, but also saves various precious primary resources. (1) Polymer-based waste composite materials: waste sand, tailings, furnace slag, pulverized coal ash and fiberglass leftovers undergo certain particle size, particle shape and surface activation treatment and then serve as reinforcing materials, while waste agricultural films, food bags, woven bags and reclaimed rubber from old tires receive a certain process treatment and then are used as base materials, for which proper additives, special interface treatment and a composite process are adopted to form composite materials based on a ball-ball, ball-fiber packaging system. (2) Silicate-based (ceramic-based) waste composite material: uses collosol including CSH resulting from hydration and bonding of active SiO_2, Al_2O_3 from waste materials and minerals from additives as matrix, and wraps some granular or fibrous waste in it to form one kind of composite material with better strength. (3) Metal-based waste composite material: adopts waste zip-top cans, toothpaste tubes

and aluminum alloy sections leftovers as matrix, and uses glass cullet, fiberglass leftovers as reinforcing materials, while the resulting high strength composite material enjoys high toughness and hardness; thus the value of renewable resources redoubles.

7.4 Development Countermeasures for Collection, Disposal and Recycling of Urban Waste

7.4.1 Principles for collection, disposal and recycling of urban waste

Internationally accepted garbage disposal mainly follows the following principles: reduce, reuse, recycle and make harmless[14-15].

1) Reduce

Input of natural resources should be minimized as far as possible in terms of production input. With respect to disposal of urban waste, reduction of waste is achieved by saving resources and reducing the resource input at source on the one hand and by disposing of and utilizing solid wastes on the other hand. Reduction of solid waste needs to start from a comprehensive development of resources and comprehensive utilization of materials during production, namely, "waste minimization" and "clean production". Disposal and utilization of solid waste belongs to the end of material production, namely, commonly understood "comprehensive utilization of waste", when various comprehensive disposal and recycling technologies are used to recycle waste so as to reduce waste.

2) Reuse

It means that the product and service utilization rate is maximized as far as possible, and products are required to be reused many times or in various ways during production and consumption to extend the service life of products and services, preventing early occurrence of waste so as to reduce waste.

3) Recycle

Solid waste is classified as a secondary resource or renewable resource, which means that articles are required at the product output end to be recycled and comprehensively utilized as renewable resources after the completion of their function. Based on improving product and service utilization efficiency, efforts are made to recycle waste into raw materials for other industries or products so as to enable resource recycling and minimize waste discharge. Recycling of solid waste is the main destination for solid waste. Recycling should follow the following principles: recycling technologies are feasible; recycling delivers relatively good economic benefits and shows a relatively strong vital force; waste should be utilized close to the discharge source as far as possible so as to save investments. Design philosophy and techniques for "cold disposal" of urban domestic waste lie in comprehensively designing one new technical proposal for waste disposal according to various advantages and disadvantages in garbage disposal, the merits and demerits of various existing technologies at home and abroad, garbage components and their characteristics, aiming at concurrently conducting full-amount disposal of wastes and resource regeneration during garbage disposal, and delivering production benefits from garbage disposal, embodying realistic social and environmental benefits.

4) Make harmless

The basic task for harmless disposal of solid waste is the engineering treatment of interfering waste to the extent of not undermining human health, not polluting the surrounding natural environment (including the primary environment and secondary environment). Currently, engineering of harmless disposal of waste has developed into one brand-new engineering technology, such as waste incineration, sanitary landfill, composting, anaerobic fermentation of faeces, thermal treatment and detoxification treatment of harmful waste, among which "high temperature rapid composting treatment process", "high temperature anaerobic fermentation treatment process" have reached a practical level in China.

7.4.2 Development countermeasures for collection, disposal and recycling of urban waste[14-15]

1) Strengthen disposal at source, promote garbage reduction

In order to solve urban garbage problems in China, it is essential to start by reducing the sources of urban domestic garbage, lowering household and commercial garbage, decreasing the use of disposable articles and packaging garbage as far as possible, reusing, repairing and using currently available articles, reducing garbage, saving energy and increasing wealth.

2) Intensify sorted collection of urban garbage

In order to achieve garbage recycling, efforts should be made to start from reinforcing management, carrying out sorted collection of garbage to lower costs for collection of waste from garbage, improving the waste collection rate and the quality of collected garbage, and promoting recycling to facilitate separate disposal of harmful matter. Various cities should proceed from their own specific conditions to put forward a garbage classification plan, and gradually popularize sorted collection of garbage. Sorted collection should be gradually implemented on the principle of starting with the easy ones and then addressing the difficult ones, and firstly coping with the simple ones and then the complicated ones. For example, street block garbage with low ash content and highly combustible content should be separately collected and incinerated, and then sorted collection and incineration of combustible matter should be gradually carried out; for cities with high ash content in garbage, ash may be separately collected for landfill disposal or directly collected and utilized; for cities where disposal modes are dominated by composting, biodegradable organic garbage may be separately collected for composting disposal.

3) Establish a garbage charging system and realize the industrialization of garbage operations

With the increasing material living standards in China, urban residents attach more and more importance to environmental quality. As the time for establishing a garbage charging system is basically ripe, residents, non-productive enterprises and public institutions, as well as productive enterprises, should be charged by population, number of employees and garbage output respectively to reduce garbage output. A garbage charge should be incorporated into local public finance and provide funding support for reforming the urban garbage system. It is necessary to act upon the requirements of garbage disposal industrialization so that domestic garbage disposal charges collected by environmental sanitation enterprises are defined as operational service charges and their charging standard be verified on the principle of compensating for garbage collection transport and disposal costs and rationally making profits. It is necessary to develop scientific calculation and charging measures and strengthen charging management.

4) Set unified standards for harmless disposal of garbage

Currently, garbage disposal facility construction standards and processes in Chinese cities are determined by the city itself according to local financial resources, technical level and existing facilities. On the one hand, local environmental sanitation departments are often incompetent in garbage disposal supervision due to local economic and technical restrictions imposed on environmental protection supervision; on the other hand, competent departments have no unified standards to regulate and supervise urban garbage disposal facilities so that currently garbage disposal levels vary in Chinese cities, and a considerable amount of garbage does not become harmless. Thus, it is imperative to develop unified standards for harmless disposal of garbage to strengthen supervision management.

7.5 Conclusions

Discussions about disposal of different kinds of urban garbage and current utilization methods show that urban waste disposal modes are diverse. With respect to method selection, we should follow the principle of "reduce, reuse, recycle and render harmless" and rationally utilize different methods and techniques to carry out collection and recycling of urban garbage. Currently, China has developed a series environmental protection laws and regulations, standards, specifications and supporting technical policies; however, it is also necessary to provide preferential industrial policies, vigorously support China's emerging garbage disposal industry. In the garbage disposal industry, people should actively open up multiple financing channels to address the shortage of funds, and adopt modern advanced diversified operation modes and operational means to enhance their own survival and profit-making ability, create good social and environmental benefits and huge economic benefits and develop the garbage disposal industry in a healthy and sustainable direction.

References

[1] Zuo Hongying, Yang Zhongzhi: "Classification, Collection and Reuse of Urban Wastes," *Productivity Research*, 2006(8):115-116.

[2] Wang Tao: Urban Garbage Recycling Technology Research and Urgent Problems, *China Environmental Protection Industry*, 2004(8):165-168.

[3] Cai Ke, Du Xichun: Discussions about Urban Garbage Recycling, *Theoretical Investigations*, 2010(7):207-208.

[4] Song Jianli, Shi Weiyong, Ni Liang, Wang Liang: "Current Situation of Urban Domestic Wastes and Research on Recycling Technology," *Journal of Hebei Agricultural Sciences*, 2009(7):58-61.

[5] Cai Lin: Sorted Collection of Garbage is the Only Way for Eradicating Garbage Pollution and Developing Circular Economy, *China Resources Comprehensive Utilization*, 2002(2):9-13.

[6] Huang Ziyi, Ge Baoyang: Discussions about Recycling of Urban Wastes, *Renewable Resources Research*, 2000(3):36-37.

[7] Fan Yuansheng: "Environmental Management of Urban Domestic Wastes in China," *Environmental Economy*, 2005(10):6-8.

[8] Chen Luyan et al: Comparisons of Components of Municipal Garbage and Disposal Strategies in Hong Kong, Guangzhou, *Foshan and Beijing, Environmental Science*, 2003(3):136-139.

[9] Tan Wanchun, Wang Yunbo: "Comprehensive Treatment and Recycling of Urban Wastes," *Journal of Changsha University of Science and Technology* (Social Science Edition), 2006(2):44-46.

[10] Zhu Hongbao, Cai Xiuping: Disposal and Recycling of Urban Garbage, *Energy Research & Utilization*, 2005(2):43-45.

[11] Chen Yang, Wang Deying, Lai Xijun: Current Situation and Prospect for Solid Waste Recycling, *Territory & Natural Resources Study*, 2003(3):69-71.

[12] Song Xiaolan: Disposal and Sustainable Development of Urban Garbage, *Journal of Changsha University*, 2001(4):2-3.

[13] Zhao Lan, Guan Yuzhuan: Disposal and Recycling of Urban Garbage, *Science & Technology Information*, 2011(7):139.

[14] Wang Yongkang: The Municipal Waste Dilemma and Its Solution Based on Reverse Logistics, *Ecological Economy*, 2010(10):172-175.

[15] He Xionglang, Zhu Xuguang: *Thinking of Theories and Practice in Development of China's Circular Economy for Urban Wastes, Contemporary Economy & Management*, 2008(30):15-18.

Chapter 8

Research on Low-carbon Technologies for Packaging Waste Disposal

Li Xiaoyong, Liu Jianwen, Zhao Xianchao

Abstract: Packaging castoffs have caused serious harm to the environment and wasted plenty of resources. Recovery utilization may produce tremendous benefits for the economy and society. Low-carbon technologies used in packaging castoffs have the characteristics of waste reduction, low-pollution, recyclability and so on. Recovery utilization technologies used in packaging castoffs with low-carbon characteristics have advanced, which benefit the promotion of the low-carbon development of the packaging industry and impel the development of the low-carbon economy.

Key Words: packaging castoff; disposal; recovery utilization; carbon reduction; technical innovation

A low-carbon economy is an economic development mode based on low energy consumption, low pollution and low emissions and represents another major progress following agricultural development and industrial development in human society. The essence of a low-carbon economy lies in efficiently utilizing energy, developing clean energy and pursuing green GDP, while its core consists in low-carbon technology innovations, industrial structure and system innovations for promoting sustainable development of human society and the economy. Development of a low-carbon economy is of great significance for boosting adjustments in China's industrial structure and transforming the economic growth mode.

With economic development, the rapid rise of the packaging industry and increasing people's income level, more and more consumer goods processed and made by the industrialized production mode enter people's life. They not only enrich and facilitate life but also cause massive packaging waste, severely pollute the environment and heavily waste resources, resulting in a vicious uncontrollable trend over time. Currently, collection, disposal and reuse of packaging wastes have become pressing matters for environmental protection in various countries and are also primary measures for controlling global environmental pollution. Packaging focusing on "low carbon" represents a development trend for the packaging industry and is an important part in the development of a low-carbon economy. Strengthening applied research on low-carbon technologies for disposal of packaging waste is an important way for saving energy, reducing emissions and protecting the environment and also one of the important means for lowering CO_2 emission intensity per unit of GDP in China by 40%.

8.1 Current Situation of Packaging Waste and Significance of their Control

8.1.1 Classification of packaging wastes and current situation of pollution

China's packaging industry started in the 1980s; with rapid development for nearly 30 years, the modern packaging industry system dominated by paper, plastics, glass, metals, package printing, packaging machinery and having certain modern technical equipment, as well as covering relatively extensive categories, has taken shape in China. Packaging has become an essential part of modern commodity production, storage, sales and social life. According to data, as of the end of 2009, the total output value of China's packaging industry surpassed that of Japan for the first time, making China the second largest packaging industry country in the world, only second to the USA, while the total industrial output value also increased to one trillion yuan. The packaging industry has been upgraded to the 14th place among 42 industries in the national economy[1]. China's packaging industry is skyrocketing by an annual 16.95% , higher than the national industrial average annual growth rate.

A rapidly developing packaging industry consumes huge resources, persistently increases packaging waste and causes increasingly severe damage to the ecological environment. Packaging waste refers to packaging containers, materials and auxiliary matter which become solid waste and are abandoned after use. Based on different materials used in packaging, packaging wastes are mainly classified into the following five categories: (1) paper packaging waste; (2) plastic packaging waste; (3) metal packaging waste; (4) glass packaging waste; (5) other packaging waste such as cloth bags and straw bags. Specifically, they mainly include paper, plastics, metals, glass, ceramics, wood and composite materials[2]. Paper waste is putrescible and can be decomposed. Main varieties of metal packaging include tin boxes, cans, steel drums and large metal storage tanks. High-capacity steel drums are collected and reused many times, while other varieties are less often collected; even aluminum zip-top cans can reach a collection rate peak of only 30%-40%. Part of glass containers and products are collected and reused and crafted glass is kept, while most of the glass is abandoned in the form of discarded bottles and cans and glass cullets. Non-degradable plastics make up the majority of plastic waste, which cannot be degraded in natural environment in less than two to three hundred years and cause "white pollution" which severely pollutes the environment[3].

Packaging waste is classified as urban solid waste and is a long-standing important environmental pollution source. If it is not disposed of or is improperly disposed of, it will exacerbate urban ecological environmental pollution. Firstly, rapidly growing waste will occupy vast cultivated land; current urban garbage has greatly accumulated in China, covering an area of more than 50,000 hectares[4], and urban garbage is expanding every year. Both garbage heap sites and garbage landfills occupy cultivated land, making China, with an excessively small cultivated area, more strained. Secondly, garbage is disposed of by simple methods such as heaping up in the open air and landfilling in natural gullies in most of the cities, which will inevitably pollute the water table and soil. Heavy metals in garbage such as As, Pb, Cd, Hg, Cu, which enter the soil will be absorbed by crops and finally endanger human health. Furthermore, these toxic substances will infiltrate into the water body through rainwater and pollute water sources and underground water resources. Secondly, common incineration methods for waste collection and disposal such as open incineration of garbage will generate Dioxin and other toxic gases, severely polluting urban air. In addition, as garbage in an urban waste tip is not quickly disposed of and some so-called useable matter such as old clothes and old injection syringes are directly reused without any disinfection and regenerative preparation, this causes secondary pollution and jeopardizes the health of consumers. Moreover, a garbage tip accommodates rats, propagates mosquitoes and flies, and it is easy to spread diseases, inflicting suffering on human beings.

8.1.2 Current situation of urban packaging waste treatment, collection and utilization

Currently, garbage has been successfully recycled in some developed countries; for example, the waste paper recycling rate reached 70% in Finland in 2007; the waste packaging recycling rate stood at 78% in Japan; the aluminum can collection rate hit 95% in the USA. This shows that impressive effects have been produced in the recycling of packaging waste, and huge economic and social benefits have been delivered in developed countries, which deserves to be imitated by China[3-5].

In China, the current accumulated garbage has reached 6.5 billion tons, and urban garbage annually grows by 8.98% on average, in which packaging waste accounted for about 1/3; packaging waste amounted to about 30 million tons in 2006, and annually soars by more than 1.6 million tons; currently, the urban garbage disposal rate is only about 50% in China[4-6]. The waste paper recycling rate is not high, only 30.4% in China, much lower than the world's average level – 47.7%. The beer bottle and metal packaging material recycling rate is relatively high, while the recycling rate of other packaging waste is considerably low, and all collected packaging products account for less than 20% of the total output of packaging products, while a considerable part of the waste which is recyclable is in turn wasted. This not only increases consumption of natural resources but also severely pollutes the environment and aggravates urban ecological environmental pollution[7-8]. In recent years, though some progress has been made in the collection of packaging waste under the guidance of relevant laws, regulations and relevant policies released by competent departments of the State and local governments, the overall situation is not optimistic. The following several problems exist in the collection and disposal of packaging waste in China:

1) Packaging waste collection system is underdeveloped

In western developed countries, packaging waste collection systems suited to national conditions have been successively established and have played an important role in controlling environmental pollution and developing a national circular economy. For example, in Germany's market-oriented Duales System Deutschland (DSD) waste collection system, a specialized DSD Company is responsible for collecting, classifying, disposing of and recycling all waste in the whole system. China's collection system is still imperfect, as evidenced by the following major problems: (1) Work on the sorted collection of packaging waste severely lags behind in China: currently work on urban waste classification is hardly carried out in China. Various packaging waste and kitchen waste are mixed together, and it is difficult to utilize their effective resources merely by landfill or incineration. The state-owned collection system has been dissolved in China where, though there are self-organized private collection systems, no professional sorting and disposal means are available. Classification of packaging waste fully relies on manual sorting and does not become accurate, making it difficult in later disposal. (2) Packaging product collection channels are chaotic in China: traditionally, previous garbage classification was supported by a collection system merely relying on a governmental administrative act in China. In recent years, the original collection systems and channels have failed to work for lack of an economic concept, while a market-supported collection system has not yet been established. Collection channels owned by commercial companies, street and civil affairs departments work separately, thus the recycling rate is very low, resources are wasted and secondary pollution becomes severe. (3) Legislation concerning packaging waste collection and disposal needs to be strengthened in China. So far, packaging waste disposal laws and regulations suited to China's national conditions have not yet been developed in China. Furthermore, legislation relating to packaging waste collection and disposal is relatively weak, and the current Law of the People's Republic of China on Prevention and Control of Solid Waste Pollution does not clearly specify the responsibilities of the producer, and the corresponding organizations and facilities for collecting, storing and disposing of packaging waste are still highly underdeveloped.

2) Production process is backward and resources are severely wasted

Though the technical level in China's packaging industry has been significantly increased in recent years during which the economy has rapidly developed, problems concerning the backward production process and severe waste of resources are still widespread compared with developed countries. For example, nearly 1/10th of 0.3 billion m^3 of timber annually consumed in China is used in various product packages. About 20 million tons of paper packaging products are annually consumed in China, among which those made with wood pulp as the main raw material account for about 40%, resulting in the consumption of more than 20 million m^3 of timber[1]. As plastic packaging is characterized by rich sources of raw materials, low price, less mass, high strength, excellent toughness and easy production, plastic packaging makes up more than 1/3rd of the total packaging materials in China. However, people generally use non-degradable ordinary plastic bags, while their waste is neither easily recycled nor naturally decomposed, thus leading to severe "white pollution".

3) Packaging waste collection rate is low

The most effective way for collecting packaging wastes lies in sorted collection and disposal of garbage. Collection and disposal of packaging waste has been conducted on a large scale and industrialized in developed countries such as Germany, Japan and the USA. Tinplate and waste corrugated paper collection rates have reached 89% and 95% respectively in Germany; collection and utilization of packaging waste has created an annual average 4 billion USD of wealth in the USA in the past ten years. In China, the annual packaging output exceeds 30 million tons, and most of the products are collected at a considerably low rate of less than 20% of total output[1]. Collection and disposal of packaging waste still significantly fails to be on a large scale in China, causing a huge waste of resources and severe environmental pollution.

4) There is no green consumption concept

"Green consumption" refers to a consumption concept and consumption mode which upholds nature, health, safety, environmental protection and energy conservation, and calls for consumers to buy products in a green package with a low degree of environmental pollution as far as possible. Green consumption is a code of conduct for consumption by the general public in a circular economy. Green consumption started later and has significantly failed to become a consumption habit in China. It is still difficult to prohibit "white" and vitalize "green" in small, medium and large cities and rural towns in China. According to a survey, 72% of people believed that developing an environmental protection industry and green products was beneficial for improving environmental conditions, while 54% of people were willing to use green products and 38% of people said that they bought green products[1]. As shown by survey data, the green consumption concept has started to take shape in China; however, not many people really take action, and the green consumption concept requires intensified publicity and institutional promotion.

8.1.3 Significance of urban packaging waste treatment, collection and utilization

China is a country where resource shortages occur and many natural resources are unrenewable, while many packages are made from unrenewable resources. Therefore, collection and utilization of packaging materials are of great importance. A considerable part of paper, plastics, glass and metals in packaging waste are completely recyclable in China, which can save enormous resources and energy, reduce environmental pollution, and deliver tremendous economic and social benefits. Five million tons of plastic garbage is annually generated in China, only about 30% of which is voluntarily recycled by individual households, while resources worth five billion yuan are wasted. Based on the current collection level, 140,000t of carton boxes are annually collected, which can save 80,000t of coal, 49 million KWH power, 23.8t of wood pulp and straw, 110,000t of caustic soda necessary to produce the paper resulting from such collected carton boxes. 1t of waste paper can remake 800kg of paper and save 4m^3 of timber, 400 kWh of power, 400kg of coal and 30t of water. Glass products account for about 1.87%, and reprocessing of 1t of glass cullet can save 1t of coal, 400 kWh of power[9]. According to statistics (see Table 8-1), when scrap iron, waste aluminum cans and waste paper are treated to remake steel, aluminum products and paper, this can bring about astonishing proportions of saved energy and reduced air and water pollution[9-10]. Garbage, especially packaging waste, has been successfully recycled to create huge economic benefits in some developed countries. For example, half of garbage has been reused in Finland; copper collected from waste accounts for 80% of national copper demand in Japan[8]; 90% of corrugated paper is made of collected waste paper in France; in the USA, three garbage disposal companies dispose of more than 15% of garbage nationwide and annually deliver more than two billion USD[8].

Table 8-1 Proportions of saved energy, reduced air and water pollution

Items	Fe	Al	Paper
Energy saving ratio (%)	65	95-97	70-75
Air pollution reduction ratio (%)	85	95	74
Decline in water pollution ratio (%)	75	97	35

Packaging waste collection and disposal are conducive to controlling environmental pollution, saving resources and energy and rationally utilizing natural resources. Under the 12th Five-Year Plan, reduction, harmless disposal and recycling of garbage have become the important topics for the government, society and industries. Scientific and efficient recycling of packaging waste is an effective way of reducing resource consumption and environmental costs, maximizing social and economic benefits. Furthermore, establishment of a new packaging waste collection system and packaging waste recycling companies will provide more employment opportunities for cities and promote China's green packaging undertaking. Therefore, we should recognize the importance and significance of packaging waste recycling and attach great importance to packaging waste recycling by carrying it out as an important project.

8.2 Applications of Low-carbon Technologies for Packaging Waste Collection and Disposal

8.2.1 Packaging waste collection system suited to China's national conditions

In order to solve problems concerning packaging waste recycling, many packaging waste collection systems with unique characteristics have been built to form a complete collection network in foreign countries, such as Germany's Duales System Deutschland (DSD), Austria's ARA waste collection system, France's ECO Emballages SA packaging waste collection system.[11]

Waste collection systems on a certain scale have not yet been developed in relevant industries in China where waste collection is limited to articles with relatively high value and relatively low collection costs, such as electronic products, plastics and paper, and the market is not standardized and collection efficiency is relatively low[8-10]. Currently, packaging waste is still not recycled by relevant government departments or companies but is uniformly collected together with other garbage by the environmental sanitation department in China. This mixed disposal mode presents various disadvantages: firstly, mixed collection of packaging waste and other domestic garbage not only increases fiscal expenditure for the government, but also makes it more difficult for garbage classification; secondly, various administrative departments are mutually independent, do not share information and often get involved in disorderly competition, which may give rise to complicated interest disputes; thirdly, a government-driven recycling system cannot arouse the enthusiasm of various market players for the recycling of packaging waste and cannot give full play to the economic potential of a packaging waste recycling system, thus hindering the market-oriented development of the recycling system. This is unfavorable for the market-oriented development of a packaging waste recycling system in the future. China should draw upon the management experience of developed countries such as Germany in establishing a packaging waste recycling system, develop non-profit recycling organizations, integrate existing garbage classification collection systems, retain garbage collection work in the original recycling system, and conduct special recycling of packaging waste, as well as building a market-oriented packaging waste recycling system.

8.2.2 Packaging waste recycling mode and process

Currently, packaging waste is disposed of mainly in the following ways at home and abroad: (1) Reuse: collect and dispose of all or part of packages after use, and then reuse them as packages; (2) Recycle: collect used packages for disposal and reprocessing, use them in different fields; (3) Resource utilization: utilize thermal energy from collected matter through incineration. The process for disposing of packaging wastes can be summarized below (Fig. 8-1):

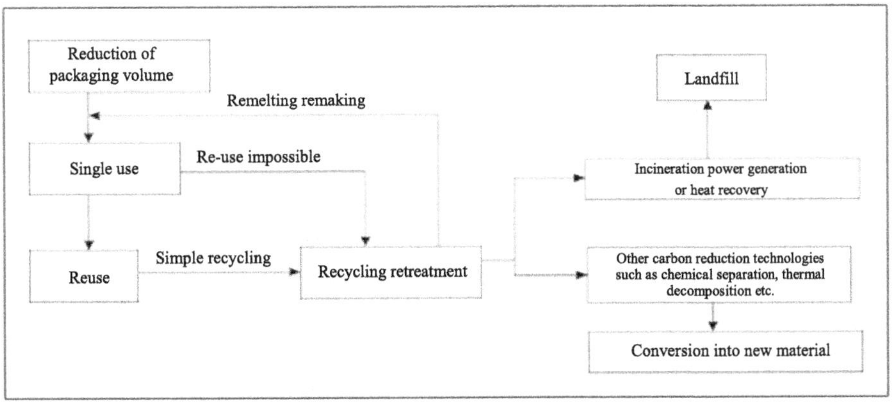

Fig. 8-1 Process for disposing of packaging waste

8.2.3 Packaging waste recycling and reuse technologies

Generally, packaging waste recycling is divided into material collection and thermal energy collection. Currently, domestic packaging waste recycling is dominated by material recycling. Material recycling technology means that the corresponding disposal technology is used to regenerate packaging waste which has been sorted (according to packaging material category) into new similar materials. Collection of thermal energy is conducted by putting mixed combustible packaging waste into an incinerator for combustion and using thermal energy from combustion as energy for power generation and heating[12].

Packaging waste is disposed of mainly by the following several common recycling technologies and methods: (1) Simple recycling technology: clean, disinfect and dry solid wastes such as plastics, glass and metals and then reuse them. (2) Incineration disposal technology: use a high temperature thermal decomposition method to oxidize and turn them into matter with small volume and less environmental harm. Its advantages are simplicity, convenience, a small volume of post-combustion residue, high density, minimal covered area and high degree of convenience at the time of landfilling, relatively high stability, easy disintegration into soil and concurrent release of huge thermal energy. Its disadvantage lies in releasing some toxic gases to pollute the environment during incineration. This disposal method is expected to be continuously used in the future for quite a long time. (3) Thermal decomposition collection technology: combustion technology designed to generate heat, steam and power; pyrolysis technology for making fuel gas, fuel oil and carbon black with medium and low heat value; gasification and pyrolysis technology for making chemical substances such as fuel gas with medium and low heat value; gasification and pyrolysis technology for making heavy oil, kerosene and gasoline.

Low-carbon technologies for recycling and reuse vary with different kinds of packaging waste:

1) Low-carbon technology for waste recycling

Plastic packaging waste is extremely detrimental to the environment and society; the plastic packaging collection process can be divided into incineration, pyrolysis and recycling. Conversion into thermal energy through plastic incineration is economically feasible for developed countries with high labor costs; however, incinerator investments are relatively large and it is easy to generate toxic gases. Plastic packaging waste recycling technology can be divided into collection-recycling-reusing, mechanical treatment and recycling, chemical treatment recycling technology.

(1) Collection-recycling-reusing: It dispenses with a processing and treatment process and is designed to make classification and selection, then clean the waste which meets basic requirements through water washing—acid pickling—alkali washing—disinfection—water washing—sodium hydrogen sulfite soaking—water washing—distillation water washing -50°C drying process before direct reuse. This method mainly focuses on some hard, smooth and easy-to-clean containers such as trays, big packaging boxes and high-capacity drink bottles.

(2) Mechanical treatment and recycling is mainly modification and recycling and is designed to improve the basic mechanical properties of regenerated materials to meet the requirements for the quality of special regenerated products. Some leftover materials, waste or cleaned single-variety plastic packages are directly smashed and mixed, and returned to the plastic processing flow. This collection method is preferred for plastic package collection; this method is also called mechanical regeneration. This is a collection method currently adopted in more cases in China. Modification regeneration is mainly divided into the following two categories: (1) Physical modification: a certain quantity of inorganic fillers are added after activation of plastic waste, and relatively good surfactants are also provided to enhance affinity between fillers and regenerated plastic materials; (2) Chemical modification: materials are modified through chemical reaction and other chain links and functional groups are introduced into the molecular chain through grafting and copolymerization, or crosslinking is carried out by a cross-linking agent, or modification is conducted by a nucleating agent, a forming agent to give waste plastics a relatively high shock resistance, excellent heat resistance and ageing resistance in order for recycling.

(3) Chemical treatment recycling technology directly decomposes packaging waste plastics through pyrolysis or a chemical reagent to obtain chemical products such as micromolecules, compounds, fuel with monomers, different polymers. The basic principle for plastic pyrolysis technology is that the original resin high polymer in a waste plastic product undergoes a relatively thorough macromolecular chain decomposition to make it return to a low molar mass state and thus obtain a product with high use value. Technologies include waste thermoplastic plastics technology, technology for making asphalt felt and plastic oil, reducing agent technology for an iron making furnace and direct oilation technology, among which oilation, gasification technologies are relatively complicated and still at a practical development stage. Chemical decomposition refers

to waste plastic hydrolysis or alcoholysis (ethanolysis, methanolysis and ethylene glycol decomposition) process in which a decomposition reaction is conducted to turn plastics into a monomeric substance or substance with low relative molecular mass and make it again become a raw material for polymer synthesis. Methods include a catalyst decomposition method and reagent decomposition method.

2) Low-carbon technologies for paper packaging waste recycling

(1) Regenerated papermaking based on paper packaging waste: regenerated papermaking based on waste paper mainly covers pulping and papermaking. Technological process for pulping: repulping, purification, screening and concentration; papermaking is carried out by delivering waste paper pulp to a paper machine, and going through a net, squeezing, drying and press polishing to make it into cylindrical paper or sheet paper.

(2) Pulp moulding technology: such methods as vacuum moulding, hydraulic moulding are adopted to rapidly, evenly sediment sundries-free collected pulp into a net model, followed by compression and drying to make a new paper packaging product.

(3) Paper-based reuse technology: hydraulic recycled pulp technology for making raw material for plastic-aluminum products or polyethylene plastics and metallic aluminum.

(4) Plastic wood technology: packaging wastes are pulverized and squeezed to directly make plastic wood products such as indoor furniture, outdoor horticultural facilities, industrial trays.

(5) HB composite board technology: a composite soft package is directly pulverized, hot-pressed to make HB composite board which is processed into a garbage bin, for example[14-15].

3) Low-carbon technology for recycling of metal packaging waste

(1) Separation technologies including gravitational separation, electrostatic separation and thermal separation

(2) Metal separation and furnace-based smelting regeneration technology: aluminum scrap is smelted in a countercurrent two-chamber reflecting furnace, an open smelting chamber reflecting furnace or other furnaces to obtain wrought aluminum alloy, cast aluminum alloy, and a deoxidizing agent for smelting steel and iron alloy. In addition, with respect to aluminum scrap, a leaching process and dry process can also be used to collect aluminum particles from scums and slags. Aluminum product packaging waste can also be used to develop a new product–aluminum polychloride mainly used as purifier for domestic water and industrial water as well as industrial wastewater.

4) Low-carbon technologies for recycling of glass packaging waste

(1) Separation technologies for glass packaging waste: gravitational separation, dense medium separation and optical separation technology

(2) Furnace-based disposal of glass packaging waste: all glass packaging containers which cannot be used, have been damaged or fragmented, can be disposed of in a furnace; furnace-based disposal not only saves raw materials and sodium carbonate for

making glass, but also acts as the fluxing agent.

(3) Glass packaging waste regeneration technologies: technologies for making glass ceramics, foam glass, fiberglass and building materials

5) Composite soft packaging waste collection technologies

(1) Thermal decomposition collection technologies: thermal decomposition collection technologies for composite soft packaging waste are mainly divided into the following four aspects: firstly, combustion technology for generating heat, steam and power; secondly, pyrolysis technology for making fuel gas, fuel oil and carbon black with medium and low heat value; thirdly, gasification and pyrolysis technology for making fuel gas with medium or low heat value or NH_3, CH_3H; fourthly, gasification and pyrolysis technology for making heavy oil, kerosene and gasoline.

(2) Chemical separation technology: currently, one method specially designed for separating polyethylene and aluminum foil composite material has been invented in China; as technical pure acetic acid and methanoic acid can be recycled and collection costs are relatively low, this separation method can concurrently deliver good economic benefits and social benefits; however, it is highly professional. Brazil's Paraguay Institute of Technology has developed a new technology for collecting composite soft packaging materials, which uses a special chemical solvent to collect three layers of different materials (paper, aluminum foil, plastics). Key technology for these chemical treatment methods lies in the treating agent.

(3) Utilization of other new technologies: a solvent separation method based on specific gravity identification, wind-driven screening, electrostatic separation, cryogenic pulverization, dissolution separation method, hydraulic separation method, X ray fluorescent separation method, infrared separation have become new trends for researching waste polymer classification recycling technologies.

With social development, people attach increasing importance to environmental protection and resource conservation, and require active improvement and innovation of packaging waste disposal technologies: (1) A recycling method is one method for the most active promotion of material recycling and is an effective recycling method for protecting resources and the ecological environment. This method can also be divided into collection-recycling-reusing, mechanical treatment and recycling, chemical treatment recycling. Chemical treatment technology is designed to turn harmful substances in plastic waste into safe and stable substances through chemical reaction, and it is considered as a harmless disposal technology, such as the method for separating polyethylene and aluminum foil composite material, with relatively low collection costs; therefore, this separation method can deliver good economic and social benefits. (2) Microbiological decomposition technology: it refers to one disposal method which relies on the role of microorganisms extensively distributed in nature to convert easily biodegradable organic components in packaging waste into humic fertilizer, methane gas or other chemical conversion products, so as to make packaging waste harmless, and it is a method with good prospects.

8.3 Development Strategies and Trends in Low-Carbon Technologies for Packaging Waste Collection and Disposal

8.3.1 Development strategies of low-carbon technologies for packaging waste collection and disposal

Wide application of packaging waste collection and disposal technologies is the effective way of reducing resource consumption and environmental costs, maximizing social and economic benefits. However, there are many problems in packaging waste recycling in China: (1) The production process is backward and resources are severely wasted; (2) There is no green consumption concept; (3) The packaging waste collection rate is low; (4) The low carbon degree of recycling technology is relatively low, for example, incineration disposal technology and thermal decomposition collection technology may consume high-carbon energy and cause new pollution such as waste gas, while saving energy or reducing pollution[3,9,15]. Therefore, it is necessary to improve the process, vigorously push forward carbon reduction or low-carbon technologies to enhance the economic value and social value of packaging waste recycling and promote low-carbon development of the packaging industry.

In order to develop carbon reduction technologies for packaging waste disposal, it is essential to firstly improve the packaging waste collection network. Actions should be taken to accelerate improvement of the packaging waste collection network, and build a roadside collection and distributed collection system for packaging waste, and encourage manufacturers and retailers to gradually establish a collection system, and apply an environmentally friendly package trademark system, and stimulate intermediaries and associations to set up a package collection organization specializing in collecting and classifying packaging waste[16]. Secondly, efforts should be made to develop one package collection disposal system suited to national conditions. Rational disposal of packaging waste is an important move for mitigating its environmental pollution and save resources. Rational disposal of packaging waste should be strengthened in design, manufacturing, circulation and consumption in a "package lifecycle": (1) Package design and manufacturing should enable reuse and easy collection of packaging containers as far as possible, cause no harmful substances at the disposal stage and prevent excessive packaging. (2) One energy and resource-saving packaging production and circulation system should be developed and, in principle, enterprises should be required to recycle and reuse their waste. (3) In commodity consumption, environmental protection awareness should be enhanced, and values and concepts should be changed, and resource-saving rational packaging should be adopted, and recycling of packaging waste after discharge should be actively supported. (4) Post-discharge disposal of packaging waste entails the establishment of one rational collection system acceptable to the general public and suited to local reuse conditions; waste should be effectively collected and transported; reusable and renewable waste packages should be collected, while non-reusable and renewable packages should be incinerated and landfilled.

8.3.2 Development trends and routes of low-carbon technologies for packaging waste recycling

A package focusing on "low carbon" represents the development trend of the packaging industry and is an important part of the development of a low-carbon economy[17]. A low-carbon circular economy requires reduction, reuse and recycling of packaging waste: (1) Reduce: subject to satisfying functions including the protection of products, facilitation of storage and transport, promotion of sales, resources not in short supply and renewable materials should be chosen for making packages as far as possible, the quantity of packaging materials should be minimized, and moderate packaging made of minimal materials should be adopted so as to reduce packaging waste at source. Excessive packaging should be opposed to decrease carbon emissions from package production and use. (2) Reuse: package design should give priority to material reuse and its ruggedness and durability, and extend the service life of a package and increase the reuse rate. Reuse of packaging waste can effectively save energy, reduce carbon emissions and resource pressure. Therefore, it is necessary to intensify technical innovations in packaging waste collection, such as chemical treatment technology and microbiological decomposition technology, and boost the collection and utilization of packaging waste, so as to effectively save resources and energy. (3) Recycle: "waste" should be reprocessed and re-treated as far as possible to make it serve in the making of new products of the same type or different type and thus re-enter the production process so as to reduce garbage. Recycling is conducted in the following two ways: firstly, primary recycling, which is the most ideal recycling mode and is designed to recycle and turn packaging waste abandoned by consumers into new products being the same as the original ones. For example, waste packaging paper bags are processed, treated and regenerated to make new paper bags; secondly, secondary recycling, which is one recycling method for using waste to produce other product raw materials different from the original. (4) Low pollution: packaging materials should satisfy conditions including less environmental pollution, cyclic utilization and self-degradation, and green package development should be intensified.

Technical innovation is the key to promoting the development of the low-carbon packaging industry and achieving a low-carbon economy. The economic development mode can be transformed into "low carbon" only when low-carbon technology innovations are continuously improved. In the development and innovation of low-carbon technologies for the packaging industry, firstly efforts should be made to strengthen independent innovation, research and develop new technologies for disposing of waste such as plastics and waste paper. It is necessary to fully utilize the talents in research institutions and universities, and actively carry out theoretical research on low-carbon technologies, and make full use of the driving power of the market to stimulate research activities. Secondly, it is essential to reinforce international cooperation. Currently, there is a big gap between China's packaging waste disposal technology and the advanced level in developed countries; in order to increase the level within a short time, it is imperative to strengthen international exchange and cooperation in addition to relying on domestic independent innovations. Enhancement of international technology cooperation and transfer can enable global sharing of technical development, greatly cushion resource and environmental pressure from packaging waste. Finally, it is necessary to give play to the role of the government and obtain policy support. Efforts should be made to guarantee input including funds, manpower and the technology platform, smooth development and the optimal allocation of social resources, and improve the price mechanism.

8.4 Conclusions

China's packaging industry is growing rapidly; however, packaging technologies lag behind those in developed countries, and packaging materials with high energy consumption and disposable packaging materials are still widely applied, and the packaging industry has not completely embarked on low-carbon development in China. Recycling of packaging waste is an important link in developing the industrial chain in the circular economy in the urban green packaging industry, and represents great significance including environmental protection, rational utilization of natural resources, coordination of ecological benefits and economic benefits. Currently, a scientific and complete waste collection disposal system has not yet been developed for packaging waste collection; processes are underdeveloped; the degree of technical innovation and low-carbon development is low. Therefore, it is suggested that packaging waste collection should be intensified; a scientific and rational collection system supported by the market should be established to design package collection, speed up the development and application of collection and carbon reduction technologies, improve recycling of packaging waste so as to make new contributions to the green packaging industry and better develop the green economy.

References

[1] National Afforestation Environmental Protection Commission and State Forestry Administration: *Broad Prospect for Development of China's Paper Packaging Industry, China Green Times,* China Forestry Publishing House, 2011.

[2] Tan Jingmei, Liang Kun, Sun Kewei: *Disposal of Packaging Wastes, China Resources Comprehensive Utilization*, 2003(4):31-34.

[3] Yang Fan: *Survey and Countermeasure Research on Current Problems in Urban Packaging Wastes, China Packaging*, 2010(2):71.

[4] Zhang Jingzhong: *Enterprise Marketing Innovations from the Perspective of Circular Economy, Ecological Economy*, 2006(1).

[5] Zhang Hongxu: *Green Packaging——New Subject for Environmental Protection, Shanghai Business*, 2000(9).

[6] Fan Yuansheng: *China's Environmental Management of Urban Domestic Garbage, Environmental Economy*, 2005(10).

[7] Zhang Jingzhong: *Pollution and Recycling of Product Packaging Wastes, Ecological Economy*, 2006(11):82-88.

[8] Tang Guohu: Current Pollution, *Recycling and Environmental Development of Packaging Wastes, Inner Mongolia Petrochemical Industry*, 2007(7):33-34.

[9] Yang Xinfang: *Collection and Reuse of Packaging Wastes, China Packaging*, 2006(4):40-42.

[10] Guo Caifeng, Xu Bo: *Recycling and Strategies of Packaging Wastes in China, China Packaging Industry*, 2007(1):32-34.

[11] Zhao Baoyuan, Shi Kaijian, Sun Bo: *Comparative Analysis of Foreign Packaging Waste Collection System and Inspirations, Ecological Economy*, 2009(3):103-106.

[12] Chen Jinghua: *Technologies for Collection, Disposal and Reuse of Plastic Packaging Wastes, Publishing & Printing*, 2002(4):38-40.

[13] Yang Wei: Overview of Recycling Technology for Composite Soft Package Wastes, *Printing Technology*, 2008(24):47-49.

[14] Cui Zhongwei: *Discussions about Technologies for Recycling of Composite Packaging Based on Waste Paper, Shanghai Papermaking*, 2009(3):62-67.

[15] Zhao Yanwei: *Research on Comprehensive Treatment of Packaging Wastes, Packaging Engineering*, 2000(6).

[16] Jiang Xiaohua: *Research on Packaging Waste System from the Perspective of Circular Economy, Northern Economy*, 2008(4):65-66.

[17] Wang Weiwei, Yang Fuxin, Hu Anhua: *Research and Application of Low-Carbon Technologies for the Packaging Industry, Packaging Journal*, 2010(4):42-45.

Chapter 9

Carbon Reduction Technologies for the Packaging Product Manufacturing Process and their Application Research

Liu Jianwen Yao Dihui

Abstract: The green and low carbon package is an important strategy that the packaging industry must use to achieve sustainable development; it must achieve green packaging using green packaging material, green packaging design and the rapid development of green packaging products in the overall industrial packaging chain. The production of packaging products is the main cause of the carbon footprint in the package. It will be important to make an effort to use carbon reduction technology and applications, energy-saving emission-reductions and applications, renewable energy sources applications in the technological innovation of package-production manufacture, carrying out a technical route for a distributed energy station providing multiple forms of energy to complement each other, based upon CWM clean fuel. Thereby this will achieve the target of utilizing energy sources and resources for packaging waste material, leading the packaging industry to achieve a low-carbon and green transformation.

Key Words: green package; production-manufacturing; carbon-reduction technology; application research

9.1 Packaging Carbon Footprint Research Methods and Progress

9.1.1 Carbon footprint and calculation methods[1]

Carbon footprint is an important indicator for measuring greenhouse gas emission and currently the internationally accepted important decision-making basis for greenhouse gas carbon emission. Carbon footprint characterizes direct and indirect greenhouse gas emission caused by human production and consumption conducts. In recent years, with vigorous implementation by relevant international organizations and developed countries, carbon footprint has been widely applied in various fields, including industrial energy conservation and emission reduction, product eco-label and international carbon trading.

In 2008, Society of Environmental Toxicology and Chemistry (SETAC), Europe Life Cycle Assessment (LCA) Steering Committee discussed the needs for carbon footprint measurement methods and relevant standards, and believed that current LCA method was fully capable of calculating product-related greenhouse gas emission according to ISO 14040 series standards, and carbon footprint can be deemed as an assessment result of global warming potential (GWP) in LCA. Currently, calculation methods mainly used for carbon footprint include Process Analysis (PA) and Economic Input-Output (EIO).

1) Process Analysis, PA

Process Analysis means that tracing starts from product end to source, and various unit processes relating to product (including resources, energy extraction, production, transport, product manufacturing, etc.) are connected, and complete lifecycle process diagram is developed, and then greenhouse gas emission data concerning various process units in process diagram are collected, and quantitative descriptions are made, and finally all greenhouse gas emissions are uniformly characterized by CO_2 as the equivalent. Conversion among different greenhouse gas factors can be conducted by making reference to greenhouse effect equivalent factors released by Intergovernmental Panel on Climate Change (IPCC) (see Table 9-1). Calculations based on complete process analysis are relatively accurate and complete process analysis is mostly used to assess carbon footprint of product or enterprise, while simplified process analysis is highly operational and is mostly used to assess carbon footprint of individual or household.

Table 9-1 Equivalent factors concerning various greenhouse emissions released by IPCC

greenhouse gas	Equivalent factor	unit
CO_2	1	CO_2 - eq
CH_4	21	CO_2 - eq
CO	310	CO_2 - eq
SF_6	23900	CO_2 - eq
HFC_S	12-11700	CO_2 - eq
PFC_S	6500-9200	CO_2 - eq

2) Economic Input-Output, EIO

Being contrary to PA, EIO starts from source (raw material extraction, crop planting, etc.) and extends backwards until final abandonment. In assessment, average data concerning various sectors at national level (mining, transportation, product manufacturing, sales, etc.) are used, and multiplication by supply chain strengths among product-related sectors is made to calculate carbon footprint of the whole system. This method involves simple collection of data and enjoys obvious advantages in analysis of macro carbon footprint and is mainly used to calculate carbon footprint at municipal or national level.

9.1.2 Application of lifecycle assessment in printing and packaging fields

1) Lifecycle assessment of printing production

With respect to printing production, lifecycle assessment about adverse effects caused by printing process on the environment may be made, or printing materials (printed material, printing ink, etc.) may be comparatively assessed to choose environmentally friendly production process and printing material.

Japanese Hiromi Uchida et al adopted LCA method to assess intaglio printing ink[2]. Data collection during assessment covered quantity of excavated resources necessary for printing ink production, energy consumed during printing ink production, printing, drying, solvent combustion and solvent recovery, emissions of CO_2 or organic solvents etc in the atmosphere. Assessment factors include resource demand, energy consumption, greenhouse effect, ozone depletion, emissions of strong photochemical oxides and toxic gasses and other environmental impact types.

Research shows that resource consumption in oil printing solvent recovery exerts the greatest environmental impact, and combustion of oil printing solvent causes a large amount of CO_2, and it has the greatest greenhouse effect environmental impact among all comparison objects, followed by oil printing solvent recovery and oil printing, among which oil printing emits organic solvents to the atmosphere at the time of printing drying, the resulting strong photochemical oxides have the heaviest impact on the environment. Water printing evaporates ethanol at the time of printing drying, while such ethanol has an environmental impact.

2) Lifecycle assessment of packaging product and production system

Product packages of many international famous enterprise groups are ecologically designed by lifecycle assessment method, such as beverage packaging container (US Coca-Cola Company, 1969), beer packaging (US Environment Protection Agency, 1974), washing powder packaging (US Procter and Gamble, 1988), fresh milk packaging (German Federal Environment Agency (UBA), 1995), reusable PET bottle and glass bottle (German Federal Environment Agency (UBA), 1996), mineral water, carbonated beverage, fruit juice and wine packaging container (German Federal Environment Agency (UBA), 2000), yogurt product delivery system (US Stonyfield Farm, 2004), etc[3].

Ren and Huo[4] conducted lifecycle assessment about corrugated paper production processes (including such board making processes as glue making, corrugation, bonding, drying and dividing and cutting; printing processes including printing and die cutting; carton forming processes including carton bonding and packaging), and found that main factors having an environmental impact included fossil energy consumption, global warming, acidification and eutrophication, among which fossil energy consumption mainly involved power and coal consumption in production processes; global warming and acidification were mainly caused by power consumption in coal-fired power generation and gas emissions from steam production by use of coal in board making processes including glue making, corrugation, bonding; eutrophication was caused by washing water discharge from starch glue maker and printing machine; power consumption in board making processes accounted for more than 50% of total power consumption, and its impact on each of fossil energy consumption and global warming accounted for 77% of total impact.

Qin [5] carried out lifecycle assessment about beer packaging, while research showed that factors with the greatest environmental impact on beer packaging system successively included eutrophication, acidification, global warming, dust and smoke, solid wastes; packaging unit consumed the largest quantity of water resources and coal resources.

Xie et al.[6] used lifecycle assessment method to research full lifecycle environmental impact of paper-plastic-aluminum composite packaging for milk and conducted comparative assessment about such impact and environmental impact of plastic packaging. Research showed that environmental impact values of paper-plastic-aluminum composite packaging and plastic packaging were 5.225 and 4.670 Pt respectively; the largest proportions of environmental impact occurred at the stage of raw material acquisition in the whole lifecycle, both were about 80%. Fossil resource consumption in plastic packaging was more than two times that in paper-plastic-aluminum composite packaging; fossil resource consumption is unrenewable, thus its environmental impact can't be decreased through relevant routes. Relatively large impact of paper-plastic-aluminum composite packaging is attributable to the fact that it has not yet been better recycled, while developing aluminum and plastic separation regeneration technology and increasing paper-plastic-aluminum composite packaging collection rate can mitigate its environmental impact.

Guo et al.[7] adopted lifecycle assessment method to assess terminal measurement model LIME, and conduct quantitative analysis of environmental impact from energy consumption and material consumption in book and periodical offset printing production in Dalian Region, Liaoning Province and their emissions, etc. Assessment analysis result showed that environmental impact from book and periodical offset printing production process mainly involved fossil energy consumption, global warming, acidification and eutrophication, etc. Power consumption in printing process accounted about 51% of power consumption in overall process and power consumption in post-printing processing process also made up 37%; power consumption in both large processes was main factor resulting in environmental impact indicator concerning fossil energy consumption in this research case. Gasses from coal-fired power generation caused such environmental impact as global warming and acidification. In addition, volatilization of glues including hot melt glue and film coating glue, etc. in post-printing processing process, collection and thermal treatment of split film and packing belt were also factors leading to environmental impact indicators concerning global warming and acidification in this research case.

Xie et al.[8] applied lifecycle assessment method to research lifecycle environmental impact of edible oil polyester (PET) packaging, and conducted comparative assessment about environmental impact from different disposal modes. Research showed that environment impact potential at the stage of PET packaging raw material acquisition made up an extremely high proportion in full lifecycle environmental impact potential and accounted for 81.8% of pre-disposal environmental impact potential. Full lifecycle environmental impact from PET packaging mainly involved fossil fuels, damage to human body caused by inorganic matters, climate change, and less dealt with carcinogenicity, ecotoxicity, acidification and eutrophication, etc. Among three main disposal modes including incineration, landfill and regeneration, incineration and landfill increased environmental impact potential before PET packaging disposal stage by 5.1% and 3.6% respectively, while regeneration decreased it by 63.9%.

3) Progress in packaging carbon footprint research

Currently, there is no real carbon footprint assessment about the packaging industry in China where only few enterprises have performed enterprise carbon footprint assessment, for example, APP (China) announced smooth completion of the first carbon footprint assessment project on March 26, 2009; it is the first enterprise for comprehensive assessment of carbon emission in domestic pulping and papermaking industry and also the second enterprise for comprehensive assessment of carbon emission in global papermaking industry (another enterprise is Kimberly-Clark).

All researches of packaging carbon footprint by domestic scholars have started from packaging lifecycle assessment method to identify main factors of process for environmental impact through lifecycle assessment of specific packaging category and estimate carbon emission by means of main factors for environmental impact so as to obtain carbon footprint of such package. Dai [9] believed that with regard to four large packaging categories including paper, plastic, glass and metal packaging, carbon footprint can be found from their energy and resource consumption. The result of lifecycle analysis of corrugated board production process shows two carbon footprints from corrugated carton: one is indirect, namely the more power consumed in coal-fired power generation is, the more indirect emission of carbon dioxide is; another one is direct, which results from steam generated by fuel as energy during production. In regard to carbon footprint from other kinds of packaging, carbon footprint from similar corrugated carton can also be found from production process from acquisition of raw materials to delivery of products from factory, which mainly stems from power and coal consumption in production.

9.2 Applications of Carbon Reduction Technologies in Packaging Product Manufacturing Process Technology Innovation

9.2.1 Applications of carbon reduction technologies in enterprises in China's packaging product manufacturing field

Packaging product manufacturing process technology innovation is an important way for package printing enterprises to expand production potential, lower production costs and enhance competition vitality and is also the subject and management direction which are worth research by contemporary package printing enterprises. Kang [10] summarized, in production practice, the following top ten process innovation improvement solutions adopted by package printing enterprises in production: die-cutting board making process innovation solution, innovative process for using letterpress printing machine for embossing, letterpress solid plate process innovation improvement solution, process innovation solution for rapid plate lock-up through electrochemical aluminum, innovation improvement solution for improving double-sheet control accuracy of printing machine, innovation improvement solution for folding endurance counter, innovation solution for rapid removal of corrugated carton die-cutting quality fault, innovation solution for solving quality problem concerning carton color fading, innovation solution for addressing paper feeding abnormality of semiautomatic folder gluer, etc.

Zhejiang Ruijiang Paper Products Co., Ltd. monthly saves coal by 18% by refitting speed regulation device for corrugated paper board production line and counter, and adopting computer program controlled and automatic deviation correction system, steam recovery unit, etc; shaftless bracket replaces telescopic paper gripping head to fully use original incompletely recovered paper; adoption of longitudinal tungsten steel cutting tool reduces troubles from replacement and saves costs by about 26%.

Shanghai Dasong Corrugating Roll Industrial Co., Ltd. is the first Chinese enterprise which launches tungsten carbide corrugating roller with the best wear resistance performance at the international level. Such corrugating roller has service life four times more than that of ordinary corrugating rollers and overcomes various defects and losses incurred to carton enterprises by easy wear of ordinary corrugating rollers.

Beijing Strongstud Electromechanical Manufacturing Factory adopts high carbon alloy steel to make corrugating roller gear sleeve material; as TC coating is sprayed onto the surface of corrugating roller gear, its wear resistance performance is three times that of the original chromeplated roll; this solution alone can save about 410t core paper (0.9 million yuan) for carton factory. Beijing Zhonghe Jiuyi Nanotechnology Co., Ltd. (a nano surface treatment company with independent intellectual property) applies nano surface treatment technology onto the corrugating roller surface to make corrugating roller hardness close to diamond, its comprehensive cost is 3/4 and 1/3 of that for chromeplated roll and thermally sprayed tungsten carbide roll respectively.

Shanghai Shenjia Machinery Industry & Trading Co., Ltd. launched a paper receiver capable of completion within 2-3s and used as the supporting equipment for high-speed corrugated paper board production line in September, 2010, which greatly improves the work efficiency of production line, reduces loss and increase benefits.

Qingdao Meiguang Machinery Co., Ltd. rolls out non-stop automatic paper receiver which can achieve completion within 0.3s, automatic tensioning and testing, semiautomatic die-cutting machine and high-speed fully automatic laminating machine, etc.

Guangdong Dongguang KaDar Computer & Datacom China (Taiwan production base) offers corrugated board (carton) packaging machinery computer and advanced carton production management ERP system.

Beijing Shouhang Wanyuan Packaging Machinery Co., Ltd. has successfully developed four-layer composite corrugated board and forming machine. This machine can meet four-layer corrugated board requirements by merely adding one special gluer (double-wall gluer) to regularly bond two core papers to make composite corrugated core paper subject to not affecting normal production of the original three, five-layer board production lines.

Zhang Shiquan, Director of Sino-Japanese Concave-convex Structure Research Institute (Jinan), has successfully developed concave-convex paperboard technology. This technology is easy to operate and can serve the intended purpose by merely substituting corrugating roller with this concave-convex structure on the existing equipment.

Jiangsu Qidong Ouxing Packaging Material Factory has successfully developed a special "corrugated" board. This concave-convex core layer different from corrugated shape results from adopting special forming technology and choosing different formulas and processes in papermaking to make various kinds of paper sheets with different applications and different gram weights; these paper sheets enjoy high expansion and contraction rate, which can either undergo punch forming or be rolled into paperboard by using single-face corrugating machine and substituting special roller; in later carton making process, the existing process equipment can be used for forming, while all of its various physical indicators are higher than that in the case of conventional corrugated carton, and it features multi-layer multi-layer composite and can replace heavy-duty carton.

Guangdong Kelon Electrical Holdings Co., Ltd. adopts flexibly printed minimally corrugated carton in which CB corrugating is changed to BE corrugating to reduce carton volume and transportation costs; costs are comprehensively decreased by 15% and comprehensive compressive strength increases by more than 10%.

Guangxi Guilin Chenguang Paper Products Co., Ltd. has successfully developed sandwich corrugated board (Wazhongwa). Its excellent structure substantially enhances board strength; shockproof compression-resistant packaging carton made of this board can reinforce composite structure packaging, etc. As this board features compact structure, and is seamless, free from nails, foldable and easy to form, it can lower comprehensive packaging costs by about 30% and enhance packaging integrity and beauty.

Henan Doyoo Industrial Co., Ltd. has successfully developed local composite technology for corrugated board and fiber reinforcement technology. Carton thickness increases while paper use level decreases; flat crush strength of carton increases by 55%; strength of main pressure-bearing surface rises by 40%; overall compressive strength goes up by 30/%; fall strength increases by 25%; it can save costs by 20%-60% compared with carton with equivalent specification. As 18.5 billion m^2 corrugated boards were produced nationwide in 2004, if this patented technology is adopted, more than two million tons base paper can be annually saved; it is called advanced technology for carton reduction, creating great prospects for carton packaging reduction with Chinese characteristics. This technology starts from changing board structure to enhance overall strength of carton while not increasing gram weight of paper and number of layers in the entire carton, thus providing technical support for carbon reduction.

These process technology innovation and improvement solutions reflect multidimensional technical innovations of enterprises in production processes and new technologies, new product development, etc, and are of paramount guiding significance for developing carbon reduction technologies in the packaging industry.

Given domestic and foreign packaging technology development trends, packaging process technology innovations which act on industry-wide technological leapfrogging mainly include cold setting corrugated board production technology, process technology for replacing EPS foam packaging by honeycomb paper packaging, Catch-Cover carton-free easy-to-carry single drug plate package, Striptabs easy-to-carry drug film package, Protektapak package.

9.2.2 Typical carbon reduction technologies for packaging product manufacturing process

1) Cold setting corrugated board production technology

This technology means corrugated paper is formed in normal temperature, and power boiler and steam heating are not required, and no pollutant emission occurs, and production costs are reduced and board strength is enhanced. Its major advantages: (1) Production line is based on low-temperature setting and low-temperature compositing; evaporation of standard-exceeding moisture is conducted by electromagnetic wave roll; startup dispenses with preheating so as to maximize energy-saving effect; (2) New generation of independently developed high-molecular binding agent is used; high temperature heating curing is not required; (3) Base paper contraction ratio is decreased to save paper sheets; paper interruption is not required in case of shutdown.

Sichuan Mianyang Xindonghua Packaging Machinery Manufacturing Co., Ltd. used its patented technology and scientific research achievements to cooperate with Sichuan Xingnan Craftworks Factory to successfully develop cold forming microwave drying corrugated board production line. This production line incorporating independently developed patent technology is designed to make corrugated board through cold forming microwave drying and is based on multiple invention patent technologies, which changes production process in which boiler (oil furnace) is not required for heating and setting corrugated board; as its design is unique and process is novel, it is the first one in China. Main technical innovations: microwave drying technology and independently developed binding agent are adopted; heat source for drying is controllable; bonding strength is high; drying is uniform and smooth; production line is rationally arranged; structure is compact; comprehensive costs are low; no boiler (oil furnace) investments are required; there is no "three wastes" discharge; it is energy-saving and environmentally friendly. This production line is the first and only corrugated board production line with proprietary intellectual property rights in China and also represents the latest process at the international level. Application of this production line will offer a new development space for the whole industry.

2) Process technology for replacing EPS foam packaging by honeycomb paper packaging

Conventional EPS foam plastics is light material made of polystyrene resin as main part by adding additives such as foaming agent, etc; it is one kind of buffer structure material most widely used at present. With respect to technology, EPS form plastics is characterized by closed-cell structure, low hygroscopicity and excellent water resisting property; furthermore, it enjoys various advantages: density is low; mechanical strength is high; buffer performance is excellent; processability is high; mold forming is easy; colorability is high; thermal adaptability is high; radioresistance is excellent, etc; however, its combustion will release styrene gas which pollutes the environment. Production and use of 1kg EPS form plastics will lead to 6.96kg carbon dioxide emission and methane emission equivalent to 0.4kg carbon dioxide, thus combined emission of carbon dioxide from 1kg EPS form plastics is 7.36kg.

Honeycomb paper packaging material based on paper-based fiber is one kind of light, high-strength structural composite material developed and produced by imitating honeycomb structure built by bees; it is made of fully recovered recycled paper as raw material; besides high strength, small mass and less use of materials, it also enjoys the following advantages: price is low and it can be 100% recycled, thus it is one kind of ideal resource-saving and environmentally friendly new material.

3) Water soluble film packaging

Water soluble film is made from polyvinyl alcohol and starch as main raw materials by adding various auxiliaries such as surfactant, plasticizer and anti-sticking agent, etc. Its process is different from conventional plastic film forming process; its process: raw materials are made into water-soluble glue with 18%-20% solid content which is spread onto mirror stainless steel band and is stripped from steel band after drying and film formation, and enters drying room for being dried to the specified moisture, followed by edge trimming and rolling to obtain product film. As one kind of novel packaging material, hi-tech water-soluble plastic packaging film is widely applied in packages for products used in water such as farm chemicals, chemical fertilizer, pigments, dyes, cleansers, water treatment agents, mineral additives, detergents, concrete additives, chemical reagents for photographing and chemical reagents for horticultural treatment etc in Europe, USA and Japan etc; its main characteristics: (1) It is environmentally friendly and safe, thoroughly degradable (100%); final products of degradation are CO_2 and H_2O; disposal of packaging wastes can be thoroughly addressed; (2) Its use is safe and convenient; direct contact with the packaged contents by users is avoided; it is suitable for serving as package for materials detrimental to human body; (3) It can be correctly measured to prevent wasting; (4) Mechanical property is excellent, and heat sealing can be performed and heat sealing strength is relatively high; (5) It has anti-fake function; it can be used as the best weapon for high-quality product anti-fake and extend service life of high-quality products.

Foreign countries (mainly Japan, USA and France, etc.) have witnessed manufacturing and sales of this kind of product by such companies as US WTP, CLLP, French GRENSOL, Japanese Nippon Synthetic Chemical Industry Co., Ltd. etc. With support from Technology Department of China National Packaging Corporation, domestic Zhuzhou Institute of Technology (currently Hunan University of Technology) and Zhaoqing Fangxing Packing Material Co., Ltd. jointly developed hi-tech water soluble film and its production equipment which has been put into production, while its products fill domestic vacancy and are entering the market.

4) Catch-Cover carton-free easy-to-carry single drug plate package

Catch-Cover carton-free easy-to-carry single drug plate package is innovative drug packaging process developed by German Romaco Siebler, which can directly integrate drug patent information to primary package for drug. Drugs are packed in elastic aluminum plastic film strips of different sizes and different shapes, with four sides bonded by thermal compression, so to as make the smallest packaging unit for drugs. With adoption of Catch-Cover packaging mode, tablet or drug film is covered by thin aluminum plastic film, while all information about this product is printed at inner side of aluminum plastic film. This packaging mode can replace conventional corrugated paper box packaging mode. Both right and reverse sides of aluminum plastic film for packaging are firmly connected by weld seams generated at the time of thermocompression bonding. In order to facilitate administration by patients, the right side of Catch-Cover package has one breach convenient for opening aluminum plastic film. Fully automatic Catch-Cover single plate package is completed by production line which integrates aluminum plastic film feeding system, thermocompression bonding equipment and trimming device. Dimensions of Catch-Cover aluminum plastic film delivered by feed mechanism are fully identical with that of strip-shaped package, and thermocompression bonding is firstly conducted according to strip-shaped package mode. On subsequent cutting sites, Catch-Cover tablets in single plate package, drug films are cut into separate packaged finished products according to their sizes, and finally delivered to counting, sorting and transport equipments. In the case of fully automatic production, maximum production capacity of this production line is 1,200 Catch-Cover single plate packages per minute.

5) Striptabs easy-to-carry single drug film package

Striptabs easy-to-carry single drug film package solution recently launched in the market is the best example for user-driven product design.

Production process of the whole production line is divided into the following three sub-processes: the first step is designed to process and treat active ingredients of drug, and mix wet and dry active ingredients of drug into homogeneous, blister-free active pharmaceutical ingredients by adopting blister-free homogeneous process technology researched and developed by Romaco Fryma Koruma. The second step aims at using coating equipment researched and developed by Optimags to make active pharmaceutical ingredients into thumb-wide, very thin flake and dry it to residual moisture less than 15%. The third step is designed for high-precision cutting. In order to cut drug film, drug film roll coated with active ingredients of drug is unfolded and delivered to thermocompression bonding equipment. Unfolded drug film roll is cut in this equipment to 20mm×30mm block, with the cutting size reaching to high precision. After accurate correction, rotary cutting tool ensures that edges of unfolded drug film are free from surplus materials.

The drug film which is as thin as skin should be buccally administered and is used as a carrier for active ingredients of drug. Drug is absorbed through sublingual gland rather than enters digestive tract so that active ingredients of drug can more quickly take effect within human body. With 60-100 micron in thickness, drug film can be dissolved within several seconds and directly enters central the nervous system through sublingual gland. Therefore, it is very suitable for hormone therapy and pain treatment. In principle, about 600 active ingredients of drug can be absorbed through skin, thus this packaging mode is widely applied.

6) Protektapak innovative corrugated package

Protektapak package is specially designed for harsh postal distribution chain environment; it is made of sturdy and durable corrugated board and can provide special protection for transport of alcohol products. As shown by testing and validation, its sturdiness is sufficient to provide 100% protection for packaged contents, and it can keep alcohol bottles safe and sound even when it falls from a 5-meter-high position.

Protektapak corrugated package features "one-size-fits-all" structural design and results from mechanical die cutting, gluing and rapid assembly of one corrugated board; its innovation is that extended end plates on both sides can replace foam plastics as buffer materials, and die-cut holes on extended end plates are mainly used for fixing packaged contents (such as wine bottles). Fixing holes with different sizes can be made through cutting in response to bottles with different sizes, thus it is widely applied. In addition, merely simple adjustment can make Protektapak corrugated package suitable for any fragile articles or articles which need high degree of protection, thus it offers very ideal packaging solution to express delivery products.

Protektapak corrugated package can replace currently common polystyrene package which is expensive and causes damage to the environment and is conducive to environmental protection and sustainable development.

7) Coated steel sheet

According to current development trends, application of coated steel sheet in packaging material has started and rapid development is shown. Coated steel sheet, namely, organic coated steel sheet refers to coated metal material obtained by applying organic coating onto the steel sheet surface. This kind of coated steel sheet dispenses with coating process after various packaging products are made in the packaging industry, thus coated steel sheet is often classified as precoated steel sheet. According to estimation made by relevant experts, coated steel sheet will be widely used in metal packaging industry and replace backward coating production process in production of current metal packaging products.

Generally, coated steel sheet is produced by metallurgical enterprise in a centralized way; as it dispenses with coating process in product making, it greatly reduces packaging manufacturing costs. According to estimation, this results in lowering costs for packaging products based on steel sheet as raw material by 5%-10%, saving energy by about 1/6-1/5, especially economizing massive investments in pretreatment and coating equipments for packaging products. Coated steel sheet enjoys advantages of both organic polymer and steel sheet; it shows high colorability, formability and excellent corrosion resistance like organic polymer, and high strength and processability like steel sheet, and is very easy to blank, bend, deeply draw and weld, etc.

Currently, there are about 474 coated steel sheet production lines in nearly 200 companies in 40 countries and regions worldwide, including about 200, more than 50, 120 and 74 coated steel sheet production lines in USA, Japan, Europe and Asia respectively. At present, annual coated steel sheet production capacity in the world is 14 million tons, including 4.5 million tons, about 4 million tons, about 2.7 million tons and about 1.2 million tons in USA, Europe, Japan and other Asian countries respectively. Combined annual production capacity in USA, Japan and Europe accounts for about 84% of global total production capacity. According to incomplete statistics, as of May, 2004, 24 coated steel sheet production lines had been built and 1,616,000t annual production capacity had taken shape in China. Moreover, more than ten production lines are being built and under preparation.

8) Stretch packaging technology

Stretch packaging technology is one modern new technology which has emerged in transport packaging field in recent years and serves as the foundation for promoting containerized transport and logistics industrialization. It adopts stretch film made by specific formula and process technology, advanced electronic technology and stretch packaging machine manufactured by exquisite mechanical manufacturing process to parcel various products with regular or irregular shapes into an integral whole so as to protect goods against scuffing, bumping, breakage, scattering, scratch and reduce oil stain and dirt, economic losses incurred by improper packaging.

Stretch packaging technology can not only change primitive, backward product packaging, but also increase unit load, enhance loading, unloading and transport operation efficiency, ensure the safety of loading and unloading personnel, transportation means; it is the foundation for developing containerized transport and logistics industry. Stretch packaging technology can also greatly reduce goods (product) packaging expenses and increase economic benefits of enterprises, which is also the key to rapid development of stretch packaging technology. Stretch packaging technology replaces various original papers and wood packages, greatly lowers consumption of such resources as timber and paper, etc. and thus plays a very important role in developing low-carbon logistics industry.

9) Innovations in corrugated board properties

Modern paper packaging is characterized by a great variety of structures and highly diverse innovation modes. As the most common paper packaging product, corrugated board has also demonstrated continuous development and innovations of its structural mode during its application; as innovative product, high-strength corrugated composite board breaks away from corrugated horizontal arrangement structure of conventional corrugated board and features corrugated vertical close arrangement structure, and it can replace heavy-duty corrugated board, honeycomb board and wood packaging and is reputed as a new type of environmentally friendly packaging material. It is particularly important for enhancing water resisting property of carton as one kind of external product package. Ordinary conventional corrugated board is relatively poor in waterproof and dampproof properties; there are many troubles and difficulties in recycling of film-coated carton. New type recyclable waterproof paperboard can be better improved in terms of waterproof and dampproof properties by adding hydrolyzable resin, and it is easily degradable in the case of recycling, thus it is an ideal process for replacing film-coated carton. Innovations in modern product packaging structure are shifting from previous simple sales package and transport package to functional packaging structure; now innovative process for corrugated board with anti-corrosion characteristic also comes out, internal paper of such carton is coated with special material so that materials within carton can be kept intact for more than five years without other anti-corrosion measures; this function originates from highly active copper element network which can effectively neutralize corrosive gases. With acceleration of China's circular economy, innovations in paper product structure and production technologies will develop in the following directions: beneficial to ecological protection, energy-saving, material-saving, consumption reduction, economical, simple and convenient; replacement of timber, plastics, glass and metals by paper will become the mainstream of commodity packaging now and in future quite a long time.

10) Innovations in paper packaging solid plate printing process technology

Solid plate is common process for letterpress printing (including flexographic printing) spot color; in the past, solid plate for conventional letterpress process printing was always copper, zinc or resin plate, while flexographic plate process was generally based on resin plate for printing, which easily caused crack and damage to printing plate, affected production efficiency and also increased production costs. Previously, zinc plate, resin plate was used to print cigarette case collar (white board bronze printing process), and clefts often horizontally occurred on printing plate during printing, which greatly affected production efficiency and production costs. With innovation inspirations drawn from offset printing process, scrapped offset printing rubber blanket was adopted to make solid relief printing plate for printing, which can not only tremendously improve product printing quality but also greatly enhance production efficiency. With the adoption of rubber blanket to replace expensive resin plate, copper plate material to make solid plate for letterpress printing (including flexographic printing), pressrun of printing plate can exceed one million so as to better reduce production costs and open up new development prospect for letterpress and flexographic printing process.

11) Technical innovations in process for making paper packaging product materials

The quality of corrugated paper exerts tremendous impact on the strength of corrugated carton; if the quality of corrugated paper is poor, quality of the resulting corrugated carton is also inferior. With addition of nano materials into corrugated paper production process, corrugated board made by special process enjoys relatively high strength, excellent dewatering performance, and is free from picking and flake-off, etc. compared with corrugated board made by ordinary process. Nano particles are very small, thus concave points and convex points on the surface of the resulting corrugated paper relatively decrease. Binding agent use level on fine and smooth paper surface is also relatively low, thus corrugating quality formed by a single facer machine is also good. Main process for producing nano corrugated paper: nano calcium carbonate material and auxiliaries with excellent performance are adopted and processed by special equipment into new type papermaking additive based on nanotechnology, while this additive is directly added into pulp with concentration being about 1% for uniform mixing, and it is then uniformly sprayed by bunch tube flowbox of paper machine onto corrugated paper forming network so that corrugated paper contains nano calcium carbonate. As nano calcium carbonate which is positive is closely bonded to negative fiber, with support from auxiliaries, about 80% nano calcium carbonate is left on the corrugated paper; after high vacuum dehydration, pressing and high temperature drying, nano corrugated paper with about 8% moisture is formed. Corrugated paper made by nano calcium carbonate material has greatly enhanced strength and can be turned into very thin corrugated paper. For example, the strength of carton made by approximately 80g/m^2 low basis weight nano corrugated paper can be comparable to that of carton made by ordinary approximately 150g/m^2 corrugated paper. Nano corrugated paper is an important production way for improving the quality of corrugated carton. In addition, innovative materials such as nano ink, nano printing plate, nano binding agent ad nano oil polish, etc. have expanded functions of printing and post-printing processing production materials, which is conducive to effectively reducing production quality faults and improving production efficiency.

12) Technical innovations in process for making paper packaging equipment parts

The corrugating roller is forming mould for corrugated board and is the most expensive important part in a single facer machine; the height of its corrugating gear determines the forming thickness and strength of corrugated board. Therefore, the produced corrugated board conforms to forming quality requirements only when corrugating roller reaches the following application effects: wear resistance is high and it is not easy to deform. Previously, after the surface of corrugating roller became worn as a result of use, under conventional repair mode, ordinary hard chrome process was adopted to treat such a surface, and its wear-resisting property was not ideal. With the adoption of corrugating roller with silicon carbide—chromium nano ceramic coating layer, its innovative processing process is shown below: the working surface of base material for corrugating roller has 0.08—0.10mm-thick silicon carbide—chromium nano ceramic coating layer which consists of the following three layers: the first layer is the ion diffusion layer firmly bonded to corrugating roller base material; the second layer is the bottom layer bonded to the first layer; the outermost layer is silicon carbide—chromium nano ceramic coating layer bonded to bottom layer. Characteristics of this process: wear resistance of such a corrugating roller is about three times higher than that of the original corrugating roller with hard chrome plating and can better increase the life of corrugating roller, shorten shutdown overhaul time and greatly decrease production costs. Anilox roller is key part for flexographic printing, and its performance has a direct bearing on product printing quality. Conventional metal anilox roller for flexographic printing machine is generally made of low-carbon steel or copper, and inking hole with electronically engraved 45° pyramid structure is made available, and then chromeplating is conducted. Plating of such a roller is thin and easily worn, especially the adoption of doctor blade type inking unit makes it easier to incur wear and reduce ink cell volume so that product printing quality is affected. Laser-engraved nano ceramic anilox roller is made of nano ceramic materials, thus its strength, toughness and wear resistance, etc. are significantly enhanced. Therefore, laser-engraved ceramic anilox roller has been widely applied on flexographic printing machine and will gradually replace metal anilox roller. Characteristics of nanometer can meet the requirements for substantially increasing ceramic hardness, toughness and wear resistance, etc, which provides technical assurance for the exquisite anilox roller and lays a solid foundation for improving carton printing quality.

9.3 Applications of Energy Conservation, Emission Reduction and Carbon Reduction Technologies for Packaging Product Manufacturing Process

9.3.1 Oxygen-enriched, oxygen-fuel combustion technologies for glassmaking

Energy conservation and consumption reduction in glass melting kiln has always been a major subject which attracts the attention of the whole industry; as nowadays energy crisis becomes increasingly worse, excessive dependence of glass melting kiln on high-quality energy has restricted development of the glass industry. Nitrogen which accounts for 78% of the air does not engage in combustion reaction during combustion in glass melting kiln, thus a large amount of nitrogen is wastefully heated and emitted into the air in high temperature, resulting in huge heat loss; nitrogen also reacts with oxygen in high temperature to generate nitric oxide, while emission of nitric oxide gas into the atmosphere is extremely easy to form acid rain, causing environmental pollution. On the other hand, with the development of high technologies, economy and society, various kinds of low-cost, high-quality glass are required, while oxygen-fuel combustion technology exactly serves as an effective means for addressing several issues concerning energy conservation, environmental protection and high melting quality, and is hailed as the second revolution for glass melting technology.

1) Pure oxygen combustion technology

Pure oxygen combustion technology was first mainly applied to the increase of production, extension of service life of kiln and reduction of nitric oxide emission; however, with the development of oxygen preparation technology and relatively stable power costs, pure oxygen combustion technology is becoming a better choice for replacing conventional air-supported combustion, which benefits from advantages of pure oxygen combustion technology in energy conservation, environmental protection, quality and investment, etc.

Application of oxygen combustion is divided into the use of oxygen-fuel combustion technology with pure oxygen combustion, pure oxygen supported combustion technology and locally oxygen-enhanced oxygen-enriched combustion technology, etc. in the whole melting part

Oxygen-fuel combustion technology enjoys the following advantages: (1) Good glass melting quality is good. At the time of oxygen-fuel combustion, glass viscosity decreases, and flame is stable, and there is no direction change, and combustion gas is available within the kiln for a long time, and pressure within the kiln is stable, this is favorable for glass melting and fining and reduces glass bubbles and stripes. (2) Energy conservation and consumption reduction. At the time of oxygen-fuel combustion, heat removed by waste gases and kiln body heat dissipation concurrently fall. Research and practice show that the melting kiln for melting ordinary sodium calcium silicon sheet glass can save energy by more than 30%. (3) Reduction of nitric oxide emission. At the time of oxygen-fuel combustion, emission of nitric oxide among waste gases from melting kiln decreases from 2200mg/Nm^3 to below 500mg/Nm^3, and dust emission and SO_2 emission decrease by about 80% and 30% respectively. (4) Improvement of combustion and melting capability of melting kiln, output in melting kiln. Where oxygen-fuel combustion is conducted within the glass melting kiln, complete fuel combustion is achieved and flame temperature is high, and mixed batch melting speed increases, and melting rate can be increased by more than 10%. (5) Low melting kiln construction costs. Oxygen-fuel combustion kiln structure is similar to the unit kiln, and it dispenses with metal heat exchanger, rangette and regenerator. Kiln body features one single structure of melting part and covers small area, and kiln construction investment costs are low. (6) Service life of melting kiln is long. Oxygen-fuel combustion can divide flame into two zones; temperature of lower part of flame rises due to injection of total oxygen, while the temperature of upper part of flame decreases somewhat so that temperature of melting kiln crown falls and burning loss to crown is mitigated; meanwhile, high-quality refractory material is used in flame space so that kiln age can exceed ten years. (7) Decreasing overall production costs.

2) Pure oxygen supported combustion technology for float glass melting kiln

According to heat transfer theory, energy for melting mixed batch within the glass melting kiln mainly comes from radiant heat from combustion flame within the kiln. Blackness of mixed batch is much higher than that of glass liquid, namely, heat absorption capacity of mixed batch is much higher than that of glass liquid so that thermal load above mixed batch is effectively increased without significant rise of melting kiln lining temperature. This is the reason why a pair of total oxygen spray guns are arranged in the $0^{\#}$ rangette position.

After a pair of total oxygen spray guns is added to the float glass melting kiln, such addition can not only increase production and enhance effects, save energy and reduce consumption, improve glass quality, but also extend the life of glass kiln to a certain extent. Specifically, there are the following advantages: (1) Glass output of glass kiln is increased by 5%-15%. (2) Thermal efficiency of kiln is improved, fuel is saved by 5%-8%. (3) Glass quality is enhanced, bubbles and stones are reduced, rate of finished products is increased by 0.5%-3%. (4) After a pair of total oxygen spray guns are added, overall erosion of kiln body by high pressure thermal current is relatively mitigated, while effective heat for melting mixed batch is significantly increased, heat which may exacerbate kiln body erosion is also decreased accordingly; rapid melting of mixed batch reduces mixed batch carry-over so as to provide guarantee for extending the service life of melting kiln. (5) Dust and scoot emission is reduced by 20%, and the possibility that regenerator checker is blocked is also decreased. (6) Pure oxygen supported combustion system and the original air combustion system are mutually independent, with flexible operation.

3) Locally oxygen-enhanced oxygen-enriched combustion technology

Locally oxygen-enhanced effort is one main application mode when oxygen-enriched air is insufficient. Ideal combustion state in the glass melting kiln: the upper part of flame is anoxic zone where crown can be protected; the middle part is ordinary combustion zone; the lower part is high temperature zone where heat can be effectively delivered to glass liquid. The key to this technology lies in introducing oxygen-enriched gas to the lower part of flame and increasing temperature in the lower part of flame (close to mixed batch and glass liquid surface) so as to change conventional flame combustion characteristics and enable gradient combustion. Temperature rise in the lower part of flame can intensify heat transfer for glass liquid by flame, which is beneficial to melting glass, reducing excess secondary air and ensuring that coefficient of excess air reaches ideal value so that oil consumption is reduced.

Temperature in the upper part of flame is lower than that in the lower part of flame in the case of being locally oxygen-enhanced, which is favorable to the life of crown and breast wall and can mitigate erosion due to decreasing thermal load in rangette and regenerator checker. Locally oxygen-enhanced oxygen-enriched combustion technology can improve fuel efficiency, reduce fuel consumption, increase production capacity, enhance glass quality, lower emission of pollutants (nitric oxide, sulfur oxide, CO_2 and particulate matters), curtail combustion waste gases, boost melting kiln maintenance capability and extend kiln lifecycle.

Locally oxygen-enhanced oxygen-enriched combustion technology input includes equipment costs, labor costs, equipment processing costs and technology use charges; locally oxygen-enhanced oxygen-enriched combustion technology can deliver approximately 4% energy conservation benefits; with calculation based on conservative energy conservation rate 3%, 500t/d float glass melting kiln consumes about 90t oil a day, daily oil saving is 2.7t, assuming that oil price is 3,000 yuan/t, daily saving amounts to 8,100 yuan, and static payback period is 200 days.

Locally oxygen-enriched combustion technology can also improve glass quality, extend furnace age, reduce emission of scoot and nitric oxide, etc, ameliorate the environment and deliver enormous social benefits.

9.3.2 Efficient energy-saving packaging equipment and packaging process automation

As packaging technology progress remains relatively defective, energy consumption and wasting are still unavoidable in industrial production involving packaging process. However, extensive implementation and use of efficient energy-saving packaging equipment, wide application of automatic control products can improve energy use efficiency and reduce energy consumption and wasting as far as possible so as to realize energy conservation, consumption reduction and green manufacturing.

Automation, high efficiency and energy conservation of packaging machinery are inexorable trends for the packaging industry to address climate change; for example, the 12th five-year development plan for food and packaging machinery industries specifies energy conservation and emissions reduction goals, and makes efforts to vigorously develop energy conservation and emissions reduction technologies and equipment for food processing and packaging, and phase out food and packaging machineries with high energy consumption and severe pollution. It encourages waste heat utilization and energy conservation of high power motor, comprehensive utilization of food processing wastes and industrial emission reduction, and active development of food and packaging machineries which are low-cost, produce good energy conservation effect and reduce emission. Energy consumption per unit of output value in food processing and water consumption per unit of industrial added value are expected to decrease by 15% and 30% respectively, national energy conservation and emission reduction requirements will be satisfied and emissions of main pollutants will conform to relevant national standards in China by 2015.

Development of major products, technologies and equipment covers substitute products, technologies and equipment for imported ones, energy conservation and emission reduction technologies and equipments, low temperature desolventizing technologies and equipment, plant protein processing technologies and equipment, high technologies and equipment for foods, aseptic packaging (filling) technologies and equipment, vacuum and gas filled packaging technologies and equipment, technologies and equipment for fresh-keeping packaging of fresh agricultural products, while automation, high efficiency and energy conservation and emission reduction performance of equipment are placed in very important positions. There will be increasingly wide application of microelectronics, computer, industrial robot, image sensing technology and new materials, etc. in packaging machineries, and packaging machineries tend to develop in the directions of automation, high efficiency and energy conservation.

9.3.3 Servo and variable frequency control energy conservation technologies

Packaging machinery development trends in various countries in the world are high efficiency of production, resource recycling, product energy conservation, high practicability of high and new technologies, commercialization of scientific research achievements, and applications of advanced technologies from other fields in packaging machinery; for example, mechanical-electrical integration technology (core is servo control), heat pipe technology, long distance remote control technology and automatic flexible compensation technology, etc. and frequency converter, PLC, touch screen and advanced sensing elements, etc. are applied in packaging production line to bring about optimal production process and greatly increase intelligentization level and reliability of system so as to save energy, reduce emission and carbon, promote technical progress in the packaging industry.

The servo control system is an automatic control system designed to ensure that controlled output variables including object position, orientation and status etc randomly vary with input target (or given value). For a long time, as an expensive component set, servo motion control has served high performance applications in which accuracy and process efficiency are required. Currently, as volume production and technical innovations are lowering costs, the servo system is frequently available in low-performance applications in which advantages of servo system are presented without any increase in overall costs.

Energy conservation principle of variable frequency governor is mainly reflected on the following several aspects:

(1) Variable frequency energy conservation. In order to ensure reliable production, certain margin is left when power drive is designed and provided for production machinery. When the motor can't operate under full load, except that power drive requirement is satisfied, excess torque increases consumption of active power and causes wasting of electric energy; where voltage is relatively high, running speed of motor may be reduced to make it save electric energy while keeping constant voltage.

When the rotating speed of motor is changed from N_1 to N_2, change relationship of its motor shaft power (P) is shown below: $P_2/P_1 = (N_2/N_1)^3$,

which indicates that the decreasing rotating speed of the motor can produce a cubic-level energy conservation effect.

(2) Energy conservation through dynamic adjustment. Maximum efficiency voltage is supplied by quickly adapting to load change. Variable frequency governor has 5,000 times/s measurement and control output function on software and always maintains highly efficient output operation of motor.

(3) Energy conservation through V/F function involving frequency conversion. Subject to guaranteeing motor output torque, V/F curve can be automatically adjusted. It reduces motor output torque and decreases input current to save energy.

(4) Energy conservation through soft start in frequency conversion. When full-voltage startup of motor is made, it is necessary to absorb 7 times of motor rated current from power grid in order to meet starting torque requirements of the motor, while large starting current means wasting of electric power and also causes much damage to voltage fluctuation at the power grid, increasing line loss and transformer loss. After soft start is adopted, starting current can change from 0 to motor rated current, which mitigates the impact of starting current on power grid, saves electric charge and also reduces rotating speed impact of starting inertia on the equipment, extends service life of the equipment.

(5) Energy conservation through increasing power factor. Motor generates torque through electromagnetic action on stator winding and rotor winding. With respect to power grid, impedance characteristic is inductive, and motor absorbs considerable reactive power during operation, causing very low power factor. With the adoption of variable frequency energy-saving governor, as its performance has changed to AC–DC–AC, load characteristic changes after rectification filtering. Impedance characteristic of variable frequency governor for power grid is resistive, power factor is very high, so the reactive loss is reduced. Variable frequency governor is one kind of excellent electronic technology product, it conducts stepless speed regulation and energy conservation for AC asynchronous motor; its application is increasingly wide.

9.4 Application of Renewable Energy in Packaging Product Manufacturing

Application of renewable energy is the most ideal carbon reduction technology in packaging product manufacturing process like other industrial production processes. Currently, technologies applied at a relatively mature level among renewable energy technologies in China include solar optothermal technology, solar photoelectric technology, wind power technology, geothermal energy utilization technology, biomass energy utilization technology; given geographical location and instability of renewable energy sources, it is difficult to ensure stable supply for the whole process system by merely adopting a kind of renewable energy; given various aspects including energy efficiency, stable supply and economic costs etc, feasible technical proposal is distributed energy station technology with multi-energy complementation.

Distributed energy supply system enjoys various advantages, including low pollution emission and high reliability, and can provide combined cooling, heating and power (CCHP); as it is close to the user side, it effectively reduces electric, thermal and cold losses resulting from long-distance delivery. Main characteristics of distributed energy supply include diversification of fuels, miniaturization and microminiaturization of equipment, combined cooling, heating and power[①]. In coordination with large power grid, distributed power supply mode can make up for the deficiency of large power grid in safety and reliability and plays an important role in improving power source structure, power supply efficiency, power supply quality and reliability, and mitigating environmental impact of the power industry and increasing economic benefits of large power grid, etc.

As many as 10,000 distributed energy stations have been put into operation in USA, Japan and Europe since the late 1970s. Development of distributed energy including combined heating and power (CHP) and renewable energy is encouraged in energy structure adjustment process in Denmark where considerable wind power, biomass power generation and CHP are developed and built in the distributed energy supply mode. CHP is not profitable in Denmark; Danish Government has taken special subsidy policy for encouraging investors to invest in and build distributed CHP and making consumers choose such an environmentally friendly power consumption mode.

In recent years, 10-20 distributed energy supply systems have been operated in Beijing, Shanghai and Guangdong, China. Fuels for actually operated systems in China are dominated by natural gas and fuel costs are relatively high; meanwhile, user loads are greatly different, and improper configuration and failure to operate in power grid lead to inefficient operation of systems in most cases.

Given current clean coal technology development level and renewable energy utilization in China, technical proposal which is easy to be recognized by the market is distributed energy station which is characterized by multi-energy complementation and is based on clean fuel-biomass coal water slurry (see Fig. 9-1). Biomass coal water slurry making process enables comprehensive utilization of high-concentration liquid waste and sludge from packaging enterprises for recycling of pollutants; clean fuel - coal water slurry provides technical guarantee for stable, low-cost operation of energy station; efficient combustion equipment can comprehensively utilize combustible solid wastes (such as packaging wastes) and organic gasses from production in packaging enterprises; renewable energy is efficiently utilized according to different geographical locations of packaging enterprises so as to build the distributed energy station characterized by multi-energy complementation, efficient and low-cost operation and low-carbon emission.

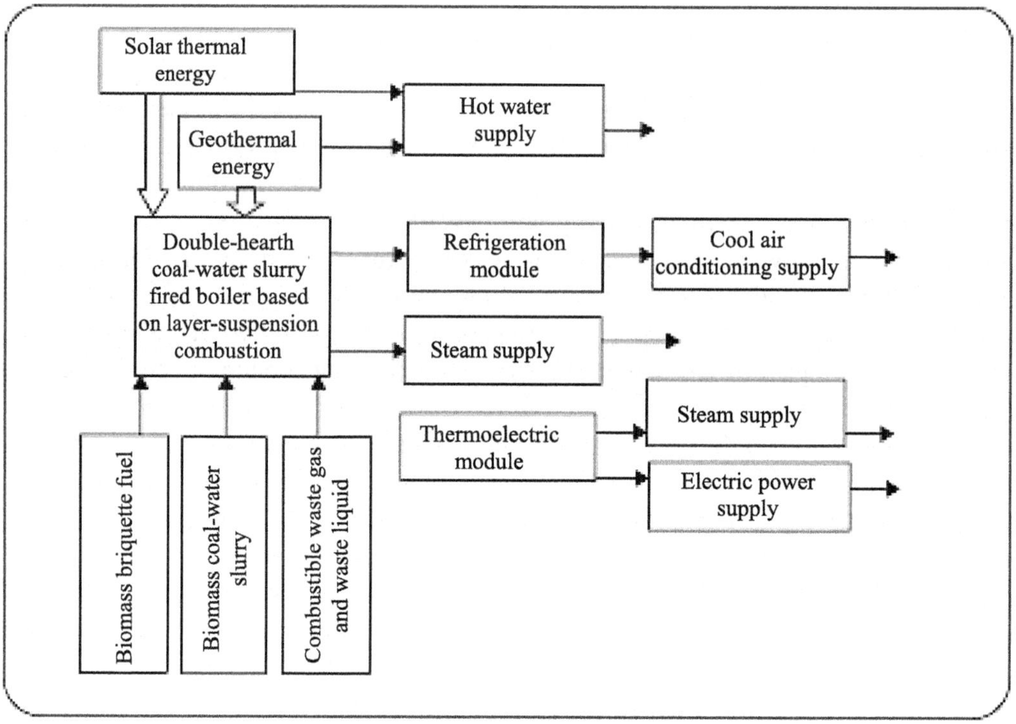

Fig. 9-1 Schematic diagram of distributed energy station which is characterized by multi-energy complementation and is based on clean fuel - biomass coal water slurry

References

[1] Nie Zuoren: *Carbon Footprint and Energy Conservation, Carbon Reduction, Materials China*, 2010(2).

[2] Hiromi Uchida: "Lifecycle Assessment of Gravure Printing Packaging Materials," *Journal of the Society of Packaging Science & Technology, Japan*, 2007(6):397-404.

[3] Zhou Tingmei: *Package and Packaging Waste Management and Environmental Economy*, Chemical Industry Press, 2007.

[4] Ren Xianshu, Huo Lijiang: "Case Study of Lifecycle Assessment of Corrugated Carton Production Process," *Packaging Engineering*, 2010(5).

[5] Qin Fengxian: *Research on Lifecycle Assessment of Beer Package*, Jilin University, October, 2006.

[6] Xie Minghui, Li Li, Huang Zechun et al: "Research on Full Lifecycle Environmental Impact Assessment of Typical Composite Package," *China Environmental Science*, 2009(7):773-779.

[7] Guo Pengying, Huo Lijiang, Ma Hailong et al: "Case Study of Lifecycle Assessment of Book & Periodical Offset Printing Production Process," *Packaging Engineering*, 2011(2).

[8] Xie Minghui, Li Li, Huang Zechun et al: Lifecycle Assessment of Edible Oil Polyester Package, *Research of Environmental Sciences*, 2010(3).

[9] Dai Hongmin: Low-carbon Economy & Green Packaging, *Packaging Engineering*, 2010(5).

[10] Kang Qilai: Multidimensional Observation of Innovative Process Technology for Packaging Production, *Printing Quality and Standardization*, 2009(12).

[11] Liu Fuqian: *Research on Application of Servo Control in Category A Packaging Machinery Control System*, Qingdao University, October, 2010.

Chapter 10

Exploration and Discussions of Carbon Reduction in Paper Packaging Material Production

Zou Yu Cai Kangheng

Abstract: With the increase in carbon dioxide emissions and in the case of global warming, reducing carbon emissions could become a dominant preoccupation in the present stage of industrial development. With regard to the production of paper packaging materials production, as both a high-energy consumption and high emission process, reducing carbon emissions highlights its importance in the production of paper packaging materials increasingly, whether for consideration of the economic benefits of enterprises or for the consideration of global warming. In this paper, by means of analyzing the cause of high-emissions in the production of paper packaging materials, the measures and methods for reducing carbon emissions in the production of paper packaging materials are discussed by optimizing the energy structure, adjusting the structure of raw materials, advancing the technological innovation, eliminating outdated industrial capacity, encouraging the optimization and reorganization of the corporation, expanding the enterprise scale and improving the management level.

Key words: paper packaging materials; reducing carbon emission; energy saving and emission reduction

According to the relevant media report, UK Carbon Trust launched on trial basis the first batch of products with carbon labels including such consumer products as chips, cheese and liquid shampoo worldwide in March, 2007. In February, 2008, Carbon Trust intensified application and popularization of carbon labels and extended them to 75 commodities from 20 manufacturers including Tesco (the largest chain department store in UK), Coca Cola and Boots. As climate change become increasingly acute, British are calculating carbon dioxide emission in the whole process of production and consumption of commodities such as foods, clothes and daily necessities etc so that commodity carbon labels enable people to consciously develop customer behaviors conducive to reducing greenhouse gas emission and mitigating global warming. This shows that human beings have attached unprecedented importance to the reduction of carbon dioxide emission. Since the industrial revolution in the middle of the 18th century, with development of human society, especially rapid industrial development, high dependence on fossil fuels has led to the entry of massive greenhouse gases such as carbon dioxide into the atmosphere and significant increase in concentration of greenhouse gases in the atmosphere, so greenhouse effect becomes increasingly severe and various natural disasters which endanger future survival and development of human society frequently occur. Carbon reduction has become an urgent task for countries due to climate change and energy shortages, etc. As a part with high energy consumption and high emission in the packaging industry, packaging material production is in particular need of intensifying energy conservation and carbon reduction. Paper, plastics, metals and glass called as four largest pillars for packaging materials are four common packaging materials in today's society. Table 10-1 shows that carbon emissions from these four packaging materials accounted for 99.8% of that from China's packaging industry in 2007, and output value from paper packaging industry was ranked first and hit 42.75% of total output value from the packaging industry, while its carbon emission made up 33.54% of total carbon emission from the packaging industry, it exerted great impact on carbon emissions from the whole packaging industry. The most effective measure for reducing carbon dioxide emission from paper packaging material manufacturing industry lies in lowering energy consumption and waste gas emission during production. The following energy conservation and emission reduction analysis proceeds from paper packaging material production process in China.

Table 10-1 Carbon emission from China's packaging industry in 2007

Category	Output value of packaging industry (billion yuan)	The proportion in gross industrial output value (%)	Carbon emissions (10^4 t CO_2)	The proportion in total carbon emissions (%)
Paper packaging	2003.22	42.75	5021.65	33.54
Plastic packaging	2000.42	42.69	7469.71	49.88
Metal packaging	504.2	10.76	1729.74	11.55
Glass packaging	162.05	3.46	722.61	4.83
Other packaging	15.53	0.33	30.58	0.2

Note: These data come from Chen Xirong special column entitled *Low-carbon Packaging Practice Methods and Countermeasures–Theory and Practice of Packaging Carbon Footprint* on www.keyin.cn.

10.1 Analysis of Energy Consumption in the Papermaking Industry

Though papermaking technology originated from China and is one of the four great scientific and technological inventions in ancient China, the development of China's modern papermaking industry still started from relatively low point since high energy consumption, high emission, high pollution and low efficiency were widespread in China's papermaking industry at one time and China's papermaking industry has lagged behind somewhat the world's advanced level. International advanced level is that comprehensive energy consumption by 1t pulp paper is 0.9-1.1t standard coal, while most of Chinese enterprises show that average comprehensive energy consumption by 1t pulp paper is 1.38t standard coal except that a few enterprises or part of production lines in China can reach such a level. In 2010, Li Ping, Section Chief of Industrial Coordination Department of National Development and Reform Commission, stated in Low-carbon Papermaking Philosophy & Practice Forum that currently domestic papermaking industry was still an industry with both high energy consumption and high pollution. As production capacity of the papermaking industry is relatively scattered, energy consumption and pollutant emission from the papermaking industry are still higher than that from other industries in terms of equivalent total output value. Packaging paper is a large category in paper products and includes paper (sheets) and paperboards for the packaging industry and package printing. Furthermore, packaging paper use level has annually grown by 17%-20% and its proportion in national paper products has increased from the original 40% to 60% in China in recent ten years, thus it really bolsters half of China's papermaking industry, and energy conservation and carbon reduction in its production process play a decisive role in energy conservation and carbon reduction in the whole papermaking industry.

High energy consumption is the main cause for high carbon emission, while high energy consumption in China's papermaking industry is mainly attributable to energy structure, raw material structure, production process, technical equipment, enterprise scale and management level, etc.

10.1.1 Energy structure

In foreign countries, fuels for the papermaking industry are dominated by heavy oil and natural gas, while in China, 70% of fuels for the papermaking industry are coal; though China abounds in coal resources, high-quality coal is available only in a small quantity, and inferior coal has lower combustion efficiency that heavy oil and natural gas and also causes severe pollution and carbon dioxide emission. Table 10-2 specifies energy consumption structure in China's papermaking industry from 2000 to 2007 and shows that coal always kept the dominant position and greatly increased year by year along with the increasing total energy consumption, especially rose by 8.78 million tons in 2004 compared with 2003.

Table 10-2 Energy consumption structure in China's papermaking industry

Energy consumption	2000	2001	2002	2003	2004	2005	2006	2007
Coal (10^4 t)	1715.9	1691.2	1747.3	1835.9	2713.9	3027.8	332.6	2279.2
Crude oil (10^4 t)	0.48	0.51	0.5	0.63	0.38	0.52	0.5	0.52
Natural gas (10^4 t)	0.3	0.26	0.27		0.37	0.54	0.64	0.74
Gasoline (10^4 t)	12.02	12.2	15.65	18.65	8.88	8.42	8.56	10.49
Kerosene (10^4 t)	3.61	3.7	2.93	1.9	0.91	0.91	0.81	0.76
Diesel (10^4 t)	22.54	22.2	29.93	35.81	25.7	26.6	25.76	22.45

(Contd.)

Energy consumption	2000	2001	2002	2003	2004	2005	2006	2007
Fuel oil (10^4 t)	19.72	21.67	21.85	22.56	26.85	28.6	32.02	32.25
Coke (10^4 t)	1.5	1.56	1.73	2	6.81	4.09	4.29	4.72
Electricity (10^8 kWh)	228.22	251.35	284.97	311.62	359.33	406.76	447.3	442.35

Note: Data come from *An Evaluation of the Effectiveness of Energy Conservation and Emission Reduction Policy of China's Paper Industry*.

10.1.2 Raw material structure

Plant fiber serves as the basic raw material for the papermaking industry and is characterized by extensive sources and low price, etc. Timber enjoys predominance in most countries. In China, the proportion of timber in use is relatively low due to fewer forest resources. According to the estimation by China Paper Association, 18.59 million tons wood pulp, 12.95 million tons non-wood pulp and 53.05 million tons waste paper pulp were used in 2010, accounting for 22%, 15.3% and 62.7% of total raw materials respectively. Waste paper has become the important resource on which the development of China's papermaking industry relies and sustainable development of China's papermaking industry has presented increasingly pronounced characteristics of circular economy. This is a raw material characteristic of China's papermaking industry. Currently, main raw material for papermaking is plant fiber at the international level, and coniferous tree or broad-leaved tree timber used accounts for about 95% of total use in some economically advanced countries.

There are the following several categories of common plant fibers in China:

(1) Straw fiber: bamboo, reed, wheat straw hood, straw, sorghum stems, Chinese alpine rush; their fibers are short and thin, and length is approximately 0.5-2mm.

(2) Wood fiber: cedar, pine tree, alamo, birch etc; their fibers are approximately 1.4-50mm long,

(3) Bast fiber: linum usitatissimum, jute, ambary, wingceltis bark, mulberry bark, cotton stalk bark, etc; their fibers are approximately 25-30mm long,

(4) Wool fiber: cotton, linters, tattered cotton cloth; their fibers are approximately 20-40mm long.

(5) Waste paper fiber

In China's papermaking industry, raw material structure is irrational since wood fiber as the raw material makes up a relatively small proportion, while non-wood fibers as raw materials make up relatively large proportion. Non-wood fibers such as straw pulp are highly consumed so that pollution is severe and the quality of paper product is not high. Waste paper recycling rate is also relatively low in China; in 2010, the world's average waste paper collection rate was 56.6%, and such collection rate generally exceeded 60% in developed countries, and that in South Korea and Germany was higher than 80%, while such figure was only 43.8% in China, lower than the world's average level.

10.1.3 New material technologies for paper package

A considerable part of paper packaging materials are used to protect products against destruction during transport. Presently, conventional corrugated boards, hardboards or foams are heavily used in China. If new technologies are adopted to make packaging materials light and decrease material use level to substantially reduce carbon value of package while guaranteeing packaging quality, undoubtedly this is also an ideal way for effectively reducing carbon. Currently, technologies for this aspect include miniature corrugated board technology, honeycomb technology and foam packaging technology, etc. Research and development of new type low-carbon packaging materials for protecting products with minimum materials not only conforms to the general situation of green low-carbon economy and but also represents one of future development trends in the packaging material industry.

The following example touches upon miniature corrugated board and is designed to reveal deficiencies in new material technologies for paper packaging in China. Miniature corrugated board is one kind of new type corrugated board which has grown up in foreign countries in recent years, such as type F, N, G etc. With in-depth research on miniaturization of corrugated board, enterprises in USA, Sweden and Germany, etc. have extended from conventional types A, B, C, E, etc. to type F (corrugating height: 0.75mm), type G (0.50mm), type N (0.46mm) and type O (0.30mm), and are stepping up exploration of more miniature corrugated paper materials, and are expanding existing miniature corrugated board production scale; currently, total output value of miniature corrugated boards accounts for 8% of corrugated boards in some countries, while in China, miniature corrugated board market is a "vacuum" one which urgently needs to be developed.

10.1.4 Enterprise scale

Currently, only a small number of large enterprise groups and backbone enterprises in China' papermaking industry enjoy international competitiveness, their influence and demonstration effect need to be further improved, and there are many small and disadvantaged enterprises, and production capacity is scattered, and industrial economy-of-scale level is low. According to statistical data from China Paper Association, there were 3,686 papermaking enterprises in China in 2009, among which more than 88% enterprises were small and medium-sized ones and only 9 enterprises reached a scale higher than one million tons/year (Table 10-3).

Table 10-3 Output and operating revenue of 9 enterprises at one million tons/year scale in China in 2009

Enterprise name	Chenming	Jiulong	Huatai	Taiyang	Zhongye	Liwen	Jindong	Zhonghua	Taige
Main business income (billion yuan)	186.2	155	151.5	139	101	110.99	82.62	61.58	64.75
Paper and paperboard production ($*10^4$ t)	300	652	155	219	1006	355	228.92	143.65	124

Note: Data come from *Top 20 Chinese Papermaking Enterprises in 2009* available on the internet.

2009 data indicate that combined output of top ten papermaking enterprises in China accounted for about 30% of total output from the whole industry, while in developed countries, total output of top ten papermaking enterprises made up about 80% of total output from the whole industry, thus China's papermaking industry presents very low degree of concentration compared with developed countries and still has much room for improvement. Small enterprises are characterized by low operation level, backward technical equipments, inefficiency of energy consumption and severe pollution. Shutting down these backward production capacities and increasing the degree of concentration in the papermaking industry are issues which urgently need to be addressed for industrial optimization and upgrading.

10.1.5 Management level

Management can be divided into two levels: one is macro level which mainly refers to national and regional policies, laws, regulations and industrial standards; the other one is micro level which mainly means management system and relevant regulations within enterprises. As supplements to hardware conditions, advanced management methods can also effectively stimulate enterprises to save energy and reduce consumption and CO_2 emission. Management level is still subject to many drawbacks at the current stage in China where industrial standards have not yet been refined and quantified, and standard execution is not tough enough. Most of the enterprises do not duly execute an energy management system; there are many deficiencies in the technical innovation system compared with foreign countries and systematic employee training mechanism is unavailable, etc.

10.2 Effective Ways for Carbon-reducing Production of Paper Packaging Materials

In order to narrow the great gap between China's paper packaging material production and international advanced level, and improve current industrial situation characterized by high energy consumption and high emission, achieve industrial transformation and upgrading, and take low-carbon, environmentally friendly, sustainable development path, we may proceed from optimizing energy structure, rationally adjusting raw material structure, actively boosting technical innovations, intensifying enterprise reorganization and integration, increasing management level, etc.

10.2.1 Optimize energy structure

1) Increase the proportion of heavy oil and natural gas in use

Heavy oil and natural gas contain high thermal content, and are fully combusted and can be utilized in a highly efficient way, from which carbon dioxide emission is much lower than that from coal combustion, so increasing the proportion of these fuels in use is beneficial to reducing carbon dioxide emission.

2) Energetically develop clean coal technology

Basic characteristics of energy resources in China (abundant coal, lean oil, less gases) determines the situation in which coal can't be abandoned in fuel structure and the papermaking industry will still need to rely on coal as main fuel for quite a long time. As there is an unchangeable objective reality that coal serves as pillar energy for the papermaking industry, undoubtedly, vigorous popularization of efficient, low-pollution clean coal technology is an effective way for carbon-reducing production in the papermaking industry.

3) Strengthen energy supply construction

Meanwhile, intensified efforts should also be made to recover residual heat, waste gases and wastes from papermaking enterprises. For example, MCC Paper Yinhe Co., Ltd. invested three million yuan in building three 500kW methane gas power generation stations; after project operation, its operating cost and daily power generation are 0.12 yuan/kWh and 3.6×10^4 kWh respectively. Based on industrial electric charge 0.65 yuan/kWh in local market, 19,080 yuan benefits can be delivered a day and more than 6.90 million yuan profit can be annually earned. Utilization of waste gases through this reconstruction has not only obtained considerable economic benefits but also reduced waste gas emission.

10.2.2 Rationally adjust raw material structure

1) Vigorously popularize forest-paper integration

The severest pollution and energy consumption in China's papermaking industry come from very backward non-wood fiber pulping enterprises. Timber and paper products are renewable and recyclable products; wood, forest, pulping and papermaking industries can expand utilization of biomass energy, reduce reliance on fossil fuels and carbon dioxide emission. Therefore, modern advanced pulping and papermaking industry with forest-paper integration is one typical low-carbon industry.

Forest-paper integration project construction was started in China's papermaking industry by making reference to foreign practice in the 1970s, which was designed to address the shortages of raw materials for papermaking and enhance wood pulp self-supply capability. With 20-30 years' development, forest-paper integration project has become increasingly perfect and standardized in terms of planting technology, operation mode and national policy and has become an important development mode for Chinese mainstream papermaking enterprises, which not only improves domestic wood pulp self-supply capability, but also gradually reduces and phases out backward non-wood fiber raw materials such as straw pulp; this raw material structure adjustment is playing a gradually prominent role in energy conservation and consumption reduction. Another important role of forest-paper integration consists in powerful function of forest for absorbing and storing carbon dioxide. Vigorous efforts in boosting development of forest-paper integration will effectively curb greenhouse effect and benefit human beings.

It is still necessary for China at the current stage to reinforce raw material forest base construction, speed up forest-paper integration, cultivate fine tree species, improve per unit yield in forest land, increase the proportion of enterprise-controlled forest bases, advocate "company + base + cooperative organization" operation mode, properly develop "company + base + farmer" operation mode, etc, enhance material supply capability of bases and change current situation of China's papermaking industry in which wood fiber proportion is relatively low.

2) Rationally utilize non-wood fiber raw materials

Non-wood fiber raw materials are mainly straw fiber raw materials; process for producing such raw materials is relatively complicated and incurs high energy consumption and severe carbon dioxide emission; however, wood fibers are insufficient, and straw fibers can't be effectively utilized, resulting in wasting, especially they are directly burnt into fertilizers in rural areas, causing more carbon dioxide emission into the atmosphere, thus use of straw fiber raw materials in the papermaking industry can't be abandoned, and non-wood resources including bamboos, reeds and agricultural wastes such as straws and bagasse should be fully utilized according to local conditions. With respect to use of non-wood fibers, enterprises may introduce advanced straw pulp production line and develop new technologies to rationally utilize non-wood fibers, phase out or decrease straw pulp under existing technologies, and reduce high energy consumption and high emission resulting from making straw pulp as planned.

3) Strengthen waste paper recycling

Recycling of waste paper can not only optimize raw material structure, alleviate shortages of raw materials, materialize cyclic development mode of the papermaking industry, but also lower environmental pollution and energy consumption resulting from pulping. According to statistics, recovery of 1t waste paper can produce approximately 0.8t paper, save 1.2t standard coal and 600kWh.

According to media report, relevant departments conducted follow-up survey on waste paper collection personnel and found that most of them adulterated sands and sprayed water into waste paper before sales to paper mills. This brought many troubles to waste paper disposal at paper mills. This phenomenon reflects that legal, specification and policy support is not sufficiently specific and duly available for waste paper recovery in China. Therefore, in order to increase waste paper recovery rate in China, the State should develop a complete set of highly operational, strictly refined and quantified legal system for recovery which gives considerations to interests of various stakeholders. Relevant departments should mandatorily require papermaking enterprises to use certain quantity of waste paper as raw material; the government should be required to provide certain subsidies to recovery enterprises to ensure that they are under normal operation and obtain reasonable economic benefits; consumers and recovery enterprises should be guided to classify and then recover waste paper; people's awareness about resource recycling should be strengthened. With regard to waste paper recycling, China should also step up research and development input in waste paper disposal technology and disposal equipment so that recovered waste paper is efficiently disposed of.

10.2.3 Actively push forward technical innovations

1) Honeycomb technology

Honeycomb technology refers to one kind of technology for turning intermediate layer of physical material into honeycomb structure to effectively reduce material consumption. Honeycomb treatment of physical material for paper packaging can greatly increase physical material utilization rate; if 20% materials are replaced by honeycomb materials at the current technical level, 20 million m^3 timber can be annually saved. According to calculation based on energy conservation and emission reduction data released by the Ministry of Science and Technology, this is equivalent to saving of five million tons standard coal and reduction of 12.86 million tons of carbon dioxide emission. Timber saving also protects our forest resources. As everyone knows, plant absorbs carbon dioxide and releases oxygen through photosynthesis, through which carbon is fixed within the plant in the form of living beings. Research shows that 1hectare forest can accommodate 120-140t carbon dioxide, equivalent to saving of 1hectare forest resources and indirect reduction of 120-140t carbon dioxide emission.

2) Foam packaging paper technology

Foam packaging paper is made by taking waste paper and flour as raw materials and firstly cutting waste paper into scrips and then grinding them into fibrous pulp, mixing it with flour at a proper proportion, injecting mixed pulp material into the extruder for extruding cylindrical particles. In the extrusion process, raw materials become foam paper by virtue of water vapour. Such foam paper particles are used as raw materials to make desired packaging materials; such packaging materials result in material saving like honeycomb materials, and production process is simple and saves energy, so carbon dioxide emission is directly and indirectly reduced.

3) Miniature corrugated paper technology

Miniature corrugated board adds new functions on the basis of inheriting functions of conventional corrugated board. Compared with similar conventional fiber presspaper, miniature corrugated board is characterized by material saving, high elasticity, strong buffering power and shock absorption, etc. For example, type G corrugated board is 30%-40% lighter than solid fiber paperboard and has strength which increases by 30%-40% (compared with paperboard of the same specification), and its weight is only half of conventional solid paperboard folding carton. Therefore, replacement of hardboard or conventional corrugated board by miniature corrugated board subject to meeting needs can save a lot of materials and indirectly reduce carbon dioxide emission from packaging materials, and contribute to carbon-reducing production of paper packaging materials.

Internationally, miniature corrugated board is deemed as mainstream in future corrugated board packaging material market thanks to its better physical characteristics, printing quality and low carbon. Development of miniature corrugated board in China entails the following several aspects:

(1) Develop suitable new corrugating roller and new models. Miniature corrugation is relatively low, the number of corrugations goes up accordingly, and initial force for paper feeding increases at the time of corrugation forming, thus it is difficult to make it by the conventional ordinary corrugated board machine.

(2) Increase input, miniature corrugation research level, technological level of machine processing.

(3) Develop more miniature corrugated paper material to make China's miniature corrugated board technology stand at the forefront of the world.

(4) Popularize miniature corrugated board application fields.

10.2.4 Intensify phase-out of backward production capacity, encourage enterprise optimization and reorganization, expand enterprise scale

The papermaking industry is capital and technology intensive, and delivers significant economy-of-scale benefits. With regard to enterprise scale, a batch of large efficient enterprises was developed through new expansion and reconstruction. In 2009, there were 30 enterprises with the annual output more than 0.4 million tons (among which 9 enterprises had the annual output more than one million tons); these enterprises are good examples basically equipped with world-class advanced technologies and equipment for pulping and papermaking and environmental protection, and give rise to the lowest greenhouse gas emission proportion among all manufacturing sectors and have improved industrial landscape to some extent. However, the situation of China's papermaking industry in which players are small and scattered remains unchanged, thus it is necessary to continue executing industrial exit mechanism, adjust and specify elimination standards, quantify elimination indicators, reinforce phase-out of backward production capacity and accelerate industrial integration. Greater efforts are made to drive enterprise reorganization, improve industrial organization form to gradually develop industrial organization structure in which large enterprises enjoy the leading position while small and medium-sized enterprises develop in a coordinated way. The government provides support to foster larger integrated pulping papermaking enterprise groups with the annual output more than one million tons and form a benign market competition pattern.

10.2.5 Effectively manage and lead low-carbon production

1) Macro management

Macro management mainly involves national and local policies, laws and regulations as well as some other control means. The State develops and promulgates relevant policies, laws and regulations for fully regulating industrial development positioning, development mode, structural adjustment, resource conservation, environmental protection, investment behavior, market access, market supervision and consumption philosophy, etc. for the purpose of creating fair, rational, satiable, orderly, healthy modern papermaking industry development environment. Enterprises are stimulated to proceed from their own advantages and disadvantages to continuously seek various ways for energy conservation and emission reduction and take low-carbon development path. For example, *Overall Work Program for Energy Conservation and Emission Reduction* released by the State in June, 2007 stressed efforts, during the 11th Five-year Plan period, to phase out backward production capacities in China's papermaking industry including straw pulp production lines below 34,000t/year, chemical pulping production lines below 17,000t/year, paper mills below 10,000t/year with waste paper as raw material and failure to reach emission standard and paperboard production lines below 20,000t/year, and boost development of China's papermaking industry in the directions of being large-scale, technology and capital intensive, low-carbon and environmentally friendly, and greatly increase modernization level of China's papermaking industry.

Energy consumption quota standard is developed; unit product energy consumption standard is not affected by such factors as market and directly reflects technical equipment, technical renovation and energy management level of an enterprise, and is one symbolic technical-economic indicator which reflects and measures enterprise energy utilization level in the most straightforward, most objective way. It can provide technical measures and management means for energy consumption control at source. Activities for reinforcing standardization, energy consumption quota for key production processes relating to products, and energy consumption management can effectively stimulate enterprises to improve process technologies, boost scientific and technological innovations, phase out backward production technologies, production modes and equipment so as to reduce energy consumption and increase energy utilization rate. Combination of energy consumption quota with industrial policies can further strengthen the role of market mechanism to inevitably increase costs for products made in the case of failure to reach standard in order to control production scale and market share of enterprises with high energy consumption, and impel comprehensive optimization and upgrading of industrial structure, and make more energy-saving enterprises and products become main players in the market and finally realize overall energy consumption reduction, successfully transform the papermaking industry towards low-carbon and environmentally friendly development. For example, Guangdong Province became the first to develop local energy consumption quota standard *Energy Consumption Quota Standard for Main Products in the Pulping and Papermaking Industry* in 2008. With the effective execution of this local standard and summarization of experience gathered in execution of this standard, a more perfect national standard is developed and executed nationwide.

2) Micro-management

Micro-management refers to management within an enterprise and serves as effective supplement and reinforcement to macro management; it covers the following aspects:

(1) Establish all-round energy management body;

(2) Develop energy conservation goal responsibility system and evaluation system;

(3) Strengthen energy statistics work;

(4) Formulate rational and effective energy management supervision system;

(5) Set up energy conservation and emission reduction research forces and develop the corresponding incentive mechanism;

(6) Carry out carbon-reducing production awareness education for employees.

References

[1] Li Weiling: *Energy Consumption Situation, Energy Conservation and Consumption Reduction Measures of China's Papermaking Industry, China Pulp & Paper*, 2011(3):61-63.

[2] Wang Wenming: *Honeycomb Technology Builds the Future for Green Low-carbon Materials*, China Hi-tech Fair the 1st Green Packaging & Low-carbon Economy Summit, November 17, 2009.

[3] Wang Wenming: *Honeycomb Paperboard Application and Market Prospect Analysis*, PACKAGING ENGINEERING Vo1.25 No.1 2004, [A] , Collected Papers at the 1st China International Packaging Summit [c] , 2001.

[4] Chen Daiguang: *Low-carbon Economy Promotes Energy Conservation and Emission Reduction in Papermaking Enterprises, Energy Conservation & Environmental Protection*, 2011(1):26-27, 33.

[5] http: //blog.sina.com.cn/s/blog_685dd39d0100iznv.html.

[6] Sun Piju: *Popularization of Honeycomb Paperboard is the Important Way for Low Carbon Emission Reduction, Material Saving and Timber Replacement, China Packaging Industry*, 2010(11): 25-28.

[7] Zhang Yihua, Wang Qiuyun: *Measures for Pulping and Papermaking Energy Conservation and Emission Reduction, Water Saving and Consumption Reduction, Paper and Paper Making*, 2008(1):38-40.

[8] Wang Yufeng, Shi Baoying: *Designate Product Energy Consumption Quota Stan-*

dard, Promote Energy Conservation and Consumption Reduction in the Papermaking Industry, Transactions of China Pulp and Paper, 2010(sup):251-253.

[9] Diao Yan, Gou Jinsheng: *Current Situation and Development Trends of Paper Packaging Materials, China Packaging Industry*, 2008(9):17-20.

[10] Cheng Peifang, Wang Qiuli: *Research and Development Trends of Paper Packaging Materials, China Packaging*, 2006(5):39-42.

[11] k Chai Xiaohui: *Intensify Enterprise Management Innovations, Boost Energy Conservation and Emission Reduction in the Papermaking Industry, China Pulp & Paper Industry*, 2007(8):17-20.

[12] Guo Yongxin: *Current Situation and Future of Papermaking Raw Materials in China, China Pulp & Paper Industry,* 2010(19):P14-17.

[13] Li Jun: *New Members of Packaging Paper Family, China Packaging Industry*, 2006(7):19.

[14] Hu Zongyuan: *Using "New Technology for Clean Pulping with Straw Raw Materials", "Agriculture, Pulp, Paper, Fertilizer Integration" New Mode Will Be Development Path for China's Papermaking Industry and Straw Pulp Enterprises——Re-discussion About Rational Utilization of Non-wood Fiber Raw Material Resources, Tianjin Paper Making*, 2009(1):2-4.

[15] Li Hongxin: *How Does China's Paper Industry Advance Before Challenges and Opportunities, China Pulp & Paper Industry,* 2010(15):8-10.

Chapter 11

Practice and Countermeasures of Low-Carbon Transportation Development of Zhuzhou City in the Urban Transformation Process

Zhao Xianchao Ding Meng

Abstract: With the increase in urban population and rapid economic growth, Zhuzhou City and other cities in China are facing the problem of traffic congestion, energy consumption and growing carbon emissions, which are incompatible with the social construction of the "two types". Low-carbon development of the transport sector has become the inevitable choice in the process of urban transformation. As an extremely important traditional old industrial base and Tan: 3+5" city group, Zhuzhou City should set an example in the field of low-carbon transport development. The paper describes the importance of developing low-carbon public transport in Zhuzhou City, summarizes the status of low-carbon transport development in the city, examines the practical experience of low-carbon transport development in the city, and finally propose several countermeasures for low-carbon transport development.

Key words: low-carbon transport; practical experience; countermeasures

On the eve of Copenhagen Conference, the Chinese Government announced that CO_2 emission per unit of GDP would decrease by 40%-45% by 2020 compared with 2005 and it was incorporated as binding indicator into national medium and long-term plan for economic and social development, and stressed efforts to speed up construction of industrial, building and transportation systems characterized by low carbon. According to a report from Intergovernmental Panel on Climate Change (IPCC), urban transportation accounted for 13.1% of global greenhouse gas emissions and was the third largest emission sector only next to energy supply and industrial production. According to Research on Direction and Route for Energy Conservation and Emission Reduction in Transportation System conducted by the Ministry of Transport, carbon dioxide emission from transportation field will hit about 1.5 billion tons and will account for 18%-20% of total urban carbon emission by 2020. In order to realize national and municipal energy conservation and emission reduction goals, it is imperative to focus on low-carbon development in such fields as industry, transportation, building etc and relying on concerted efforts from all walks of life in society.

In response to the current grim situation of increasing energy consumption and carbon emission from urban transportation, more or less low-carbon transportation development practice has been carried out in part of Chinese cities. In 2011, the Ministry of Transport determined ten cities including Tianjin, Chongqing, Xiamen, Hangzhou, Nanchang, Guiyang, Baoding, Wuhan and Wuxi to launch pilot work on building a low-carbon transportation system, which marks a new step forward in the development of urban low-carbon transportation in China. In addition, Policy research on low-carbon transportation is also being conducted in Shanghai and Beijing, etc.

11.1 Significance of Development of Low-carbon Public Transportation in Zhuzhou City

Changsha-Zhuzhou-Xiangtan city group is one of the national comprehensive reform pilot areas for building resource-saving and environmentally friendly society as approved by the State Council and is the main city carrier for implementing national strategy and ambitious goal for rise and development of Central China, and core growth pole for economic development of Hunan Province. As Zhuzhou City is an important pole in Changsha-Zhuzhou-Xiangtan city group, low-carbon transformation and development of transportation in Zhuzhou City is not only beneficial to easing current urban traffic congestion but also of great practical significance for Zhuzhou City to realize low-carbon transformation in the new historic opportunity period.

Firstly, development of low-carbon transportation in Zhuzhou City is the inevitable choice for relieving resource, energy and environmental pressure. The transport industry is a key oil consumer and also one of the important sources for greenhouse gases and air pollution, especially motor vehicle exhaust gas emission from the transport industry accounts for about 60% of air pollutants. All 627 in urban areas of Zhuzhou City have been changed to hybrid electric buses, with average oil saving more than 15%. Development of low-carbon transportation in Zhuzhou City can effectively mitigate huge pressure on urban resource, energy and environment to a certain extent.

Secondly, development of low-carbon transportation in Zhuzhou City is the important means for making steady headway in building resource-saving and environmentally friendly society. According to estimation, total oil consumption from transportation accounted for 29.8% of total oil consumption in the whole society in China in 2005. As the second largest oil consumer only next to the manufacturing industry in the whole society, the transport industry is one of the major fields for resource saving; furthermore, exhaust gas emission caused by rapidly growing city cars also increasingly affects air quality. In recent three years, all 627 in urban areas of Zhuzhou City have been changed to hybrid electric buses, and total operating distance exceeded 45.88 million km, 115.54 million people-times were transported and 2.2 million liters oil was annually saved, and emission of various harmful substances such as carbon dioxide etc was 14,730t.

Development of low-carbon transportation in Zhuzhou City can effectively promote the construction of resource-saving and environmentally friendly society.

Thirdly, development of low-carbon transportation in Zhuzhou City is an urgent need for timely alleviating urban traffic congestion. Currently, with the increasing income of urban residents in Zhuzhou City, the number of private cars has soared and is approaching 0.6 million. More and more urban residents choose private car as transport means for daily travelling, which causes serious traffic congestion on typical roads including Shifu Road and Hetai Road during peak hours to some extent. Therefore, development of low-carbon transportation in Zhuzhou City is a pressing need for mitigating urban traffic congestion.

Finally, development of low-carbon transportation in Zhuzhou City is the important guarantee for effectively improving physical and psychological health of residents. At present, increasing urban motor vehicles result in decreasing urban air quality, especially growing motor vehicle exhaust gas emissions lead to frequent haze weather and are extremely easy to cause lung cancer. As urban residents excessively choose private car as transport means and refrain from daily exercise, this is also very easy to increase occupational diseases such as lumbar spondylosis, etc. Thus development of low-carbon transportation in Zhuzhou City is the important guarantee for effectively improving physical and psychological health of residents.

11.2 Overview of Current Situation of Transportation Development in Zhuzhou City

Zhuzhou was called Jianning in ancient times and is an important growth pole in national comprehensive reform pilot areas for building resource-saving and environment-friendly society in Changsha-Zhuzhou-Xiangtan city group and is also included in the first batch of national key industrial cities. Located in eastern Hunan Province by north and the lower reaches of the Xiangjiang River, currently it has jurisdiction over five districts including Tianyuan, Lusong, Hetang, Shifeng, Yunlong and county-level Liling City, Zhuzhou County, Youxian County, Chaling County and Yanling County. In recent years, urbanization has been continuously accelerated and the transport undertaking has flourished in Zhuzhou City, mainly reflected in the following several aspects.

11.2.1 Continuous increase in road mileage

As of August, 2011, existing road mileage was 813.06km in built-up areas in Zhuzhou City, including 38.05km express ways, 80.46km main trunk roads, 113.75km secondary trunk roads and 580.8km branch roads.

11.2.2 Skyrocketing motor vehicles

Currently, there are more than 525,000 motor vehicles in Zhuzhou City, up nearly 250,000 compared with the end of 2005 (276,000). Among existing motor vehicles, there are more than 175,000 cars and more than 348,000 motorcycles, while private cars account for 23.3% of total motor vehicles.

11.2.3 Remarkable effects have been produced on transportation management

In 2009, more than 50 million yuan was invested and an intelligent transportation management system for roads was completed in Zhuzhou City, especially the successful establishment of national transportation management model city marks a new step forward in Zhuzhou City's transportation management.

11.2.4 Public transportation undertaking is relatively developed

In July, 2009, a three-year action plan for electric buses was officially initiated in Zhuzhou City. Currently, 627 hybrid electric buses are available in urban areas, and the rate of buses available for every one million urban residents exceeds ten standard busses; the ticket price for all air-conditioned buses in urban areas was adjusted to 1 yuan/people-time (student 0.8 yuan/people-time) as from January, 2011; it has been included in the first batch of national pilot cities for demonstration popularization of energy saving and new energy vehicles. In 2011, it became the first to launch a bicycle leasing system. Presently, barrier-free facility renovations at main and secondary trunk roads and construction of special bicycle lanes in urban areas have been basically completed, and city-wide coverage by public bicycles has been basically achieved, and approval has been obtained from the Ministry of Housing and Urban-Rural Development, and it has been included in the second batch of national candidate cities for "Urban Pedestrian and Bicycle Transportation System Demonstration Project". License plate is prohibited for diesel taxicabs and dual fuel taxicabs are popularized in Zhuzhou City as from 2010. So far, 550 among 1,955 taxicabs have been refitted into dual fuel taxicabs in the whole city, and nearly 200 new dual fuel taxicabs are purchased.

11.3 Main Practical Experience in Development of Low-carbon Transportation in Zhuzhou City

11.3.1 Receive high attention from government leaders and actively strengthen the leading role of planning

Development of urban low-carbon transportation especially public transportation can't do without correct guidance from the government and planning.

In recent years, under the correct guidance of Zhuzhou Municipal Party Committee and Municipal Government, transportation work in Zhuzhou City has been closely aligned with strategic deployment made by Municipal Party Committee and Municipal Government for implementing "four large strategies" and building "four Zhuzhou Cities" and remarkable progress has been made in low-carbon transportation development. Especially with high attention from Municipal Party Committee and Municipal Government, Zhuzhou City successively established Zhuzhou Municipal Road Transportation Safety Committee and Zhuzhou Municipal Urban Transportation Work Comprehensive Coordination Committee, and developed a regular joint meeting system, timely carried out coordination and research on urban road transportation issues to gradually form an integrated operation mechanism for urban transportation planning, construction and management.

Based on preparation, revision and improvement of *Overall Plan for Zhuzhou City* (2005-2020), various urban transportation plans including *Overall Transportation System Plan for Zhuzhou City, Special Plan for Public Transportation of Zhuzhou City, Road Transportation Management Plan for Zhuzhou City, Road Transportation Safety Management Plan for Zhuzhou City, Parking Lot Plan for Zhuzhou City and Slow Transportation System Plan for Zhuzhou City* have been prepared and revised by investing more than six million yuan since 2011. With successive implementation, these plans help gradually improve the urban transportation planning system, further give play to the leading role of low-carbon transportation development and provide scientific basis and work guidance for near, medium and long-term urban transportation development of Zhuzhou City.

11.3.2 Strengthen active participation from enterprises and actively give scope to roles of the market

Practice proves that merely the leading role of government is much insufficient for the development of urban low-carbon transportation, and it is essential to reinforce active participation from enterprises and energetically give play to roles of the market.

Under the guidance of the government, transportation enterprises in Zhuzhou City including Zhuzhou Public Transportation Co., Ltd., Hunan Zhuzhou Xiangyun Group Co., Ltd., Zhuzhou Nonstop Commercial Express Bus Co., Ltd., Zhuzhou Hailian Taxi Co., Ltd. and Zhuzhou Ping'an Taxi Co., Ltd. vigorously participated in various theme activities including "Safe Zhuzhou", "Low-carbon Transportation" and "Energy Conservation & Emission Reduction". Meanwhile, city-wide enterprises such as CSR Zhuzhou Electric Locomotive Co., Ltd. (CSR Zhuzhou Locomotive), CSR Zhuzhou Institute Co., Ltd. (CSR Zhuzhou Institute) and CSR Zhuzhou Electric Motor Co., Ltd. (CSR Zhuzhou Electric Motor) etc. have proceeded from their own advantages to actively make technological breakthroughs and develop multiple low-carbon transportation technologies so as to provide technical support for urban low-carbon transportation development.

A public bicycle leasing system was launched in Zhuzhou City in 2011. This public bicycle leasing system is operated in the mode in which the government plays the leading role, market-oriented operation is performed and management is conducted by enterprises. With enterprise-oriented operation, bicycles and all parts supplied for this leading system are specially customized, and manufacturers offer maintenance services immediately after bicycles become worn and torn somewhat. This mode stresses efforts to actively give play to roles of the market subject to the leading role of government. Take public bicycle leasing system for example, the advantage in actively giving play to roles of the market is that public bicycles in Zhuzhou City have started shifting from initial leading system construction to intensive industrialization development. Currently, a new industry which integrates public bicycle production, research and development, operation and management has gradually taken shape in Zhuzhou City. Zhuzhou has won the bidding in single source procurement in Yantian District, Shenzhen and will witness construction of 6,000 locking pillars and launch of 5,000 bicycles. In addition, Xi'an City has signed pilot agreement, while more than ten cities including Jining in Shandong Province, Xinyu in Jiangxi Province and Linwu in Hunan Province are actively discussing cooperation intention with Zhuzhou. Bicycle "Zhuzhou Mode" has become the object imitated by many cities.

11.3.3 Firstly make breakthroughs in key fields, actively carry out pilot work and gradually popularize them

Remarkable achievements have been made in key fields including construction of public bicycle leasing system, operation of electric buses and application of dual fuel taxicabs in Zhuzhou City.

1) Priority development of public transportation: electric buses bolster green Zhuzhou

In 2007, *Opinions of Zhuzhou City Concerning Implementation of Priority Development of Urban Public Transportation* ((ZZF 2007) No.21 Document), *Administrative Measures of Zhuzhou City for Public Transportation Management* were developed and issued in Zhuzhou City, which are designed to vigorously implement strategy for priority development of public transportation. Meanwhile, public transportation facilities are built according to national technical standards and public transportation facility construction land is allocated by the government; passenger vehicles for public transportation are dealt with by quota subsidies and low-ticket price policy; urban public transportation lines are not used for other purposes, and are not contracted or subcontracted by individuals, etc.

In order to vigorously implement low-carbon public transportation strategy, Zhuzhou has become the first to initiate Three-year (2009-2011) Action Plan for Electric Buses, under which more than 400 million yuan would be invested and 627 buses in urban areas would be changed to electric or hybrid buses to create the first electric bus city worldwide in 2009-2011. By the end of 2009, there were 627 electric buses in Zhuzhou City, making it included in the first batch of national pilot cities for demonstration popularization of energy saving and new energy vehicles. In September, 2011, all buses were turned into electric ones in Zhuzhou City ahead of schedule, making it become the first "electric bus city" in China (Fig. 11-1).

Fig. 11-1 Electric buses in Zhuzhou City

2) Low-carbon practice in taxicabs: new world for dual fuel taxicabs

Dual fuel taxicab (Fig. 11-2) based on both natural gas and gasoline is another green low-carbon travel mode launched in Zhuzhou City following electric bus, public bicycle leasing system. In July, 2009, Three-year (2009-2011) Action Plan of Zhuzhou City for Electric Buses was specially released in Zhuzhou City, which states that license plate is prohibited for diesel taxicabs and dual fuel taxicabs are popularized from January 1, 2010. So far, 550 among 1,955 taxicabs had been refitted into dual fuel taxicabs in the whole city, and nearly 200 new dual fuel taxicabs were purchased (as of August, 2011). According to low-carbon transportation plan of Zhuzhou Municipal Transportation Bureau, all diesel taxicabs in this city will be replaced within 4-5 years and non-diesel taxicabs will be basically available in the whole city by about 2015.

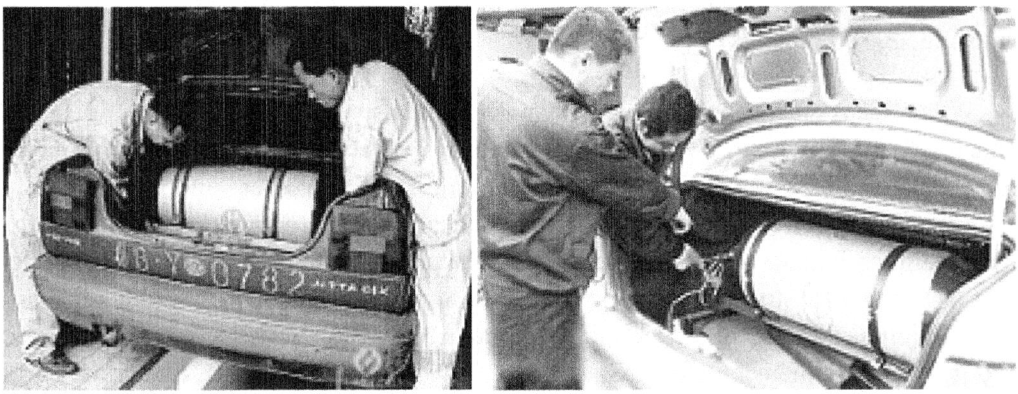

Fig. 11-2 Dual fuel taxicab in Zhuzhou City

3) Low-carbon practice in bicycle: the bicycle leasing system is being improved

Launching ceremony for public bicycle leasing system was held on Shennongcheng Plaza in Zhuzhou City on May 6, 2011. The public bicycle leasing system stimulates residents to travel through green low-carbon means such as bus and bicycle, etc. and 400 leasing sites plan to be built in residential areas, main and secondary trunk roads, transportation hub, public areas, park scenic spots, etc, and 10,000 bicycles plan to be put into operation to form a city-wide network system. Currently, 279 stations, 2,000 public bicycles in Phase I project have come into service (Fig. 11-3). In July, 2011, Zhuzhou City was approved by the Ministry of Housing and Urban-Rural Development to be included in the second batch of national candidate cities for "Urban Pedestrian and Bicycle Transportation System Demonstration Project" following six cities including Hangzhou, Chongqing and Kunming which obtained such approval in 2010.

Fig. 11-3 Public bicycles in Zhuzhou City

Currently, barrier-free facility renovations on main and secondary trunk roads and construction of special bicycle lanes in urban areas have been fully completed. 5,102 public bicycles have been put into use and "city-wide coverage" by public bicycles has been basically achieved.

11.3.4 Improve the quality of urban transportation, actively seek pilot work from the State

With a focus on enhancing the quality of urban transportation and developing "green Zhuzhou" and "low-carbon Zhuzhou", Municipal Party Committee and Municipal Government have jointly implemented "Smooth Traffic Project" and have actively applied for "National Model City for Transportation Management", "Pilot City for Demonstration Popularization of Energy Saving and New Energy Vehicles" and "Model City for Urban Pedestrian and Bicycle Transportation System".

Municipal Party Committee and Municipal Government have made concerted efforts to fully implement urban "Smooth Traffic Project" since 2000. Municipal Party Committee and Municipal Government set work goal for seeking "Smooth Traffic

Project" first-class management level (namely, National Model City for Transportation Management) at the beginning of 2008 and incorporated it into 2008 and 2009 Government Work Report. In July, 2010, the Ministry of Public Security and the Ministry of Housing and Urban-Rural Development jointly released document which marks Zhuzhou's success in developing into National Model City for Transportation Management.

With focused implementation of Three-year Action Plan for Electric Buses, Zhuzhou was successfully included in the first batch of pilot cities for demonstration popularization of energy saving and new energy vehicles in 2009. Zhuzhou was successfully included in the second batch of national candidate cities for "Urban Pedestrian and Bicycle Transportation System Demonstration Project" in 2011.

11.3.5 Speed up construction of transportation facilities, energetically improve the road system

Improvement of transportation facilities and road system is the important guarantee for effectively boosting low-carbon development of urban transportation.

From 2007 to 2009, 5,438 million yuan investment was made in road transportation infrastructure in urban areas of Zhuzhou City, accounting for 4% of gross national product in urban areas; 47 municipal infrastructure projects were fully implemented in urban areas; sidewalk and non-motor vehicle lane maintenance and management measures were developed; infrastructure was complete and allocation was rational; roads in built-up areas were provided with barrier-free access systems; urban road network was further optimized.

In 2008, urban "five-reconstruction" projects (reconstruction of concrete roads into asphalt roads, side street and by-lane reconstruction, sidewalk slab reconstruction, main trunk road side building appearance improvement reconstruction project, buried electric wire project) were vigorously carried out and urban road transportation infrastructure was improved in an all-round way.

In 2009, based on quickening construction of Shisong Avenue, Xiangjiang River No.5 Bridge, Zhuzhou-Changsha Highway Tianxin Flyover, additional 2,024.19 million yuan was invested in newly building (reconstructing, expanding) four backbone roads including Hongyi Avenue, Yingbin Avenue, Zhuzhou Avenue and Yandi Avenue, and Phase I project for urban road intelligent transportation monitoring system which is ranked 1 in Hunan Province and first-class nationwide was completed.

11.3.6 Pay attention to low-carbon transportation management, actively solve difficult problems concerning transportation

In recent years, road traffic order has been strongly rectified and an intelligent road transportation management system has been built in Zhuzhou City.

Road traffic order has been vigorously rectified. Special rectification was conducted to address main trunk road traffic order, and series centralized rectification activities were launched to cope with serious illegal traffic acts, illegal acts relating to two-wheeled and right three-wheeled motorcycles, special vehicles (bus, taxicab, dangerous chemical transport vehicle, school bus), rural road transportation safety, pedestrian and non-motor vehicle illegal acts, etc.

In 2009, more than 50 million yuan was invested in completing intelligent road transportation management system of Zhuzhou City. This system exercises all-around, all-weather, multi-means transportation monitoring of roads in urban areas through 90 television surveillance points, 40 traffic flow collection points, 83 signal control points, 42 automatic recording systems for red light running, 4 stationary overspeed monitoring systems and 15 LED guidance display screens in the whole city.

11.3.7 Mobilize all residents for participation, vigorously launch low carbon publicity

Universal and active participation by residents is the important guarantee for urban low-carbon transportation development. In recent years, various kinds of extensive vivid low-carbon transportation theme publicity activities have been held to gradually make "harmonious transportation, civilized transportation, green transportation" philosophies deeply rooted among people in Zhuzhou City (Fig. 11-4).

Fig. 11-4 Green transportation publicity activities in Zhuzhou City

In recent years, special columns dedicated for Zhuzhou City were launched in mainstream media including *Zhuzhou Daily, Zhuzhou Evening News*, TV station, broadcasting station to carry out in-depth, high-density, multidimensional publicity about traffic laws, regulations and common knowledge relating to safety, and advocate civilized travel by residents.

Meanwhile, theme activity publicity and signing activities were unveiled among primary and secondary school students on memorial days to significantly enhance their awareness about low-carbon transportation.

11.3.8 Attach importance to low-carbon technology innovations, proactively conduct low-carbon technology research and development

In recent years, technical innovation support has been intensified and preferential policies have been issued to actively support enterprises to engage in low-carbon technology research and development and jointly make technical breakthroughs in Zhuzhou City. 86 patents relating to low-carbon technologies were approved in the whole city from 2008 to 2009, among which 17 patents were related to electric vehicle and 24 patents involved energy conservation and emission reduction.

For example, hybrid electric and pure electric urban buses made by CSR Times reach advanced level in terms of energy conservation and emission reduction in China, which can save oil by 30% and reduce exhaust gas emission by more than 60%.

11.4 Main Countermeasures for Sustainable Development of Low-carbon Transportation in Zhuzhou City

In recent years, Zhuzhou Municipal Party Committee and Municipal Government have attached great importance to low-carbon transportation construction, and remarkable achievements have been made in urban road planning, electric public transportation development, dual fuel taxicab application and popularization, public bicycle leasing, etc, especially it has successively become the city for application and demonstration popularization of new energy vehicles and the first electric bus city, making low-carbon transportation development of Zhuzhou City rise to a relatively advanced level in China.

However, overall, some problems still exist in low-carbon transportation development of Zhuzhou City. Currently, rail transit and bus rapid transit (BRT) have not yet been launched in Zhuzhou City; intercity railway is under construction; urban low-carbon transportation is also increasingly subject to severe problems including urban road congestion and increasing transportation-induced exhaust gas emissions. In addition, there is a salient problem: carbon dioxide emission from transportation is skyrocketing. According to estimation, carbon dioxide emission from transportation in Zhuzhou City was about 936,100t in 2005, while such figure was about 1,401,000t in 2008; overall growth of carbon dioxide emission was as high as 49.66% with annual average growth standing at 14.39% from 2005 to 2008. With the rapid development of road transportation and rapid entry of cars into households in Zhuzhou City, carbon dioxide emission from road transportation means is rapidly increasing, and such growth is expected to last for a period of time.

Low-carbon transportation development is a systematic project. Against such a background, in order to accelerate low-carbon transportation development of Zhuzhou

City, a plan should be developed in advance and the following countermeasures are recommended in the light of actual situation of transportation development in Zhuzhou by making reference to domestic and foreign low-carbon transportation development practice.

11.4.1 Prepare a special low-carbon transportation plan, actively build an integrated transportation hub

On one hand, the *12th Five-year Overall Plan for Urban Transportation Development* has been formulated in Zhuzhou City. This Plan is a comprehensive and systematic one; however, it less touches upon low-carbon urban transportation development and only introduces transportation energy conservation and environmental protection, it is difficult to effectively guide low-carbon urban transportation development. Given the needs for future low-carbon urban transportation development, in order to fundamentally optimize urban function layout and foster coordinated development of urban transportation, land utilization and spatial structure, active actions should be taken to specially develop programmatic guiding measures including *Urban Low-carbon Transportation Development Plan*, the corresponding *Action Plan for Low-carbon Transportation Development of Zhuzhou City (2011-2015) and Opinions Concerning Implementation of Low-carbon Transportation Development of Zhuzhou City*, etc, and based on alignment with existing overall urban plan and special transportation plan, scientific planning and construction are conducted to increase public transit network density and station coverage rate; lines and stations should be extended to urban function areas including residential quarters, commercial districts, school clustering areas, so as to improve public transit accessibility.

On the other hand, the fact that the proportion of urban residents who choose walking and public transport to travel is not very high is attributable to excessive transfer time. Therefore, how to combine integrated transport transfer hub construction with road transportation planning design is the key to the current urban low-carbon transportation development. For this purpose, it is necessary to not only build urban spaces for compact development but also develop integrated transport transfer hub, and give overall considerations to population and industrial layout, integrate spatial functions to form group-based mixed land use layout pattern, and gradually shift land use from "two-dimensional" expansion to "three-dimensional" effect enhancement, and focus on its comprehensive utilization and development, and increase urban density, boosting unity between land use and transportation supply and strike transportation supply and demand balance. Attention should be paid to proceeding from long-term perspective and considering urban function division to optimize urban transportation layout network, and stressing connections among multiple transportation means including railway and road etc to build seamless connection among multiple transportation means so as to form a green integrated urban transportation system in which travel mode is dominated by public transport.

11.4.2 Give priority to developing urban public transportation, quicken improvement of bicycle leasing system

Given that subway and bus rapid transit (BRT) have not yet been built in Zhuzhou City and the public transportation system is also being continuously optimized, the government should further give priority to public transportation in urban transportation, and accelerate construction of large capacity rapid public transportation including rail transit and bus rapid transit, etc, and build special bus lanes, choose transportation-oriented development mode, and implement public transportation-oriented land development mode, and guide orderly expansion and renewal of urban land use, and promote continuous optimization of urban spatial layout. Backbone network for public transportation should be improved and public transportation sharing rate should be increased. Public transportation enterprises should energetically improve public transportation operation services and jointly guide benign development of urban transportation.

The bicycle is the most low-carbon transportation means among current transportation means. As one of the first cities for launching "public bicycle leasing" project in China, Zhuzhou City has witnessed certain achievements in development of public bicycle transportation. Firstly, further perfect special bicycle road system and pedestrian system, and improve pedestrian network facilities, foster a comfortable walking environment, and strengthen network-focused construction of pedestrian system connected to commercial zones, cultivate and develop good pedestrian traffic environment; secondly, increase input, further improve public bicycle leasing system, perfect leasing outlet layout and supporting service management facilities; thirdly, actively research and develop bicycle self-lock and surrounding parking lot unavailability early warning and management techniques, etc.

11.4.3 Strengthen urban transportation operation management, improve urban transportation management

Firstly, propel construction of intelligent public transportation system. Actively research, develop and popularize comprehensive dispatching system for intelligent transportation, build integrated dispatching platform for intelligent transportation, perform intelligent road and vehicle operation management, digital management, enhance modern management of urban public transportation; build and popularize intelligent dispatching information platform for taxicabs, and set up unified stations for taxicabs, conduct differentiated operation.

Secondly, reinforce real-time sharing of transportation information resources. Make travelers informed of real-time road information, choose proper travel mode or timely change lines to ease traffic congestion, improve overall operating efficiency of urban public transportation, reduce energy consumption and carbon dioxide emission.

Thirdly, exercise coordinated traffic signal control. Actively push forward research, development and construction of coordinated traffic signal control system commensurate with mixed urban transportation, and adjust coordinated control timing parameters on real-time basis, effectively reduce vehicle delay, stop and acceleration times, shorten the queuing time.

11.4.4 Intensify scientific and technical innovations, vigorously popularize and apply new energy vehicles

As one of the important energy saving and new energy vehicle research and development bases in China, Zhuzhou City has paid great attention to the development of the new energy vehicle industry in recent years. With generous support from relevant departments of the State, efforts have been actively organized to implement major special science and technology projects for electric vehicles, break technical bottlenecks, and make great progress in research and development of full electric vehicle and key components, industrialization demonstration popularization; core technologies for energy saving and new energy vehicles have been basically mastered; relatively complete key component supporting system has preliminarily taken shape; one batch of hybrid and electric vehicle products have been developed. Currently, there are two new energy vehicle manufacturers in Zhuzhou City, and new energy vehicle industrialization has been basically achieved; it is the only area in which both series and parallel hybrid power technologies are concurrently mastered and batch commercial operation of full vehicles is performed in China.

Firstly, actively boost innovations and applications of substitute energy technologies. Give considerations to demonstration and popularization of energy saving and new energy vehicles, and pilot projects for low-carbon urban transportation construction, proceed from local conditions to spread alternative fuels such as natural gas and alcohol fuels, etc. and oil replacement technologies, and encourage applications of alternative fuels in urban public transportation and taxicab fields; establish access mechanism for energy saving and new energy vehicles, and reinforce demonstration and popularization of hybrid and pure electric vehicles; form a guarantee system for new energy public transportation and taxicab operation.

Secondly, make continuous, active efforts to seek funds from the Ministry of Science and Technology, National Development and Reform Commission, provincial new industrialization, technical innovation projects, major special science and technology projects, etc, provide more support for development of new energy industry in Zhuzhou City.

Thirdly, actively guide new energy vehicle market. According to "10-city & 1,000-vehicle" Action Plan of the Ministry of Science and Technology, the State will popularize and apply more than 60,000 energy saving and new energy vehicles in bus, taxi, public service, municipal, postal fields etc in future 3-4 years. All existing buses in Zhuzhou City will be replaced by new energy buses made in the city within three years, and replacement of public vehicles including taxicabs and postal vehicles, etc. by new energy vehicles made in the city and complying with requirements is encouraged and guided; mini electric vehicles will be popularized in communities and scenic spots in the city.

11.4.5 Improve policy system, guarantee low-carbon urban transportation development

Excessively rapid growth of private cars in Zhuzhou City has led to serious problems including traffic congestion and environmental pollution. The carbon emission (kg/(capita·km)) from cars is dozens of times that from rail transit and ground public transportation, while walking and bicycles basically do not incur carbon emission. From this perspective, the key to low-carbon urban transportation lies in decreasing the proportion of cars in urban transportation. For this purpose, besides correct guidance at the philosophical level, accelerated efforts should be made to improve the policy system beneficial to the development of low-carbon transportation in Zhuzhou City in the light of actual regional situation.

Table 11-1 Comparisons of carbon emissions among various travel modes

Travel mode	Rail transit	Bus	Car	Electric vehicle	Bicycle	Walking
Carbon emissions (kg/(capita·km))	0.049	0.075	0.135	0.017	0	0

Firstly, carry out differentiated parking charge policy based on area, time slot and standard in urban areas. Different parking charges are applied to different areas, different time slots so as to regulate spatial and temporal distribution of road traffic loads; high charge and temporally progressive rate policy is adopted in areas where supply and demand contradictions are acute such as central business district and parking lot etc to encourage short-time parking; rational pricing is conducted for on-street and off-street parking lots, and existing parking facility resources are fully utilized; park-and-ride hub parking lots are free of charge or are available low charge.

Secondly, develop and improve an energy conservation planning system for the transport industry, and perfect fixed asset investment project energy conservation evaluation and examination system, energy conservation law, regulation and standard system, energy conservation supervision system, energy conservation statistics system and energy conservation testing system for the transport industry, and enhance energy conservation supervision capability.

Thirdly, establish and improve transportation energy conservation investment and financing mechanism, and gradually form transportation energy conservation input mechanism in which national and local capital serves as guidance and enterprise capital acts as main part, and establish special fund for energy conservation to encourage and support construction of energy conservation supervision system, and research, develop, popularize and apply new technologies and products for energy conservation.

11.4.6 Intensify publicity, encourage residents to travel in a low-carbon, green way

According to the result of online opinion survey conducted by Zhuzhou Public Transportation Co., Ltd. (http://zzggjt.com/), 483, 49 and 60 voting residents traveled through public transportation, bicycle and walking respectively, accounting for 52.6%, 5.3% and 6.5%. This survey result shows that Zhuzhou residents have relatively good green travel awareness.

However, from another perspective, the proportion of Zhuzhou residents who chose low-carbon transportation means was still not high; for example, only 49 voting residents chose bicycle, accounting for 5.3%, less than that of residents who chose private car (6.4%), suggesting that the awareness of residents about travel based on low-carbon green transportation still needs to be further improved.

For this purpose, action should be taken to intensify publicity, improve publicity means and methods, and enhance residents' awareness about low-carbon green travel.

Firstly, fully utilize public media including network, TV and newspaper, etc. and public service advertising to actively publicize "low-carbon transportation, green travel" philosophy, encourage the general public to choose "green transportation" travel modes including public transportation and bicycle, and decrease dependence of the general public on private transportation modes such as cars, etc, make enterprises and residents foster low-carbon transportation outlook and develop strong atmosphere for low-carbon transportation development.

Secondly, carry out all-people activities such as "Low-carbon Travel Month", "No Car Day", "Green Travel Month" etc in a deep-going way, continuously enhance residents' consciousness about giving priority to low-carbon transportation modes such as bicycle and walking, etc.

Thirdly, vigorously launch activities for appraising and choosing advanced individuals, advanced households and advanced collectives in low-carbon transportation, develop citizens' awareness about green travel, and create a good social atmosphere for everyone's involvement in building low-carbon transportation city.

References

[1] Xu Jianmin: "Analysis of China's Low-carbon Transportation and Its Promotion Measures," *Urban Insight*, 2010(4):13-18.

[2] Sun Dehong: "Urban Low-carbon Transportation Development Mode and

Measures," *Environmental Protection in Transportation*, 2011(1):43-45.

[3] Xie Xiao'an, Lu Xiaozhi, *Liu Yang:* "Thinking and Countermeasures about Low-carbon Transportation in Shijiazhuang City," *Journal of Shijiazhuang University of Economics*, 2011(1):96-98.

[4] Wang Guangrong: "Research on Urban Low-carbon Transportation System Construction," *Forward Position*, 2011(13):126-129.

[5] Feng Liguang, Zhangwei, Zhang Haozhi: "Thinking of Construction of China's Urban Low-carbon Transportation System," *Highways & Automotive Applications* 2011(1):36-39.

[6] Li Zhenyu: "Low-carbon Urban Transportation Mode and Development Strategy," *Engineering Research*, 2011(2):106-111.

[7] Zhang Taoxin, Zhou Yueyun, Zhao Xianchao: "Analysis of Current Situation and Route of China's Urban Low-carbon Transportation Construction," *Urban Studies*, 2011(1):69-75.

[8] Anonymity: *Report on Efforts in Building Zhuzhou into National Model City for Transportation Management, http://www.0731.cn/news/html/100324/0M5JZ10324112059_1.html.*

Chapter 12

Discussion of Typical Foreign and Domestic Urban Low-Carbon Development Modes

Teng Fei, Liang Benfan

Abstract: Cities, as the accumulation areas of human production and life, are the main sites of energy consumption. Therefore, building low-carbon cities plays a vital role in responding to climate change and the energy crisis. Low-carbon cities started earlier in foreign countries, and the developed countries such as the United Kingdom, the United States and Japan have formed different developmental models. However, throughout the world, there are also numerous improper practices such as blindness and disorderliness in the building of low-carbon cites. Cities need to choose the appropriate low-carbon development model based on their own characteristics. Therefore, all the different experiences and the successful development models needed to be summarized and analyzed. This chapter, based on the domestic and international cases of low-carbon city construction, systematically analyzes and summarizes the development models, the main practices, the relevant policies and measures of the low-carbon cities at home and abroad so that other cities can learn from them in the building of their own low-carbon cities.

Key Words: low-carbon cities; typical cities; development models

12.1 Introduction

With the year-by-year increase in the gap between global supply and demand for fossil energy such as coal, oil and natural gas, and the increasingly severe global warming trend, sustainable development of human society is subject to grave threats. In 2003, the UK became the first to put forward a "low-carbon economy" in a white paper *Our Energy Future*[1]; after the Copenhagen Climate Change Conference was held in 2009, a "low-carbon economy" development mode based on low energy consumption, low pollution and low emissions has become a consensus reached by various countries in the world. Various countries have declared their emission reduction goals; China has also announced the goal for reducing carbon dioxide emissions per unit of GDP by 40%-45% by 2020 compared with 2005. Development of a low-carbon economy has become a global inexorable trend. As cities are areas where population, buildings, transport, industry and logistics are concentrated and also incur high intensity energy consumption, cities inevitably become hot spots and key areas involving greenhouse gas emissions and play a leading role in addressing global climate change and breaking the energy shortage predicament[2]. Currently, more than half of the world population is living in cities. Their total energy consumption accounts for 60%-80% of the world's total energy consumption. Meanwhile, cities are also the foremost CO_2 emission sources in the world[3] (see Fig. 12-1). Therefore, construction of low-carbon cities will affect the world's energy demand and greenhouse gas emissions to a considerable extent and is the important element in building an action system for addressing climate change[4].

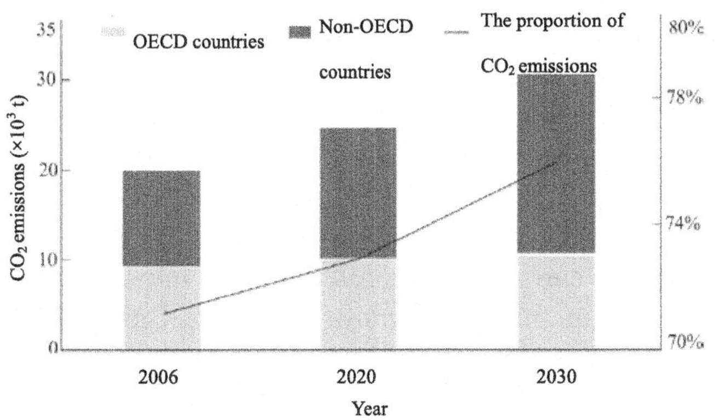

Fig. 12-1 Current situation and forecast of CO_2 emissions from global cities[6]
Source: International Energy Agency.

Currently, low-carbon city construction practice has been started and valuable experience has been gathered in many cities in foreign countries, while in China the National Development and Reform Commission and World Wildlife Fund (WWF) officially launched the "China Low-Carbon Development Project" and determined Shanghai and Baoding in the first batch of pilot cities on January 28, 2008. In 2010, the National Development and Reform Commission also vowed to carry out low-carbon city construction in China's five provinces and eight cities; afterwards, low-carbon city construction attempts were made in many domestic cities[5]. As low-carbon city construction is progressing in various countries, different low-carbon city development modes have been fostered according to their own specific problems and favorable conditions. However, presently there is still a lack of systematic research and summarization of these different low-carbon city development modes. In this paper, based on domestic and foreign experience and achievements in low-carbon city construction, low-carbon city development strategies and relevant policies will be summarized, and typical low-carbon city development modes will be organically clarified with the aim of providing certain references for low-carbon city construction through summarization and the thinking behind domestic and foreign examples.

12.2 Foreign Low-Carbon City Development Modes

12.2.1 Overview of foreign low-carbon city development

As cities are gathering places for politics, the economy and people, and are aggregates of energy consumption and they play an important role in the development of a low-carbon economy, not a few countries have recognized the economic and social problems involved in energy issues and climate change. Currently, global low-carbon city planning and construction are in full swing. Many world-class megacities such as London, Tokyo, Paris and New York have specified low-carbon city construction goals and have developed relevant plans or action plans. Per-capita emissions in most of the key cities in the world are higher than those in their rural areas; however, with active low-carbon city planning and development, emissions in many cities have substantially decreased, and per-capita greenhouse gas emissions in some cities have been lower than the level in their rural areas; for example, per-capita greenhouse gas emissions in New York City in 2008 had decreased to 6.4t carbon dioxide equivalent[6].

Internationally, case cities which can serve as a reference for low-carbon city construction at the current stage are mainly members of the Large Cities Climate Leadership Group; these cities have entered the stage of implementation of low-carbon city construction goals, including London, New York, Los Angeles, Copenhagen, Tokyo, Toronto, Portland, Amsterdam, Austin, Chicago, Madrid, Stockholm and Seattle. Most of the case cities have developed indicators for substantial, quantifiable reductions of carbon dioxide emissions; for example, London has vowed to reduce carbon dioxide emissions by 60% by 2050 compared with 1990[7]; Los Angeles has set an emission reduction goal for 2030, namely a reduction of 35% compared with 1990[8]; Madrid has announced the goals for reducing emissions by 20% and 50% by 2020 and 2050 respectively[9], compared with 2004 standards; Stockholm has declared that carbon dioxide emissions will decrease by 60%-80% by 2050 compared with 1990 and it will become a zero emission city[10].

It's worth noting that most of these cities are at the post-industrial urban development stage and have stood at the forefront of the world in terms of energy renewal and environmental protection long ago, and are subject to less energy, resource and environmental pressure compared with China. They enjoy a relatively large operational space for urban emission reduction, thus they have inborn advantages in building low-carbon cities[11].

12.2.2 Foreign typical low-carbon city development modes

1) British mode–take London for example

As a country which first put forward the "low-carbon economy" in the world, undoubtedly the UK is the forerunner in low-carbon city planning and practice.

According to statistics, the annual carbon emissions just in London account for 8% of carbon emissions throughout UK; such a figure may increase to 15% by 2025[12]. In order to stimulate the UK to transform towards a low-carbon economy as early as possible, the UK Government established a Carbon Trust in 2001 to work with enterprises and the public sector in the form of a private institution and draw support from various organizations to reduce carbon emissions while developing low-carbon technologies. The Carbon Trust and Low Carbon Cities Programme (LCCP) have jointly promoted the UK's low-carbon city project, while respective low-carbon city plans for the first batch of three model cities including Bristol, Leeds and Manchester have been developed under expert and technical support provided by the LCCP[13]. Meanwhile, a series of low-carbon London actions plans have been formulated by London City. *The Mayor's Climate Change Action Plan* released by London City in 2007 plays a very critical role in London's low-carbon action plans[14]. Table 12-1 lists the major means for London to build a low-carbon city.

Table 12-1 Major means for London to address climate change and build a low-carbon city

Major fields for emission reduction	The proportion of carbon emissions (%)	Realization approaches and specific measures	Carbon emission reduction goals ($\times 10^4$ t) (until 2025)
Residential houses	40	Green Homes Programme	770
Commercial and public buildings	33	Green Organizations Programme	700
New development projects		Distributed energy supply system; Strengthen the requirements for energy saving	100
Energy supply		Transition to decentralized and sustainable energy supply; Encouraging waste generation and its application; Incentive to build renewable energy power station	720
Transportation	22	Changing the travel way of the people of London, increasing investment in public transport, walking and bicycle systems; Encouraging low carbon transport and energy; charging for carbon emissions in transportation	430

Source: Greater London Authority, "Action Today to Protect Tomorrow: The Mayor's Climate Change Action Plan", February 2007, 2008-05-28.

Main characteristics of London's efforts in addressing climate change and building a low-carbon city[15]:

(1) Quantitative indicators were specified. The low-carbon action plan of London City explicitly stated that carbon emissions from 2007 to 2025 would be controlled to

within 600 million tons, namely, annual carbon emissions would be reduced by 4%.

As shown in Table 12-2, building and transportation are two major fields for emission reduction. Carbon emissions from residential houses accounts for 38% of total carbon emissions in London City, while those from commercial and public buildings and from transportation make up 33% and 22% respectively; they become major fields for reducing carbon emissions. The UK has developed the corresponding renovation measures according to different carbon sources. Firstly, the "Green Home Plan" was implemented for residential houses in stock, such as insulation renovation subsidies for attics and walls, household energy conservation and recycling information, social residential house energy conservation renovation. Secondly, the "Green Institution Plan" including a building renovation partnership plan and green building label system was carried out for commercial and public buildings in stock. Finally, with respect to ground transportation, efforts were dominated by the change in residents' travel modes, and more government investment was made in public transportation, a walking and bicycle system. Low-carbon transportation is encouraged and charges are collected for transportation carbon emissions[16].

Table 12-2 UK's carbon emission and emission reduction goals

Major fields for emission reduction	The proportion of carbon emissions (%)	Carbon emission reduction goals ($\times 10^4$ t) (until 2025)
Residential houses	38	770
Commercial and public buildings	33	700
Transportation	22	430

(2) New energy is developed. Clean energy is applied and energy utilization efficiency is improved to realize low-carbon city goals and reduce carbon by 7.2 million tons by 2025. Main methods: encourage garbage power and application, localize renewable energy, build large renewable energy power stations, stimulate renewable energy power generation through new policies and plans, encourage carbon storage, develop new energy conservation projects. The Government actively encourages and guides low-carbon consumption and low-carbon city construction. Fiscal input is comprehensively utilized, and publicity is energetically launched among the general public; enterprises are encouraged to participate in low-carbon city construction, and considerations are given to the actual situation of the city; low-carbon city development is driven through major projects.

(3) New technology, government policies and public governance are concurrently implemented. The Government actively advocates construction of a low-carbon city and encourages enterprises and the general public to participate in this process; feasible policies are developed and high and new technological means are adopted to vigorously develop low-carbon energy such as solar energy; meanwhile, private carbon trusts are established to contact enterprises and government⑯.

2) American mode——take New York for example

Though the USA refused to join the *Kyoto Protocol*, all walks of life in the USA neither passively regard climate change nor give up the exploration of low-carbon development. The U.S. Senate proposed a Low-Carbon Economy Act on July 11,2007, suggesting that the USA may take the development of a low-carbon economy as its future important strategic choice[17]. In January 2009, Obama announced the American Recovery and Reinvestment Act which stressed development of new energy as an investment priority and stated that 150 billion USD was planned to be invested and it would take three years to double the output of new energy in the USA, and the proportion of new energy power generation in total energy power generation would be increased by 10% and 25% by 2012 and 2025 respectively[18]. On February 15,2009, the USA officially promulgated the *American Recovery Reinvestment Act* in which total investment reached 787 billion USD and development of new energy was listed as the important content, including development of efficient batteries, smart power grid, carbon storage and carbon capture, renewable energy such as wind energy and solar energy[19]. Meanwhile, actions have been taken in many American cities to build low-carbon cities, and remarkable achievements have been made. As an international metropolis, New York has the largest urban area and population in the USA and plays an important role in the world's economy. New York has developed a good organizational structure and process in cleaning the city in the pursuit of addressing global climate change and exploring low-carbon city development.

New York is taken as an example because this city shares similarities with many megacities in China: (1) Growth pressure and low-carbon emission requirements coexist. New York has the highest population density in the USA, with the urban population continuously growing at an average annual rate of 14%. The urban population of New York was nearly 8.4 million in 2009 and will increase by one million by 2030 on the existing basis. Meanwhile, New York City has also specified very clear carbon emission reduction indicators: carbon emissions will be reduced by 48.7 million tons by 2030 compared with that estimated under current trends, and carbon emissions in 2030 will decline to only about 70% of carbon emissions in 2005[20]. (2) The municipal infrastructure is old and obsolete and finds it hard to cope with population growth and the needs for efficient city operation. (3) New York City has a high residential density and is a public transportation-oriented city. Meanwhile, the car utilization rate is also very high in this city; its living mode and travel mode are similar to those in many large cities in China[21]. In recent years, though the urban population has continuously risen in New York City, the carbon emission level in the whole city has constantly decreased (see Fig. 12-2); the carbon emission level declined by about 13% in New York City from 2005 to 2009.

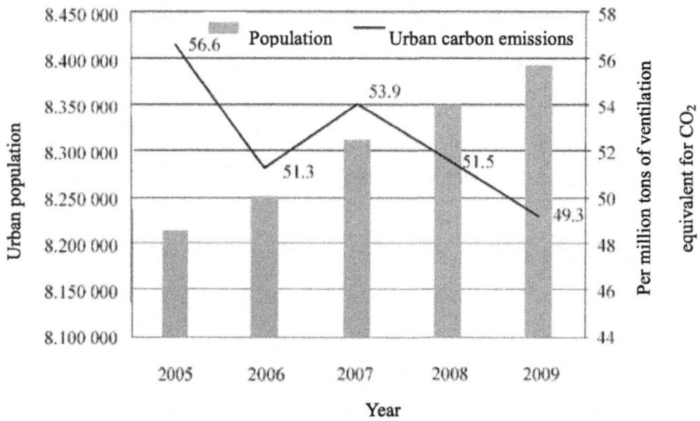

Fig. 12-2 Carbon emissions and population in New York City (2005-2009)

These mainly benefited from the following three measures taken in New York:

(1) Vigorously carry out a "green building" emission reduction plan

Major emission sources in New York are building and transportation. Emissions from buildings make up the largest proportion in total emissions; nearly 80% of carbon emissions come from buildings; change in their emission level can directly affect the change in total emissions. New York City has put forward a "green building" emission reduction plan, while specific measures include: (1) Lighting upgrading and independent electric meters. Space in a non-residential building area of more than 929m^2 is required to be provided with independent electric meters, and owners are required to regularly deliver a power consumption report each month. (2) Building emission benchmark test: Owners of large buildings are required to analyze power consumption and emissions from their buildings by means of an analysis tool provided by government so that owners and potential buyers can provide building energy conservation data. (3) Funding support for green buildings: A loan fund was established for providing funds for owners who have difficulties in building renovation and upgrading, and commercialization of financial support is actively advocated to consistently provide impetus for projects. An Environmentally Friendly Product Procurement Act has been executed in New York City since January 2007. People in New York can obtain a government subsidy when they are building or renovating houses if the design complies with the green building standard. In addition, education was launched among the general public through government projects such as "GreeNYC" in New York City where a low-carbon lifestyle is advocated and qualified energy auditors and technical personnel necessary for low-carbon development are provided through training[22].

(2) Optimize layout of urban public transportation system

Emissions from the transportation sector accounted for about 1/5 of total carbon emissions in New York, and such a proportion continuously increased from 18.28% in 2005 to 20.27% in 2009. Much work has been done in transportation planning, popularization and application of urban public transportation in New York. Compared with other large cities in the USA, public transportation system utilization efficiency is extremely high and non-motor vehicle transportation is nearly 80% in New York (see Fig. 12-3). The New York City subway is one of the largest subway systems in the world. Currently, an extensive network with 26 subway lines, 468 stations and more than 1,300km in total length has been developed. New York City public transportation daily delivers about five million passengers. There are more than 5,900 buses and more than 230 operating lines in New York. The developed public transportation system better satisfies people's travel needs so as to decrease the purchase and use of self-drive cars. The proportion of no-car households in New York is much higher than that in other cities, thus urban per-capita carbon emissions are greatly reduced.

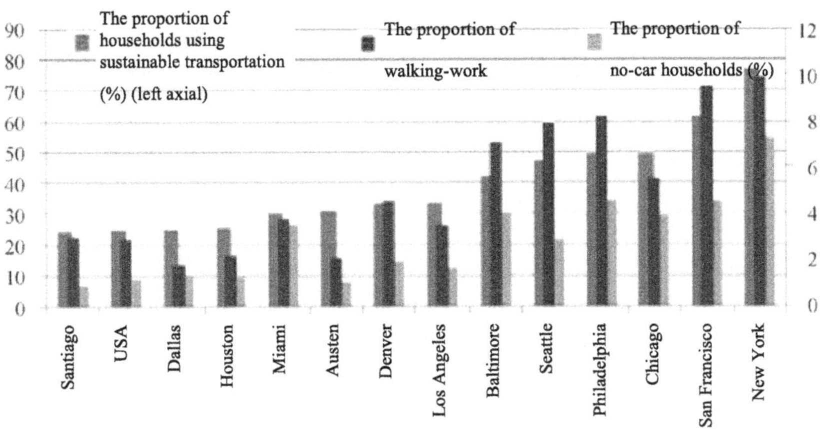

Fig. 12-3 Proportions of transportation means used and proportions of no-car households in large cities in the USA

Note: Sustainable transportation means refer to non-motor vehicle transportation means.

(3) Adopt a clean energy supply

(1) A medium and long-term energy development plan (2007-2030) for New York City has been developed, which is mainly designed to reduce energy demand, increase the new clean energy supply and promote modernization of the energy infrastructure. According to this plan, a clean, safe and affordable power supply will be greatly increased by 2015. (2)In order to ensure that the energy supply side and demand side cooperate to jointly seek safe development of urban energy, New York City has set up an Energy Planning Committee. (3) Efforts are made to speed up the development of renewable energy and new energy, expand the clean energy supply. Series policies have been released in New York City to develop renewable energy and new energy including solar energy, wind energy, tidal energy. These policies include: firstly, accelerating power restructuring, construction of clean power plants and power grid. Secondly, expanding a clean distributed generation project (Clean DG). The efficiency of distributed generation is two times higher than that of conventional power generation; if distributed power generation equipment is used in building energy efficiency renovation, costs can be

recovered within 3-5 years, while it takes 5-7 years to recover costs in the case of conventional energy efficiency renovation. Thirdly, the renewable energy market is actively cultivated. For example, a tax reduction is given for installation of solar panels by offsetting 35% of the installation costs in the first three years, 20% in the 4th and 5th years. (4)Modernization of energy transmission facilities is boosted, and the safety of the urban power grid is improved. Firstly, the power grid is protected and power grid modernization is promoted through legislation and polices. Secondly, construction of the natural gas infrastructure is strengthened[23].

3) Japanese mode–take Tokyo for example

Given the special geographical conditions in Japan, the adverse impact which climate warming exerts on Japan's agriculture, fisheries and environment is much higher than that in other countries in the world; for this reason, the Japanese Government is actively coping with climate change, advocated and initiated in the Kyoto Protocol, as well as playing a leading role in building a low-carbon society[15].

In July 2008, the Japanese Cabinet adopted an *Action Plan for Building a Low-Carbon Society* and made it public. In July 2008, the Japanese Government chose six local cities where practicable and effective measures were vigorously taken to prevent the greenhouse effect, known as "Environmental Model Cities". In April 2009, Japan's Ministry of the Environment announced a draft policy entitled *Green Economy and Social Change*, which was designed to reinforce Japan's low-carbon economy by taking measures for reducing greenhouse gas emissions. Japan believes that a "low-carbon society" serves as the foundation for a "low-carbon economy" and its low-carbon society follows the following three principles: reduce carbon emissions; advocate a frugality spirit and shift from a high-consumption society to a high-quality society; harmoniously coexist with nature.

Since 2006, the Tokyo Metropolitan Government has released a "Tokyo After 10 Years" plan which lists the specific emission reduction goal: carbon emissions in Tokyo will be reduced by 25% by 2020 compared with 2000, from which the construction of a low-carbon society kicked off. *Tokyo Climate Change Strategy——Basic Policy for the "10-Year Project for a Carbon-Minus Tokyo"* was unveiled in June 2007, which detailed the medium and long-term strategy of the Tokyo Government for addressing climate change. Its basic policies covered various aspects: stimulate enterprises to reduce emissions, lower residential waste, save energy in government facilities and cut carbon dioxide emissions from transportation[24]. Meanwhile, four basic policies for low-carbon Tokyo were defined: (1) Help enterprises reduce carbon dioxide emissions, offer emission reduction routes to enterprises, establish a special fund for encouraging enterprises to adopt energy saving technologies. (2) Residents are encouraged to take the household as a unit and adopt a low-carbon lifestyle to reduce energy consumption, energetically advocate energy saving lamps; residents are encouraged to take energy saving measures for houses and add thermal insulation windows. (3) CO_2 emissions from urban construction are reduced and the newly built infrastructure complies with energy conservation regulations; energy consumption requirements for new building facilities are higher than current standards. (4)CO_2 emissions from the transportation system are lowered and new energy vehicles are popularized.

Main practices for Tokyo to develop a low-carbon city are shown below[25]:

(1) Low carbon starts from clean energy

Low-carbon energy is the basic guarantee for building a low-carbon city. Tokyo has been vigorously researching, developing and utilizing new technologies and new processes for green low-carbon energy including solar energy, biomass energy, wind power and hydropower. The proportion of primary energy used increased, and use of carbon-based energy (oil, natural gas, coal) was reduced, and utilization of clean energy such as nuclear energy and hydropower also declined from 1998 to 2008, as shown in Fig. 12-4. The overall proportion of carbon-based energy used consistently fell, suggesting that Tokyo's energy structure was becoming increasingly clean.

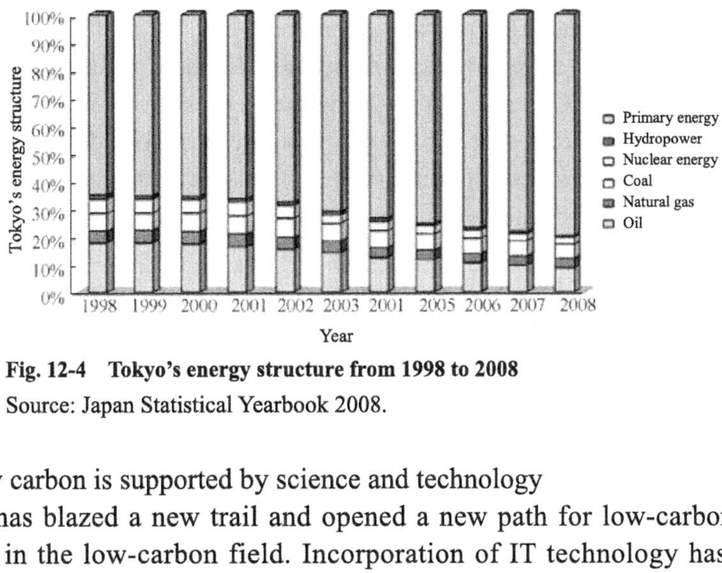

Fig. 12-4 Tokyo's energy structure from 1998 to 2008

Source: Japan Statistical Yearbook 2008.

(2) Low carbon is supported by science and technology

Tokyo has blazed a new trail and opened a new path for low-carbon science and technology in the low-carbon field. Incorporation of IT technology has helped open a new era for more information-based, more people-oriented low-carbon science and technology. In 2007, the Tokyo Municipal Government worked with other functional departments to successfully apply the Internet of Things in the whole city[26].

(3) Low carbon leads ecological transportation

According to relevant information, CO_2 emissions from Tokyo's transportation sector hit 14.66 million tons and accounted for 26.2% of total emissions in 2006. In order to reduce carbon emissions from the transportation sector, Tokyo adopted various measures including energy conservation and emission reduction; for example, in order to vigorously advocate low-pollution low-consumption vehicles, Tokyo Metropolitan Government provided certain fiscal subsidies for buyers; in order to boost application of biodiesel, biodiesel was introduced to public buses within urban areas of Tokyo, and a feasibility study concerning the application of the second generation of biodiesel in public buses within urban areas was conducted .

(4) Low carbon is injected into green building

According to statistics, 60% of energy consumption in the whole of Tokyo came from buildings in 2005. Thus the Tokyo Metropolitan Government released *Tokyo Green Building Plan, Green Label Plan, 2007 Tokyo Energy Conservation Charter, 2008 Overall Environmental Plan for Tokyo*[8]. Energy conservation measures have been extensively adopted in Tokyo municipal government agencies, which have played a demonstration role in popularizing energy conservation philosophy and energy conservation technologies in the whole of society. The Tokyo Metropolitan Government has required that an environmental report must be delivered to the Government in the case of a new building covering more than 10,000m^2 so that the building's owners can conduct a sustainable low-carbon design according to the requirements of Tokyo Metropolitan Government.

4) Other modes–take Copenhagen, Berlin and Seoul as examples

(1) Copenhagen, Denmark–low-carbon community

A typical representative for low-carbon city development in Denmark is the low-carbon community, while Copenhagen is its typical representative city. A low-carbon community mainly starts from a national energy policy goal for addressing the impact from global climate change and reducing carbon emissions; local governments play a pioneering role in energy conservation applications; community energy conservation construction is performed with low-carbon energy saving demonstration projects as guide in most cases. Generally, a low-carbon community follows the following ten principles: zero carbon, zero waste, sustainable transportation, sustainability and local materials, local foods, low water consumption, animal and plant protection, cultural heritage conservation, fair trading and happy, healthy lifestyle.

In Copenhagen, the power supply mostly relies on a zero carbon mode, and wind and biomass power generation is used, completely white modern windmills can be seen everywhere, and there is the largest offshore wind power plant in the world. In 2008, Copenhagen topped the list of the world's 20 best cities compiled by British lifestyle magazine MONOCLE for a high life quality and great attention to environmental protection. In 2009, Copenhagen, Denmark announced that it was expected to become the first carbon neutral city in the world by 2025.

(2) Berlin, Germany–low-carbon construction of buildings

Low-carbon construction of buildings has been actively practiced in Berlin, Germany. Low-carbon construction of buildings has been pursued through an "Energy Savings Partnership" and Berlin building modernization renovation project, throughout Berlin. Currently, in the most successful building emission reduction mode "Energy Savings Partnership" in Berlin, energy conservation renovation of 1,300 public buildings has been completed and 0.6 million tons of carbon emission reductions have been made; about 60 million EUROS of private investments have been attracted, and nearly 24 million EUROS of energy costs have been saved[27].

(3) Seoul, South Korea–exquisite combination of mitigation policy and adaptation policy

Seoul, South Korea has subtly combined a mitigation policy with adaptation policy. In 2009, Seoul Metropolitan Government unveiled a Master Plan for Low Carbon Green Growth, which integrated three large policies including climate change mitigation, adaptation and green economy growth and enabled direct investment worth 2 billion USD into research and development in ten green technology fields including hydrogen energy cells, solar batteries, a smart power grid and climate change adaptation technologies, for comprehensive climate change mitigation and adaptation[27].

12.3 Domestic low-carbon city development modes

12.3.1 Overview of domestic low-carbon city development

There are more than 600 cities and more than 20,000 towns in China, among which 287 cities above prefecture level produce energy consumption and carbon dioxide emissions accounting for 55.48% and 58.84% of total energy consumption and total emissions in China respectively. If the remaining cities and towns are considered, energy consumption and carbon emissions make up more than 80% of total energy consumption by society[28]; therefore, construction of a low-carbon city is the important content for energy conservation and emission reduction. China's economic, environmental and energy situation greatly differs from foreign developed countries, thus it is necessary to not only make reference to foreign low-carbon city development modes, but also consider China's actual conditions, so as to take a low-carbon economic development path with Chinese characteristics.

In January 2008, the World Wildlife Fund (WWF) chose Shanghai and Baoding as pilot low-carbon cities; thus China has entered the stage of developing low-carbon cities. A "China Electricity Valley · Low-Carbon Baoding" goal has been set for Baoding, while a "Low-Carbon Expo" was practiced in Shanghai during the Shanghai World Expo. In the same year, the Ministry of Science and Technology initiated a "10-City 1,000-Vehicle Plan" which was designed to popularize and apply new energy and electric vehicles on a large scale in cities. This Plan was an important move for developing low-carbon cities in China. In August 2010, the National Development and Reform Commission launched pilot work on national low-carbon provinces and low-carbon cities and determined five provinces including Guangdong, Liaoning, Hubei, Shaanxi, Yunnan and eight cites including Tianjin, Chongqing, Shenzhen, Xiamen, Hangzhou, Nanchang, Guiyang and Baoding as pilot low-carbon provinces and cities, which marked official incorporation of China's low-carbon city development into a national project.

12.3.2 Domestic typical low-carbon city development modes

1) Low-carbon city construction practice and mode in Shanghai

Since reform and opening up, rapid GDP growth has been maintained in Shanghai on the one hand and energy consumption per 10,000 yuan GDP has annually declined by 4%. However, Shanghai's economic structure features an excessively high proportion of secondary industry and its industry is a heavy one. Thus Shanghai's per-capita carbon emissions have been still relatively high. In 2008, Shanghai's per-capita carbon dioxide emissions were nearly 13t, while the world's per-capita carbon dioxide emissions and China's per-capita carbon dioxide emissions were 4t and about 5t respectively, suggesting that Shanghai's per-capita carbon dioxide emissions much surpassed both the world's per-capita level and China's per-capita level. In terms of per-capita carbon emissions, Shanghai also topped the list of large cities with a population of 10 million in the world. Per-capita carbon emissions in New York, London, Paris and Tokyo were only 30%-50% that in Shanghai. In 2007, Shanghai's carbon dioxide emissions were 239 million tons, out of which those from industry, transport and buildings accounted for 58.21%, 18.77% and 16.21% respectively. This indicates that Shanghai's industrial carbon emissions are in a dominant position, while emissions in large cities with populations of 10 million in developed countries are basically dominated by emissions from buildings or transport, such as New York, London and Tokyo[30].

Bright spots in building Shanghai into low-carbon ecological city concentrate on Chongming Eco-island, Lingang New City and Hongqiao Business District, among which Chongming Eco-island is positioned as "ecological new town" to be built with "low ecological footprint" philosophy, with planning content including: build ecological function zone, develop green transportation, fully utilize renewable energy, pay attention to combining urban morphology with ecological function, and apply environmental protection and energy conservation technologies for buildings[31]. Low-carbon technologies are applied in such fields as buildings, transportation, energy and resources recycling, etc. to build low-carbon community. In Lingang New City, construction focuses on building local areas including low-carbon community and low-carbon industrial park to boost application of low-carbon technologies, and solar power generation is conducted as unique feature and low-carbon industrial park is established to vigorously develop low-carbon industries including high-end manufacturing and port service etc and propel integrated application of low-carbon technologies. As the first low-carbon business district in Shanghai[31], Hongqiao Business District is developed with core zone fully characterized by green buildings above national standard one-star level among which more than 50% of green buildings are ones at two-star level and six green buildings are at three-star level.

Meanwhile, core philosophy of low-carbon and sustainable development was actively advocated by virtue of 2010 Shanghai World Expo so as to greatly boost construction of Shanghai as low-carbon city. 2010 Shanghai World Expo was the first world expo for which "low-carbon expo" philosophy was officially put forward; low-carbon philosophy was implemented at various levels and various stages from park planning to airport construction, from demonstration application to technology display, so as to successfully turn world expo park into demonstration zone for disseminating, popularizing and practicing low-carbon economy philosophy, which exerted important impact on driving construction of Shanghai as low-carbon city. For example, more than 90% of pavilions within 5.28 km^2 work expo park are energy-saving and environmentally friendly, recyclable green buildings. Utilization of large social activity for boosting low-carbon city construction provides beneficial model for low-carbon city construction practice.

2) Baoding's low-carbon city construction practice and mode

Baoding is a typical city with industry-led low-carbon city construction. Baoding is marked as "China Electricity Valley" with respect to industrial development and strives to become "City of Solar Energy" in terms of social development. Opinions Concerning Construction of Low-carbon City (Trial Implementation) were released in Baoding at the end of 2008, stating that construction is mainly based on "China Electricity Valley" and "City of Solar Energy" plan, and their construction proceeds from development of new energy and renewable energy industry, comprehensive application of new energy, energy conservation and emission reduction; 21 million kWh power can be saved and 17,000t carbon dioxide emission can be made through utilization of new energy throughout the city each year[32]. Baoding is based on new energy industry as the leading means for low-carbon city construction, while new energy has become the industry with fastest growth in Baoding. With several years' development, more than 170 backbone enterprises specializing in solar energy, wind energy, power transmission and transformation, energy storage equipment have been built in China Electricity Valley. In 2008, Baoding was awarded "National Hi-tech Industry Base for New Energy Industry" by National Development and Reform Commission.

3) Tianjin low-carbon city construction practice and mode

"Carbon finance" has been actively exploring for building unique low-carbon city in Tianjin. In Tianjin, "carbon finance" as an emerging financing activity is chosen as low-carbon city development goal to develop favorable platform for financing and urban environment. Sino-US Low-carbon Finance Research Center was established in Tianjin in September, 2009; Tianjin Municipal Government and the United Nations signed memorandum in October, 2009, under which support is provided to build Tianjin into low-carbon city and the United Nations will set up a low-carbon economy center in Tianjin[33].

4) Development practice and mode in Yanqing County, Beijing

(1) Functional positioning of Yanqing County in Beijing[34]

According to the *Main Functional Zone Plan for Beijing*, Beijing's functions are divided into four categories: capital function core zone, urban function expansion zone, new urban development zone and ecological conservation development zone. Yanqing County falls within ecological conservation development zone, in which main tasks lie in strengthening protection and construction of ecological environment, guiding relative aggregation of population, and guiding rational development and utilization of natural resources, developing eco-friendly industries and turning it into ecological shelter zone for the Capital and ideal leisure and recreation space for residents.

(2)Measures for building Yanqing County into low-carbon ecological city[35]

Extensive development of forestry carbon sink. Forestry carbon sink measure is taken to absorb and fix greenhouse gases including carbon dioxide, which is currently the most economical, most realistic and most effective means for addressing climate warming. Presently, in Yanqing County, woods greening rate reaches 72%, and forest land covers an area of 146,000 hectares and 987,000t carbon dioxide is annually absorbed, accounting for 13.9% of that in Beijing. Meanwhile, carbon sequestration is also conducted by protecting and restoring wetland in Yanqing County. Yanqing has become the real carbon sink base for Beijing and serves as one green ecological shelter zone in northwestern Beijing.

Low-carbon construction of agriculture. Shift from "high carbon" to "low carbon" entails adjustment of industrial structure. Construction of bases for nuisance free agricultural products, green foods and organic agricultural products was initiated in Yanqing in 2006. Currently, an urban ecological agricultural system uniquely characterized by organic agriculture, featured agriculture and recycling agriculture has been developed in Yanqing County.

Low-carbon development of the tourism industry. In Yanqing County, based on functional positioning of the Capital, the tourism industry is actively developed, and international tourism leisure business district is energetically fostered, and major projects including TangleWood Music Valley, Huihuang International Conference & Resort District, Badaling Great Scenic Spot and Sleepless Longqing Gorge are made; traditional scenic spots are vigorously upgraded and renovated; high-end leisure business functions and cultural creative industry are developed; characteristic folk tourism brand is built; overall quality of the tourism undertaking and development of comprehensive capability are promoted in an all-round way.

Development and utilization of new energy. Solar lamps, solar greenhouses and sunrooms are passionately popularized by making use of Yanqing's advantages characterized by high sea level elevation and abundant sunlight; meanwhile, shallow geothermal energy and deep terrestrial heat are developed for refrigeration in summer and heating in winter so that abundant geothermal resources in this region are effectively used. Currently, large biomass gasification stations, medium and large-scale biogas digesters are built to provide clean cooking energy for farmers in Yanqing, resulting in annually saving fuels by 50% and reducing carbon dioxide emission by 3,200t. Deqingyuan Chicken Farm Power Station generates electricity by using biogas from chicken manure, which not only annually supplies 14 million kWh green electricity to power grid, but also generates afterheat equivalent to 4,500t standard coal for heating and reduces greenhouse gas emission by more than 80,000t; it is listed as "the largest biogas power generation technology demonstration project in the world" by the United Nations. Currently, geothermal heating area in Yanqing reaches 0.59 million m^2, accounting for 14% of heating area in the whole county, saving coal by 12,000t. In addition, a lot of work has also been done in wind power generation, new energy vehicle application and construction of smart power grid demonstration zone, etc. in Yanqing County. New energy developed in Yanqing County was equivalent to only 10,000t standard coal, accounting for only 2.2% of total energy consumption; such figure reached 60,000t standard coal in 2009, accounting for 15% of total energy consumption in the whole county, being nearly seven times that in 2005.

5) Low-carbon construction practice and exploration in other cities

Rizhao City and Wenjiang District in Chengdu City have joined UNEP's carbon neutral network, with commitment to disclose greenhouse gas emission information, endeavor to cut greenhouse gas emission, and realize low-carbon development and carbon neutralization. Rizhao City was granted the first "World Clean Energy Award" in 2007, while its low-carbon city development strategy is development of "City of Solar Energy". Qingdao City, Beijing, Hangzhou City and Dezhou City have signed *Daegu Declaration* to join International Solar Cities Initiative and have undertaken to actively take actions to control CO_2 emission; Dezhou City also obtained the qualification for holding 2010 ISCI World Congress. Furthermore, Nanchang, Wuhan, Changsha, Shenyang, Zhuhai, Jilin, Xiamen, Hangzhou, Guiyang, Wuxi and Chongqing, etc. have also specified low-carbon city goals by developing and releasing low-carbon city development plans (see Table 12-3).

Table 12-3 Summary of low-carbon city construction in China

City	Goals	Action measures or development plans
Hangzhou	Low-carbon industry, low-carbon city	In the first implementation of the "low carbon new deal" in China: public bicycle rental scheme, new energy; Low-Carbon Technologies Museum
Nanchang	Economic priority area of low carbon	Low carbon industry positioning around solar, LED, service outsourcing, new energy vehicles, etc.; To build three major economic demonstration zone
Zhuhai	Low carbon economic demonstration area	Developmental Strategy of New Energy
Suzhou	Low Carbon demonstration Industry Park	Industrial upgrading with energy saving and environmental protection as the core
Chongqing	Low Carbon Industry Park	Application of geothermal energy; To build a low carbon Research Institute
Beijing	low-carbon city, low carbon business district	CBD expansion to east; Green energy utilization; Low carbon standard for construction; Development of circular tram; Creating an international financial and cultural media center
Jilin	Low carbon demonstration area	Case study pilot city of low carbon economic zone; Research on the industrial structure adjustment of heavy industrial city

(Contd.)

City	Goals	Action measures or development plans
Dezhou	Low-carbon industry	Development of wind power equipment; Biomass power generation; "China Sun valley"
Wuxi	low-carbon city	Low carbon urban development research center
Guiyang	ecocity	Ecological low carbon summer community
Xiamen	low-carbon city	LED lighting, solar building; Energy Museum
Rizhao	tourist city	Scale application of solar energy

Source: public data.

12.4 Summary of Low-carbon Development Modes in Domestic and Foreign Cities

Resource endowment, industrial base and development stages of various cities and development strategies of their countries or territories are different, thus diverse low-carbon city development modes are developed. Overall, low-carbon city construction is relatively mature and more effective low-carbon development modes have been identified in foreign countries, while in China, though beneficial attempts and preliminary exploration have been made in many cities, it is at the initial stage and explicit development modes have not yet been found. Given various kinds of development practice at home and abroad, current low-carbon city development modes can be summarized as the following four modes[36]:

12.4.1 Integrated "low-carbon society" mode

We consider modes in which low-carbon transformation has been successfully achieved in cities in UK, Japan and Demark, etc. as integrated "low-carbon society" mode. Such mode almost covers every aspect of urban economic development and various fields from energy supply to energy consumption, including new energy development and utilization, green building, environmental protection and transportation, low-carbon consumption, etc. This kind of cities is mostly at later stage of industrialization and has good foundation for economic transformation, such as London, Tokyo and Copenhagen.

12.4.2 Low-carbon industry pull mode

Low-carbon industry refers to the industry which realizes economic output with less greenhouse gas emission relative to energy-intensive industry, and mostly means knowledge-intensive and technology-intensive industry. Low-carbon transformation in part of international cities is pulled by low-carbon industry, namely urban development is fostered by focusing on development of a or a kind of low-carbon industry (industries) and gradually weakening development of other industries and finally forming low-carbon development mode with relatively single industrial structure. Typical examples are Birmingham and Boston, among which the former focuses on development of cultural industry or creative industry and the latter stresses development of low-carbon hi-tech industries; low-carbon development in both cities is achieved by building knowledge-based city.

12.4.3 "From particular point to extensive area" development mode

At the initial stage of low-carbon transformation in urban economic development, most of cities follow the following "from point to area" mode: demonstration zone is firstly built to explore advanced development philosophy and transformation experience, and then drive low-carbon development of the whole city. A typical city is Masdar Ecological Low-carbon City under construction in the United Arab Emirates, which is a "zero carbon emission" ecological park explored to be built. Such a mode is also one mode universally attempted in domestic cities; for example, the low-carbon industrial park is built in Chongqing; the Ministry of Science and Technology establishes "low-carbon science and technology demonstration zone", serving as the effective way for exploring low-carbon technology development and diffusion.

12.4.4 "Low-carbon supporting industry" development mode

We define industries which provide support for development of low-carbon industries as low-carbon supporting industries. Such a kind of industries may not be low-carbon themselves and may actually incur high energy consumption, such as wind generator manufacturing, manufacturing of polycrystalline silicon and photovoltaic equipment necessary for solar energy utilization, which are essential for development and utilization of the above renewable energy. From the perspective of national even global low-carbon economic development, though such a kind of industries are energy-consuming, they make important contributions to the development of low-carbon economy and are also the indispensable and important part in the development of low-carbon economy. Therefore, we also consider development of cities with "low-carbon supporting industries" as one transition mode for development of low-carbon city, especially at the initial stage of exploring low-carbon city development nationwide, such a kind of cities play a crucial role. Development of such kind of industries also lays a certain foundation for urban economic growth, economic structure adjustment and transformation of economic growth mode.

In addition, there are also other low-carbon development attempts in many cities; however, high attention is paid to an aspect of low-carbon development in most cases, such as vigorous development and utilization of renewable energy, improvement of transportation system, development of low-carbon buildings etc. This paper defines such kind of cities as the initial stage of low-carbon development at which relatively full-fledged low-carbon development system has not yet been formed. Though low-carbon city construction is in full swing in China, most of cities are only at the exploration stage and it is still necessary to explore a rational development mode suited to own characteristics.

References

[1] Department of Trade and Industry (DTI), UK Energy White Paper: Our Energy Future-creating A Low Carbon Economy. London: TSO, 2003:1-142.

[2] Sustainable Development Strategy Research Group of Chinese Academy of Sciences: *2009 China Sustainable Development Strategy Report——Exploring Low-carbon Path with Chinese Characteristics*, Science Press, 2009.

[3] United Nations. World Urbanization Prospects: The 2005 Revision. New York: United Nations, 2005: 1-196.

[4] Xia Chaoran: *Research on International Low-carbon City Planning Strategies and Governance Modes, Beijing Planning Review*, 2011(5):24-27.

[5] Dai Yixin: *Analysis of Necessity and Governance Mode for China's Low-carbon City Development, China Population Resources and Environment*, 2009(3):12-17.

[6] Cai Bofeng, Cao Dong: *China's Low-carbon City Development and Planning, Environmental Economy*, 2010(12):33-38.

[7] Mayor of London. The Mayor's Climate Change Action Plan. http://www.london.gov .uk/mayor/environment/climate-change/ccap/index.jsp. 2007-02-26.

[8] Green LA Action Plan. http://www.ladwp.com/ladwp/cms/ladwp010314.

[9] City of Madrid Plan for the Sustainable Use of Energy and Climate Change Prevention 2008. www.c40cities.org/docs/ccap-madrid-110909.pdf.

[10] Government of Stockholm. Stockholm Fossil Fuel Free City 2050 [EB/OL]. http://www.stockholm.se/KlimatMiljo/Klimat/Stockholms-Action-Programmeon-Climate-Change/Stockholm-Fossil-Fuel-Free-City-2050/.2009-08-13

[11] Li Chaosu, Ma Zhenbang, Zheng Qi et al: *Comparative Study of Domestic and Foreign Low-carbon City Construction Cases, Urban Development Studies*, 2011(1):31-35.

[12] "Climate Change In London" [EB/OL]. http://london.gov.uk/mayor/environment/climate-change/london.jsp. [2008-05-28].

[13] Wee-Bean Jong. Energy Saving Trust, and the Carbon Trust. Low-Carbon Cities Programmer. [EB/OL]. http://www.frdata.co.uk/CcsPres2007/RichardRugg.pdf [2008-5-28].

[14] Great London Authority. Action Today to Protect Tomorrow: The Mayor's Climate Change Action Plan. February2007. [EB/OL]. http://www/london/gov.UK/mayor/environment/climate-change/docs/ccap-fullreport.pdf. [2008-5-28].

[15] Yuan Yi: *Research on China's Low-carbon City Development Modes*, Master's Thesis, Agricultural University of Hebei, 2011.

[16] *International Comparative Study of Low-carbon City Development Modes.*

Journal of Jixi University, 2011(8).

[17] Zhou Peijiang: *Develop Low-carbon Economy, Promote Green Development , http://www.cjbd.com.cn/bdxc/2010-09/30/cms400106 article.shtml. [2010-09-30].*

[18] Li Ya: *Obama Calls for Two Parties to Pass American Recovery and Reinvestment Plan as Soon as Possible, http://business.sohu.com/20090105/n261570091.shtml, 2009-01-05.*

[19] Wu Min, Hu Yanni: *Content of American Recovery and Reinvestment Act and Its Impact Analysis [EB/OL],http: //map.infocio.cn:8899/stock. [2009-02-17].*

[20] Sun Yufei: *Research on Urban Low-carbon Development Strategies and Measures——Take New York for Example, Foreign Investment in China*, 2011(4):2-4.

[21] Song Yan, Peng Ke: *Discussions of the Route for Promoting Realization of Low-carbon City Through Urban Master Planning——Take New York City, USA for Example, Planner*, 2011(4):94-99.

[22] Sun Yufei: *Research on Urban Low-carbon Development Strategies and Measures——Take New York for Example, Foreign Investment in China*, 2011(4):2-4

[23] Wang Wei, Ke Wanzhi: *New Energy Policy of New York City and Its Inspirations for Beijing, Expanding Horizons*, 2009(5):87-89.

[24] Tokyo Metropolitan Government, Tokyo Climate Change Strategy: A Basic Policy for the 10 year Plan for a Carbon-minus Tokyo, June 2007.

[25] Zhang Wanlu, Zeng Yunmin: *Tokyo's Low-carbon City Development——Experience and Inspirations*, Economic Development Mode Transformation and Independent Innovation——the 12th Annual Conference of China Association for Science and Technology, 2010, 1:1-6.

[26] Tang Dingding: *Inspirations from Development of Low-carbon Economy in Japan, World Environment*, 2009(5):62-64.

[27] *Successful Practice and Beneficial Inspirations in Domestic and Foreign Low-carbon City Construction, Industrial & Science Tribune*, 2011(8):

[28] Qiu Baoxing: *Overall Thinking of China's Low-carbon Eco-city Development, Construction Science & Technology*, 2009(15):12-14.

[29] Chen Yongming: *Policy Suggestions Concerning Development of Low-carbon Economy in Shanghai, Party & Government Forum*, 2010(3):9-11.

[30] Chinese Society for Urban Studies, China Association of City Planning, etc: *China City Planning Development Report 2008-2009*, China Architecture & Building Press, 2009.

[31] Gu Chaolin, Tan Zongbo et al: *Research Progress in Climate Change, Carbon Emission and Low-carbon City Planning, Urban Planning Forum*, 2009(3):38-45.

[32] Baoding Municipal People's Government: *Opinions of Baoding Municipal Government Concerning Construction of Low-carbon City (Trial Implementation)*, 2009.

[33] *Successful Practice and Beneficial Inspirations in Domestic and Foreign Low-carbon City Construction.*

[34] China News Network http:// www.chinanews.com/NY/2010/08-03/2442865.

shtml.

[35] Science & Technology Committee of Yanqing County: *Low-carbon Construction of Yanqing as Ecological City, Science and Gulture*, 2011(3):18-21.

[36] Liu Wenling, Wang Can: *Low-carbon City Development Practice and Development Mode, China Population Resources and Environment*, 2010(4):17-22.

Chapter 13
Transformation Modes and Reference for Domestic and Foreign Industrial Low-carbon Development

Chu Chengshan Liang Benfan

Abstract: Industry is very important since it supplies powerful materials and equipment for human beings. At the same time, however, industry consumes too much energy and too many resources, and becomes the main factor causing environmental pollution. Industrial transformation to low-carbon development can reduce fossil energy consumption, dissolve the bottleneck of energy and resource supplies, strengthen urban comprehensive competitiveness, and promote the sustainable development of a society based on economy. This paper makes a thorough analysis of the models of industrial transformation towards low-carbon development at home and abroad, which will provide useful experience and a reference for industrial transformation in China.

Key Words: industrial transformation; low carbon development; models

13.1 Introduction

Industry is a very important industrial sector in any country; industry provides powerful material wealth and technical equipments for social development and makes important contributions to the progress of human civilization. Industry consumes enormous energy and resources and is also the important factor which causes environmental pollution.

Firstly, industry is the largest energy-consuming sector. Energy consumption in industrial field accounts for about 1/3 of total energy consumption in the whole society in developed countries. In China, with industrialization and urban advancement, infrastructure construction and rapid development of various industries, quick increase in people's living standard, various factors drive significant growth of energy consumption in China. Industrial energy consumption makes up overwhelmingly large proportion in total energy consumption in China. According to analysis of relevant data from 2005-2010 China Energy Statistical Yearbook, industrial terminal energy consumption accounted for more than 65% of total energy consumption in the whole society in China. Meanwhile, industrial power consumption accounted for more than 70% of total power consumption in the whole society.

Secondly, industry is the largest pollutant emission sector. Industry causes huge greenhouse gas emission and pollutant emission during fossil energy consumption and industrial production, such as carbon dioxide emission from lime and cement production; sulfur dioxide, soot and nitric oxide emissions from thermal power plant; toxic heavy metals from nonferrous metal production and processing, etc. According to data from *The First National Pollution Source Census Communique* (based on 2007) released by the Ministry of Environmental Protection in February, 2010, emissions of various pollutants from industry made up an important proportion in national total emission. Sulfur dioxide and nitric oxide from industrial sector accounted for 91.37% and 66.11% of total emission in the whole society respectively, while theses pollutants were exactly main causes for ecological environment deterioration.

Energy conservation and emission reduction in industrial field and industrial transformation are of great significance for reducing fossil energy consumption, and resolving energy and resource bottlenecks, enhancing urban competitiveness, addressing climate change and boosting sustainable development of economy and society.

Series problems caused by industrial overdevelopment make it necessary and inevitable for industrial low-carbon transformation. Currently, industrial transformation modes with respective unique features have been adopted according to the characteristics and advantages in some foreign cities, while these modes serve as great reference for industrial low-carbon transformation of Chinese cities.

13.2 Foreign Classic Modes

13.2.1 Chicago carbon emission trading operation mode

Founded in 2003, as one pioneering organization for voluntary participation in greenhouse gas emission reduction trading and legally binding emission reduction which is the first one in the world and the only one in North America, Chicago Climate Exchange (CCX) launched electronic trading for greenhouse gas emission permit and offset program as from December 12, 2003 and has become the first remission registration, emission reduction and trading system for all greenhouse gases specified in *Kyoto Protocol*.

CCX has a set of trading rules designed, governed and voluntarily observed by members; any CCX members must voluntarily and legally make joint commitment to reduce greenhouse gas emission. These members may honor commitment through internal emission reduction, permit purchase or purchase of credits generated by emission reduction projects which satisfy specified criteria. CCX establishes emission reduction indicator trading platform and adopt market mechanism to ensure that enterprises realize energy conservation and emission reduction indicators so as to achieve multi-win outcome between enterprises, between enterprise and country.

Such market-oriented emission reduction mode can enable optimal allocation of resources so as to ensure that emission reduction and low-carbon development are completed with the lowest cost in the most convenient way and promote industrial energy efficiency improvement and development of new energy industry.

13.2.2 Manchester's service industry development mode

Britain is the birthplace of the world's industrial revolution and kept ahead of other nations in the world in industrialization and urbanization level over a certain period of time. Meanwhile, it was also the first to be subject to series industrial, social and environmental problems after industrialization. The urban industrial transformation path in Manchester, the important city in north Britain, serves as the important reference for developing countries to develop low-carbon economy and undergo industrial transformation.

Manchester had an important position in British industrial development history. In the 19th century, Manchester became the important cotton wool textile and textile machinery manufacturing center in Britain. With industrialization development, Manchester gradually developed into an important industrial city with multiple industries among which textile, machine tool, general machinery, motor, food processing and chemical engineering were particularly prominent. In the early 1960s, the manufacturing industry accounted for about 70% of Manchester's economy; rapid and high industrial development brought not a few "byproducts" to Manchester, including severe environmental pollution and energy consumption.

As industrial and economic development incurred series conflicts, Manchester's industrial economy gradually transformed to service economy, as directly evidenced by evolution from the emphasis on the manufacturing industry to further rationalization of industrial structure. In the early 1990s, proportion of the manufacturing industry in Manchester's economy drastically declined to about 20%, while the proportion of employed population in the service industry in total employed population soared to 84% and proportion of employed population in the manufacturing industry sharply decreased to 11%. In recent years, Manchester's economic transformation has still accelerated. In particular, employed populations in such industries as finance, education and tourism have skyrocketed. For example, the number of practitioners in the financial industry rapidly increased to 68,200 in 2007, up nearly a half compared with 1997. Notably, Manchester is rapidly developing in creative industry and cultural industry. Now Manchester has become the collecting and distributing center for creative industry in central and northwestern Britain and is home to many high education, cultural and media creation organizations and sports brands which future boost the development of local economy. According to the plan developed by Manchester government departments, more efforts will be made to improve transportation, save energy, develop green cities, renovate the environment and address climate change in Manchester so that this city more adapts to local economic development trends.

Rise and development of the service industry in Manchester lays a foundation for smooth transformation of the manufacturing industry and sustainable urban development while providing more and more employment opportunities and gradually turning Manchester into commercial, financial, insurance and transport center in northwest Britain. Industrial transformation of Manchester stimulates tremendous development of urban economy and is also beneficial for forming new knowledge-based economic system in Manchester. These experience and practices can serve as great reference for promoting industrial transformation and upgrading of heavy industry cities in China.

13.2.3 Birmingham and Boston's low-carbon industry pull mode

Low-carbon industry refers to the industry which realizes economic output with less greenhouse gas emission relative to energy-intensive industry, and mostly means knowledge-intensive and technology-intensive industry. Low-carbon transformation in part of international cities is pulled by low-carbon industry, namely urban development is fostered by focusing on development of a or a kind of low-carbon industry (industries) and gradually weakening development of other industries and finally forming low-carbon development mode with relatively single industrial structure. Typical examples are Birmingham and Boston, among which the former focuses on development of cultural industry or creative industry and the latter stresses the development of low-carbon hi-tech industries; low-carbon development in both cities is achieved by building knowledge-based city.

13.3 Major Modes in China

13.3.1 Wujin low-carbon ecological demonstration zone mode

With an overall goal for building "three-in-one" national low-carbon ecological demonstration zone which encompasses low-carbon manufacturing clustering zone, low-carbon modern service industry clustering zone, low-carbon demonstration application clustering zone, etc, Wujin Hi-tech Industrial Zone focuses on development of low-carbon industry with support from low-carbon research and development organizations, research and development personnel, in which construction of low-carbon communities is taken as the carrier, and low-carbon environment is fostered to seek guarantee, and new energy, new technologies and new processes are used to vigorously develop production and lifestyle with low-carbon economy philosophy, which are designed to rapidly develop Wujin Hi-tech Industrial Zone into low-carbon industrial park which is a leading one in the whole province, becomes first-class nationwide and internationally famous. For this purpose, relevant work is carried out from the following aspects:

Accelerate the development of the low-carbon manufacturing industry. Emphasis is placed on developing semiconductor lighting industry, wind power equipment industry, photovoltaic industry, new energy vehicle and power battery industry, energy-saving environmentally friendly building material and equipment industry and smart power grid equipment industry.

Speed up construction of enterprise energy conservation project. More support is provided for enterprises to perform clean production; all key enterprises complete clean production audit within three years; accelerated actions are taken to phase out backward production equipment and backward production processes, and enthusiasm and initiative of enterprises are aroused to push forward technical renovation for energy conservation.

Quicken the cultivation of low-carbon service industries. Priority is given to cultivating and developing low-carbon service industries including low-carbon research and development, incubation, testing and finance, etc, and building seven large service industry clustering zones including low-carbon industry research and development incubation zone, low-carbon international overseas student incubation park, LED modern service clustering zone, information industry park, industrial design park, low-carbon information center and headquarters economy park.

Accelerate application of renewable energy. Energy wasting is eradicated at source and energy utilization efficiency is improved; solar energy is popularized and solar photo-thermal utilization equipments are installed in new buildings not more than 12 storeys; solar photovoltaic technologies are popularized for construction and renovation of municipal lighting including street lamps, landscape lighting and bus station platforms etc; use of geothermal energy is explored, and shallow strata geothermal resources are developed and utilized.

In addition, efforts are also made to quicken construction of low-carbon transportation and low-carbon buildings, resource-saving, environmentally friendly, business-friendly, livable environment; low-carbon lifestyle is advocated and low-carbon living habit is developed.

With construction of low-carbon demonstration zone, Wujin County will gradually foster low-carbon transformation development of industries.

13.3.2 Baotou wind power new energy industry mode

In Baotou City, action is taken to energetically develop green, environmentally friendly energy, actively introduce domestic and foreign wind power generation equipment manufacturers to develop wind power generation equipment manufacturing industry and make mechanical machinery industry in Baotou City larger and stronger. For this purpose, 12 domestic wind power equipment manufacturers including Goldwind Science & Technology, Sinovel Wind and China Water Investment Group Corporation are introduced to build a complete wind power equipment manufacturing system covering wind power host, tower, blades and gear case, etc. Wind power equipment is expected to reach four million KW in Baotou City during the 12th Five-year Plan period.

Meanwhile, wind resource advantages of Baotou City are utilized to vigorously develop wind power generation. With the goal for building Baotou City into two million KW wind power city, on the principle of "developing large based, incorporating into large power grid", emphasis is placed on large-scale, cluster-oriented development and construction of wind power and expansion of wind power industry. Currently, wind farms planned to be built in Baotou City are ten wind farms including Bayin Wind Farm, Baiyun Ebo Wind Farm, Bailingmiao Wind Farm and Zhaohe Wind Farm, etc, with the planning area of 3,446km^2 and total installed capacity of 12.1 million KW. Measures will be taken to continue accelerating construction of Damao, Guyang and Baiyun large wind farms, popularizing and applying large wind power equipment, expanding installed and grid-connected scale, and improving power generation efficiency and building Autonomous Region's important wind power base in Baotou City during the 12th Five-year Plan period.

New energy industry represented by wind power is developed to boost low-carbon development and transformation of industries in Baotou City.

13.3.3 New energy equipment manufacturing industry modes in Baoding and Dezhou

1) Baoding

Baoding is situated in the triangular hinterland of Beijing, Tianjin and Shijiazhuang, while its downtown is 140km, 145km and 125km away from Beijing on the north, Tianjin on the east and Shijiazhuang, the capital of Hebei Province on the southwest respectively. Baoding City covers an area of 21,100km^2 and has a population of nearly 11 million, it is the most populous city in Hebei Province. Baoding City is a "National Science and Technology Demonstration City for Comprehensive Utilization of Solar Energy" and the first "National Renewable Energy Industrialization Base" as well as is included in the first batch of pilot cities for China low-carbon city development project as determined by World Wildlife Fund.

Baoding has strong foundation for new energy industry and energy equipment manufacturing industry. In Baoding City, "China Electricity Valley" concept was put forward by making reference to "Valley" development mode in California, USA, and with central task focusing on new energy and energy equipment manufacturing industry, all-round efforts are made to build "China Electricity Valley" and further improve six large industry systems including solar photovoltaic power generation, wind power generation, efficient power saving, new type energy storage, power transmission, transformation and automation, and cultivate and expand low-carbon industry clusters at a certain scale. With construction of key project "China Electricity Valley", persistent work will be done to build industry parks including solar photovoltaic power generation, wind power, efficient power saving, new type energy storage, power electronic device, power transmission, transformation and automation, etc, develop internationally influential internationalized new energy and energy equipment manufacturing base with annual sales revenue more than 100 billion yuan within about ten years.

With the development of new energy industry, one low-emission, low-pollution, low-consumption, ecological economic growth solution is established through exploration and circular, saving, sustainable low-carbon city development path is taken in Baoding.

2) Dezhou

In Dezhou, acceleration of new energy industry development is taken as the important acting point and breakthrough point for city-wide emerging industry transformation and upgrading in the future, especially the 12th Five-year Plan period.

Solar energy utilization field. With a focus on two large industries including solar photo-thermal industry and photovoltaic industry, backbone enterprises including Himin, Ecoosolar, Zhongli, Tenglong, Jingweite, Yuying are relied upon to extend industry chain, quicken popularization of mature suitable photothermal products, vigorously research and develop photovoltaic technology products, enable coordinated development of photothermal and photovoltaic industries. With respect to solar thermal utilization, large scale, integration and intelligentization represent development directions, and emphasis is placed on developing heat pipe type heat collector, flat-plate collector, built-in metal runner glass evacuated collector tube, vacuum tube solar water heater, etc. Output of solar water heaters will reach 8 million (sets) and heat collection area will be 16 million m^2 through this city by 2015. Efforts are stepped up to make breakthroughs in development of high-end technologies including solar high-temperature power generation, solar sea water desalting equipment, environmentally friendly solar drying equipment, etc, and actions are taken to accelerate the development and construction of solar medium and high temperature thermal power generation demonstration projects. With regard to solar photovoltaic industry, independent research and development of high efficiency monocrystalline (polycrystalline) silicon production, application and auxiliary product production technologies is encouraged, and priority is given to developing monocrystalline (polycrystalline) silicon slices, solar battery pieces, amorphous silicon membrane solar cell, solar cell modules for buildings, photovoltaic power generation control system, solar air conditioners, solar lamps and quartz crucibles, etc. A batch of ground solar power stations, photoelectric building integration and landscape lighting demonstration projects are built. Output of photovoltaic crystalline silicon cells and amorphous silicon membrane solar cells will exceed 500MW throughout this city by 2015.

Wind energy utilization field. Enterprises including Tongyu, Century Energy, Zhongnan are depended on to actively introduce advanced wind power equipment manufacturing technologies, enhance the capability for making auxiliary products including wind power main shaft, tower, blades, gear case; stress is laid on developing large capacity wind turbine generator system equal to and above 1.5MW and high-end products including auxiliary wind generator, mechanical drive, operation control and frequency converter, and continuously increasing wind power manufacturing level, and manufacture complete wind power generation unit in a localized way as soon as possible. Auxiliary production capacity for complete wind power generation unit equal to and above 1.5MW will reach 750 (sets) throughout this city by 2015.

New energy vehicle field. Enterprises including Qilu Bus, Fulu Motor, Fuxing Electric Car are encouraged and supported to research, develop and innovate key technologies for pure electric and hybrid electric vehicles, and quicken renovation and transformation of conventional vehicle technologies, and intensify support for application and demonstration popularization of new energy vehicles, and foster a batch of self-owned brands for full vehicle and auxiliary industries, and gradually nurture technical advantages, brand advantages and market advantages. New energy vehicle production capacity will reach 20,000 throughout this city by 2015.

New energy industry is developed to promote energy conservation and emission reduction, thus boosting economic development mode transformation and industrial structural optimization and upgrading in Dezhou City.

13.3.4 Wuhan's mode in which smart industries take the lead

As Wuhan is included into the first batch of pilot cities for smart city construction under overall planning of China 863 Program, it will take ten years to conduct comprehensive integration of complicated large system, technology and business integration, comprehensive research and development involving multiple specialties and multiple disciplines in Wuhan so as to develop smart brain for social development of Wuhan. Smart services provided in the smart city include smart application systems covering the following five large fields: urban management and services, services for urban industries and economy, urban social people's livelihood services, infrastructure services, resource and environmental management and services. These are designed to accelerate enhancement of Wuhan's comprehensive competitiveness, and boost a new round of scientific and technical innovation so as to create brighter city life for Wuhan's residents.

Therefore, Wuhan centers on Future Sci-tech City to develop smart industries for leading economic development and transformation. Covering an area of 66.8km^2, Wuhan's Future Sci-tech City is situated in east Wuhan City, within East Lake National Self-dependent Innovation Demonstration Area, and among East Lake National Scenic Tourist Area, Jiufeng National Forest Park and scenic Liangzihou Lake. With respect to scientific research and industrial direction, Wuhan's Future Sci-tech City is positioned as the one which covers the following five leading industrial fields: optoelectronics information, biological medicine, energy and environmental protection, high-end equipment manufacturing and modern service industry, with the emphasis on developing optoelectronics information, energy and environmental protection, modern equipment manufacturing industry, and with efforts to concentrate on boosting agglomeration development of such industries as new materials, Internet of Things, new energy and cultural creative industry, etc. Future Sci-tech City will be turned into world-famous new sci-tech city, global top-level talents gathering place, China's strategic emerging industry birthplace, and innovation engine for driving rise of Central China. According to the plan, by 2020, in Wuhan's Future Sci-tech City, enterprise revenue will reach 300 billion yuan; there will be 500 enterprises above designated size, 2,000 entrepreneurial enterprises; 2,000 high-level entrepreneurial teams for scientific and technological innovation will be introduced and cultivated; the total number of science and technology working personnel will exceed 100,000; 200 domestic and foreign famous enterprises and 200 scientific research institutions will be rooted in this Park; 50 world-class research and development institutions are cultivated; 50 world-class original achievements will be researched and developed; 50 major application-oriented research achievements will be developed and converted; 1-2 emerging hi-tech industries having significant impact on economic and social development and people's lifestyle will be cultivated.

13.3.5 Chengdu's mode in which low-carbon city leads industrial transformation

Chengdu Municipal Government deliberated and adopted *Work Plan of Chengdu City for Building Low-carbon City* at the 56th executive meeting on December 25, 2009, which officially took effect on January 14, 2010. According to this Plan, efforts will be made to develop low-carbon economy, strengthen low-carbon management and advocate low-carbon life, and explore one low-carbon city construction and development path suited to Chengdu's actual conditions, and by the end of the 12th Five-year Plan period, non-fossil energy consumption will account for more than 30% of total energy consumption in the whole city; forest coverage rate will be increased and exceed 38%; carbon dioxide emission per 10,000-yuan GDP during the 12th Five-year Plan period will be decreased and lower than 1.15t (such figure was 1.4t in 2010) so as to ensure that it will be at the advanced level in central and western regions. For this purpose, this Plan specifies the following major tasks:

(1) Promote structural adjustment, develop low-carbon economy. Vigorously develop low-carbon emission industries, build new energy industry base, boost utilization of renewable energy, and work on straw power generation project and strengthen technical innovations.

(2) Reinforce energy conservation and emission reduction, reduce carbon consumption. Intensify energy conservation and emission reduction at industrial enterprises, push forward one package service concerning energy audit, and boost building energy conservation, and strengthen energy conservation and emission reduction in transportation field, and intensify energy conservation at public institutions, further popularize energy-saving lighting products, improve energy conservation monitoring system, actively carry out contract energy management.

(3) Foster low-carbon philosophy, build low-carbon society. Deepen publicity about

low-carbon philosophy, and hold China Low-carbon Economy Development (Chengdu) Forum, propel urban low-carbon construction, and carry forward forest planting activity and advocate low-carbon life.

(4) Proceed with experiment and demonstration, drive low-carbon development. Establish low-carbon demonstration district (city) county, build zero carbon agricultural industry demonstration park and zero carbon tourism industry demonstration park, implement green lighting project, apply for renewable energy building application demonstration city and take "free bicycle" action

(5) Innovate system and mechanism, promote construction of low-carbon city. Improve energy consumption statistics monitoring system of Chengdu City, accelerate implementation of evaluation, reward and punishment measures of Chengdu City for energy consumption per unit of GDP and drainage basin pollutant emission, and put in practice compensation measures for public welfare forest, and establish Southwest Environmental Exchange.

13.4 Conclusions

Industrial low-carbon transformation is related to resource endowment of a city, its industrial positioning, industrial base, economic growth stage and decision-makers' philosophy, etc, thus different industrial transformation modes are fostered. Various Chinese cities have their own characteristics, however, industrial transformation modes available in domestic and foreign cities especially transformation modes in developed countries can serve as an important reference for industrial transformation and sustainable development of Chinese cities.

With the analysis of domestic and foreign low-carbon industrial development modes, industrial low-carbon transformation development of these cities or regions can be summarized as follows:

(1) Carry out structural transformation, namely increase the proportion of the tertiary industry in primary, secondary and tertiary industrial structure of national economy, and decrease the proportion of the secondary industry accordingly. Meanwhile, optimize and adjust different subsectors within industrial sectors, lower the proportion of heavy industries with high energy consumption and high emission, and increase the proportion of industries with low energy consumption and high added value.

(2) Phase out backward production capacity and process equipment, develop large units and suppress small ones, and upgrade and reconstruct existing industries.

(3) Speed up the development of emerging industries, give prominence to developing strategic emerging industries dominated by high-end equipment manufacturing, new energy, energy conservation and environmental protection, electric vehicle, new medicine, new materials, biological breeding and information industry.

(4) Develop low-carbon industry park to drive industrial low-carbon transformation

and development.

(5) Expand use of renewable energy.

(6) Conduct energy and carbon emission trading, actively practice contract energy management, utilize market mechanism to carry through energy conservation and emission reduction. Simulate industrial transformation and upgrading.

Based on these industrial transformation modes, proceed from characteristics of specific cities to choose suitable industrial transformation path so as to bring about low-carbon development of economy and society.

References

[1] Fan Yujia: *Strategic Significance of Manchester's Experience for Urban Industrial Transformation of Shaoxing at the Middle Stage of Industrialization, Journal of Shaoxing University*, 2008(2):58-60.

[2] Chinese Society for Urban Studies: *China Low-carbon Ecological City Development Strategy*, China City Press, 2009.

[3] Chen Hongbo, Chu Chengshan: *Xinyu City as Comprehensive Demonstration City for Fiscal Policy Concerning Energy Conservation and Emission Reduction——Industrial Low-carbon Implementation Plan*, November, 2010.

[4] *Opinions of Dezhou Municipal People's Government Concerning Acceleration of New Energy Industry Development*, Dezhou Municipal People's Government, November, 2010.

[5] Liu Wenling, Wang *Can: Low-carbon City Development Needs "Chinese Characteristics", China Urban Economy*, 2011(8):18-19.

[6] *Build "Two Million KW Wind Power City" in Inner Mongolia, Rare Earth Information*, 2010, the 1st issue

[7] *New Thoughts for Chengdu's Low-carbon Economic Development and Environmental Protection Planning during the 12th Five-year Plan Period*, Chengdu Social Science Online: http: //www.cdss.gov.cn/yanjiu/ZZJJ/5608.htm.

[8] *The First Batch of Pilot Cities for Smart City Construction under Overall Planning of China 863 Program*, China Daily : http: //www.chinadaily.com.cn/hqsj/shbt/2011-10-26/content_4179588.html.

Appendix I The Top-ten Cities of Low-carbon Development for 110 Selected Chinese Cities in 2009

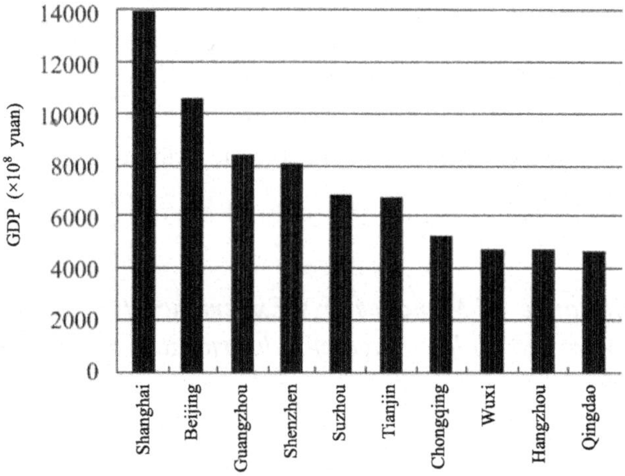

Fig. A1-1 Top ten cities in terms of GDP (constant price in 2005)

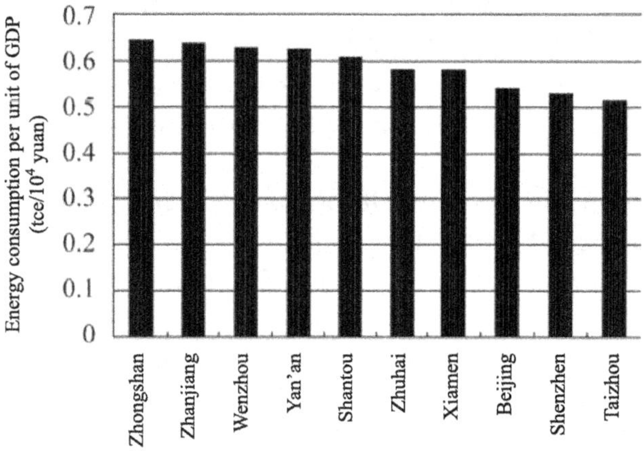

Fig. A1-2 Top ten cities in terms of energy consumption per unit of GDP (inverse)

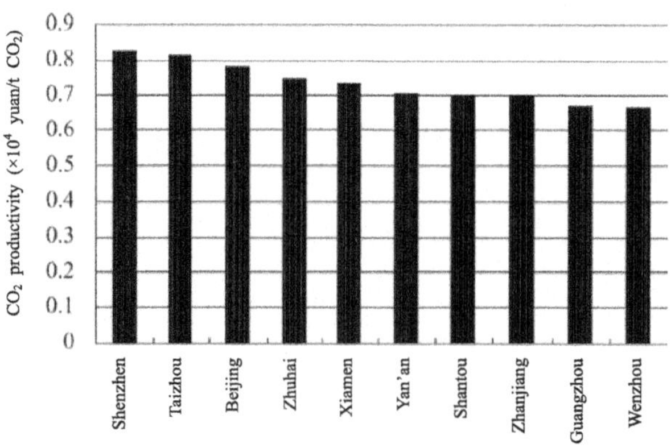

Fig. A1-3 Top ten cities in terms of CO_2 productivity

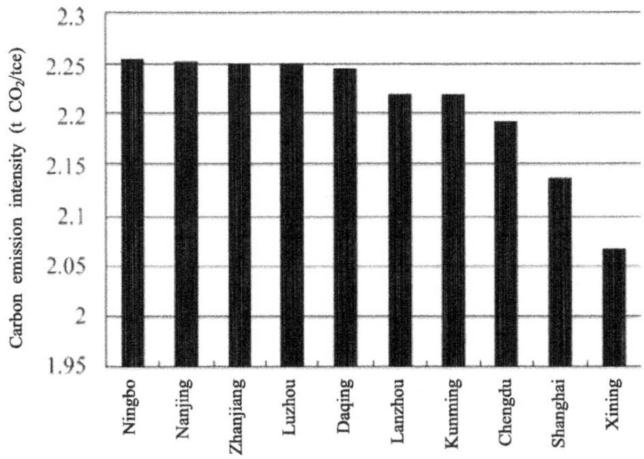

Fig. A1-4 Top ten cities in terms of carbon emission intensity (inverse)

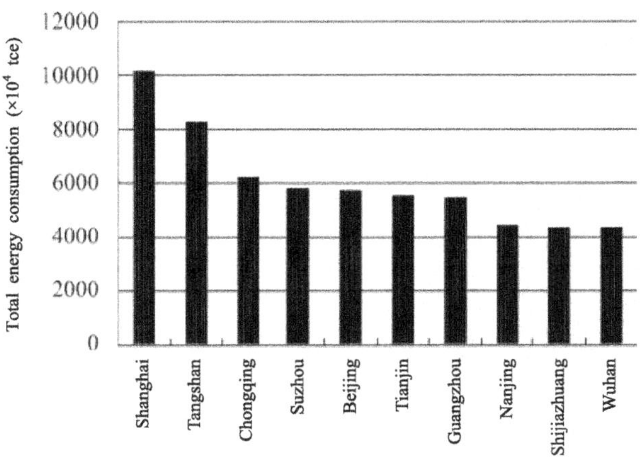

Fig. A1-5 Top ten cities in terms of total energy consumption (inverse)

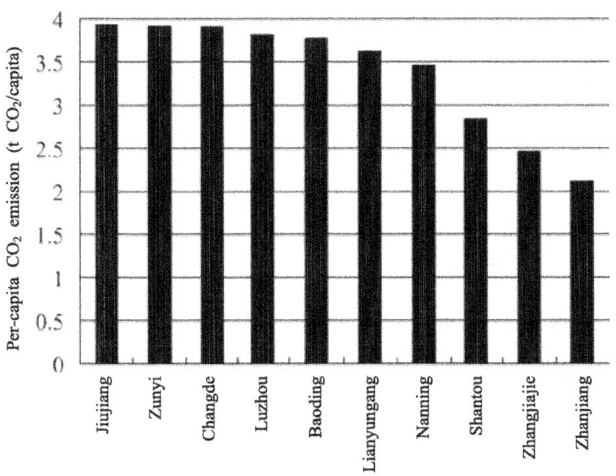

Fig. A1-6 Top ten cities in terms of per-capita CO_2 emission (inverse)

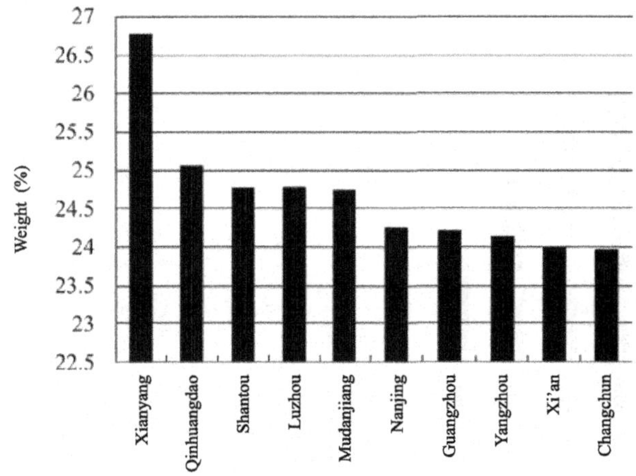

Fig. A1-7 Top ten cities in terms of residents' low-carbon consumption

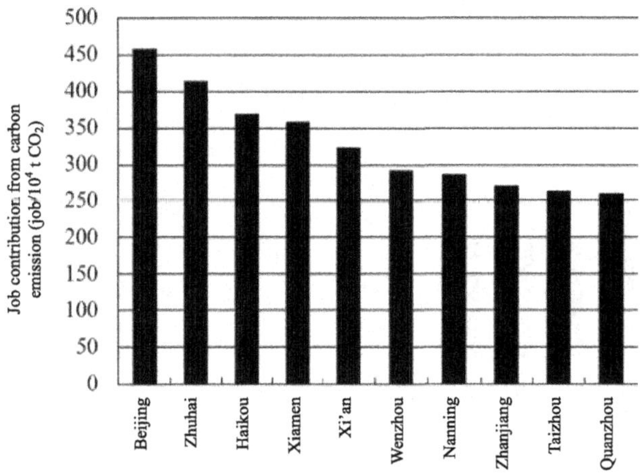

Fig. A1-8 Top ten cities in terms of job contribution from carbon emission

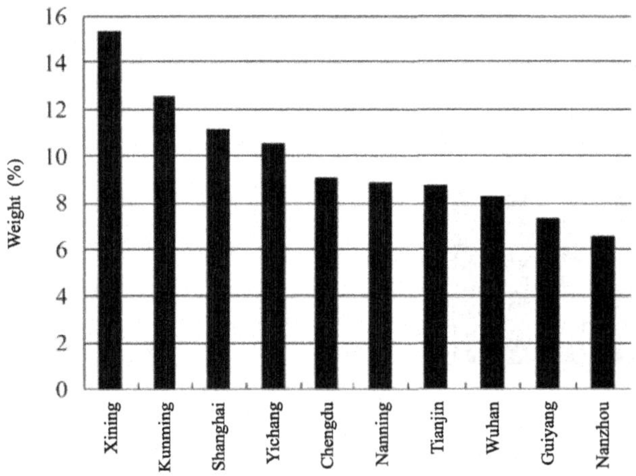

Fig. A1-9 Top ten cities in terms of non-fossil energy

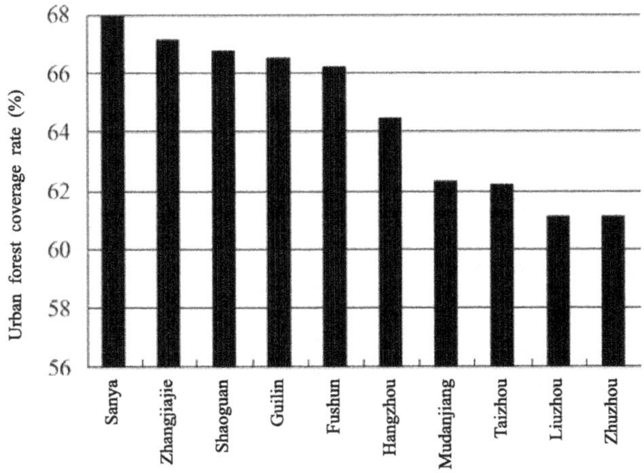

Fig. A1-10 Top ten cities in terms of urban forest coverage rate

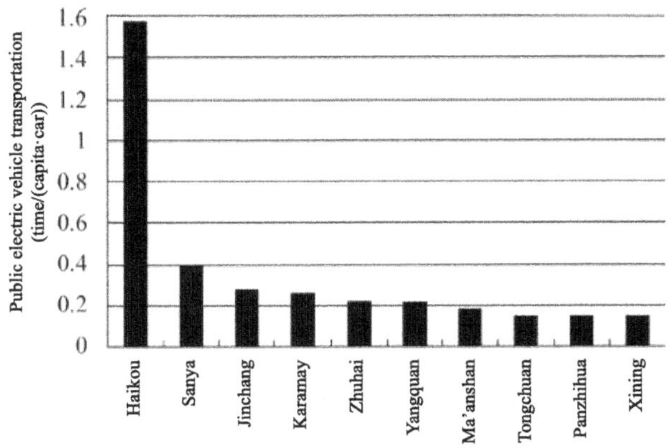

Fig. A1-11 Top ten cities in terms of public electric vehicle transportation

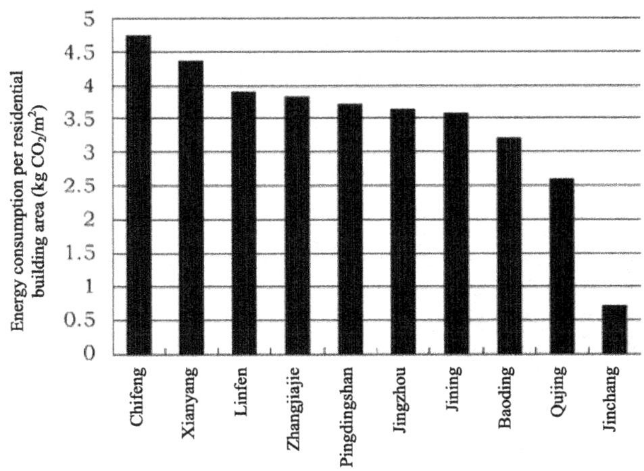

Fig. A1-12 Top ten cities in terms of energy consumption per residential building area (inverse)

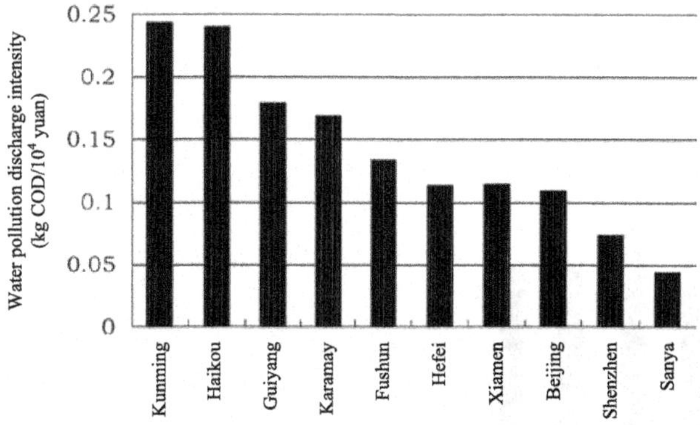

Fig. A1-13 Top ten cities in terms of water pollution discharge intensity (inverse)

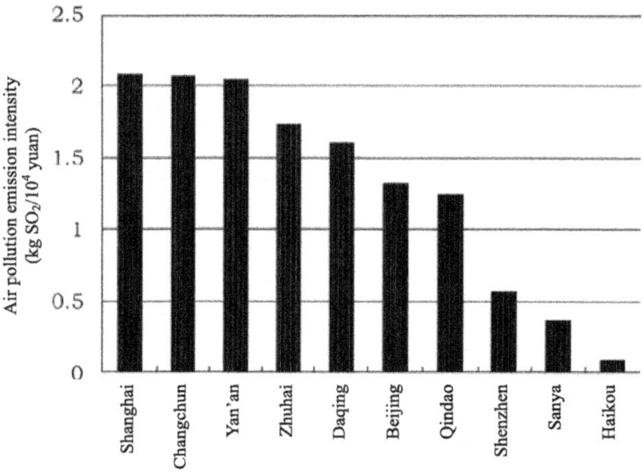

Fig. A1-14 Top ten cities in terms of air pollution emission intensity (inverse)

Appendix II The Bottom-ten Cities of Low-carbon Development for 110 Selected Chinese Cities in 2009

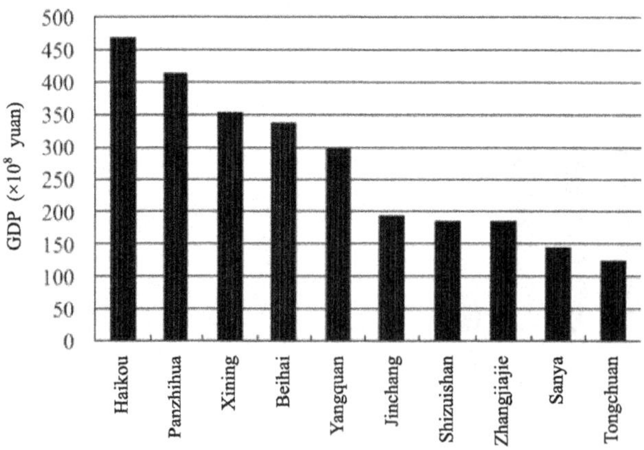

Fig. A2-1 Urban GDP (constant price in 2005)

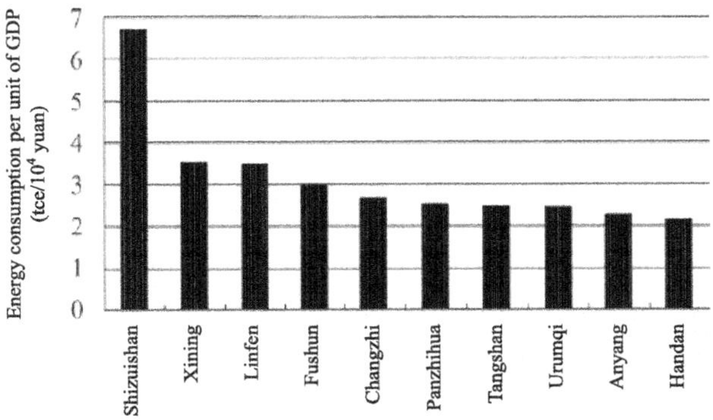

Fig. A2-2 Urban energy consumption per unit of GDP (inverse)

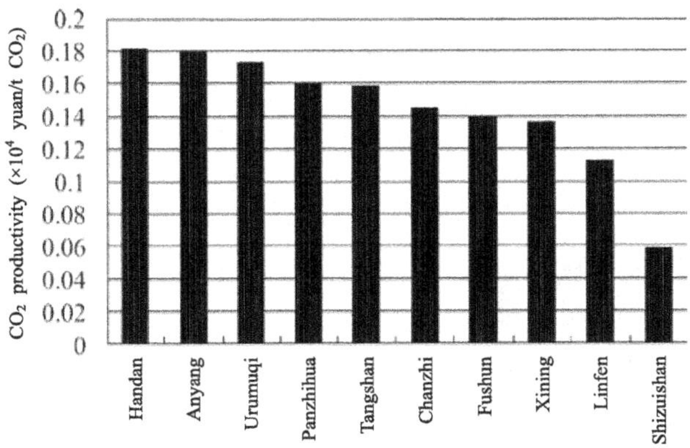

Fig. A2-3 Urban CO_2 productivity

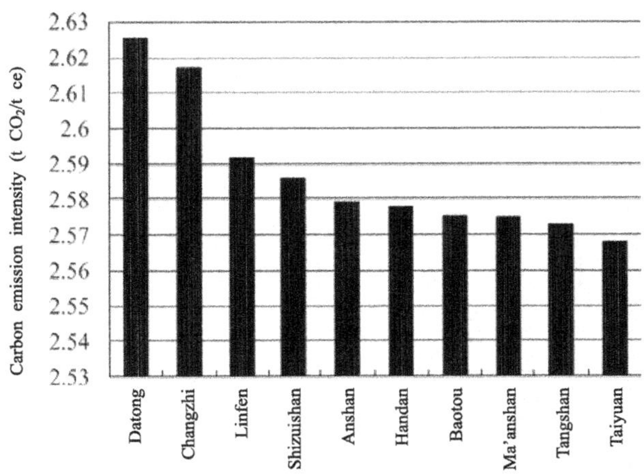

Fig. A2-4 Carbon emission intensity in urban energy consumption (inverse indicator)

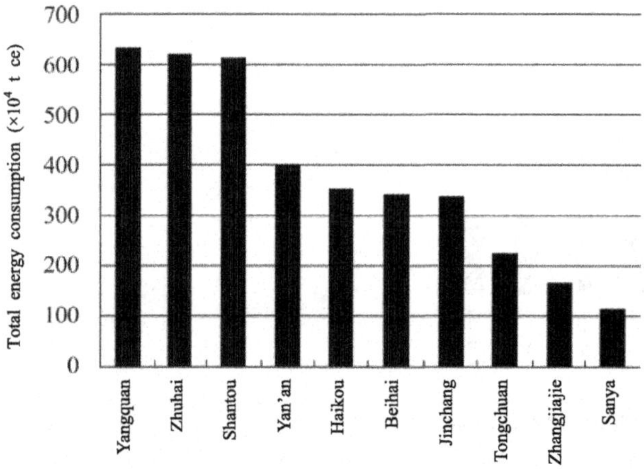

Fig. A2-5 Urban total energy consumption (inverse indicator)

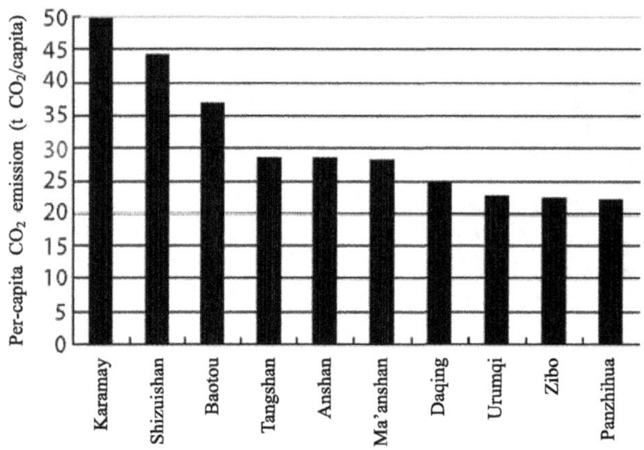

Fig. A2-6 Urban per-capita CO_2 emission (inverse indicator)

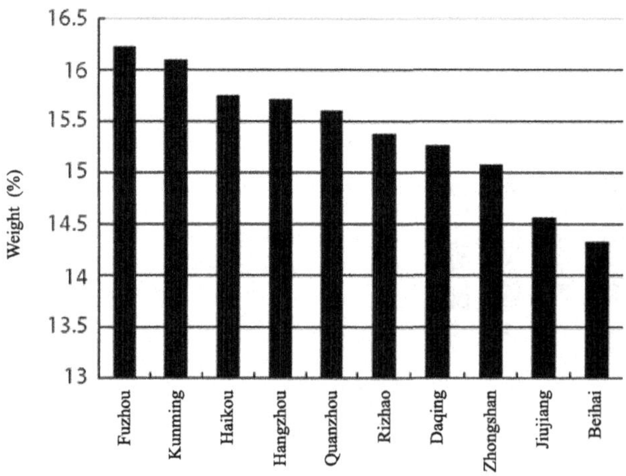

Fig. A2-7 Urban residents' low-carbon consumption coefficient

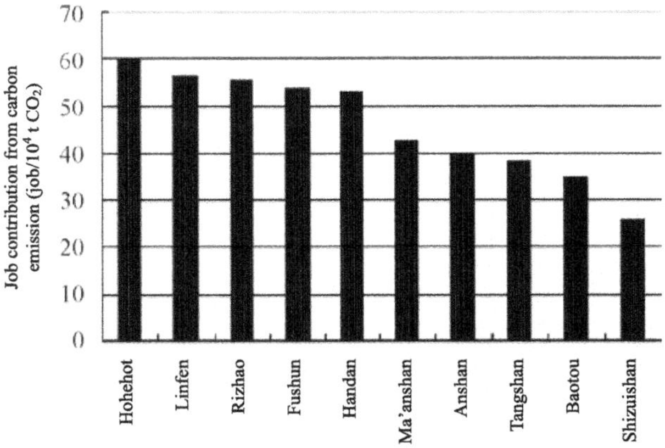

Fig. A2-8 Job contribution from urban unit carbon emission

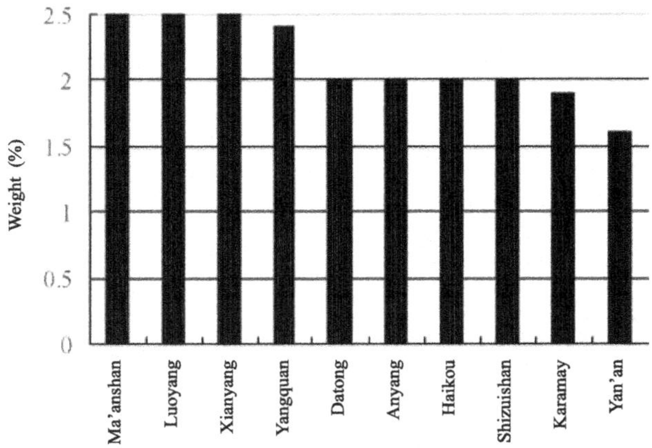

Fig. A2-9 Proportion of urban non-fossil energy consumption

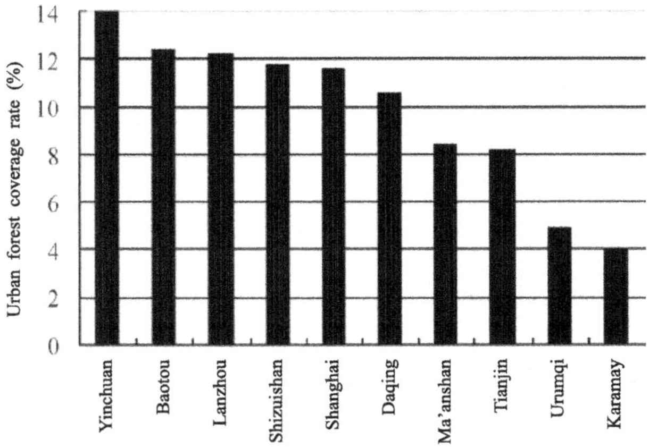

Fig. A2-10 Urban forest coverage rate

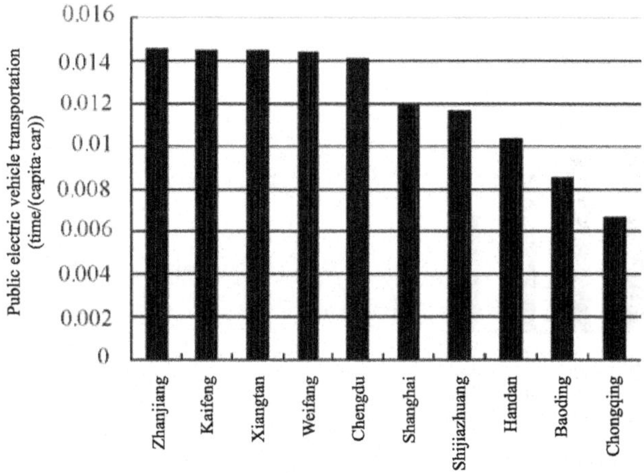

Fig. A2-11 Per-capital riding number per urban public electric vehicle

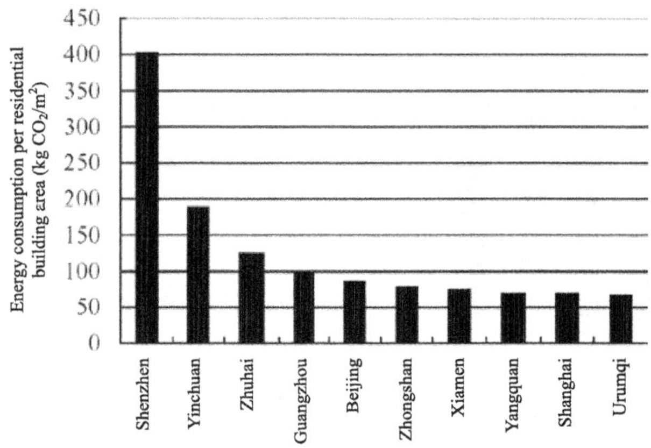

Fig. A2-12 Urban energy consumption level per living area (inverse indicator)

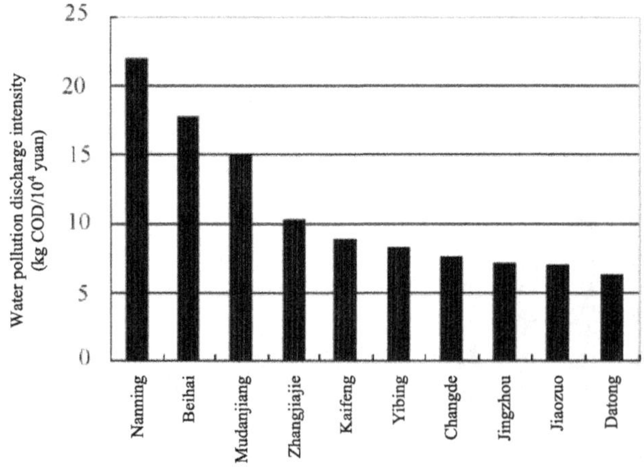

Fig. A2-13 Urban water pollutant discharge intensity (inverse indicator)

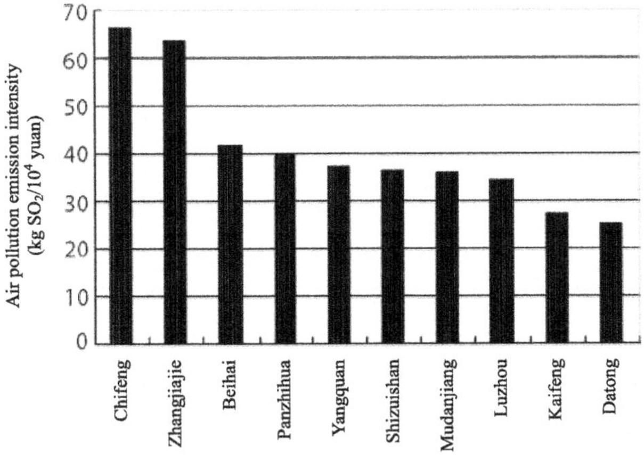

Fig. A2-14 Urban air pollutant emission intensity (inverse indicator)

Appendix III The Data of Low-carbon Development Indicators for 110 Selected Chinese Cities in 2008

Table A3-1 Summary of low-carbon development data concerning 110 Chinese cities in 2009

City	Economic indicator		Social indicator		Resource indicator		Facility indicator		Environmental indicator	
	Low carbon production index	Low carbon consumption index	Low carbon population index	Low carbon employment index	Clean energy index	Forest carbon sink index	Low carbon traffic index	Low carbon building index	Water environment index	Atmospheric environment index
	GDP output per unit of carbon dioxide emissions (10000 yuan/t) (Constant price)	Low carbon consumption expenditure of residents (%)	Per capita carbon emissions (t/capita)	Jobs per unit of carbon emissions (Jobs/t)	Proportion of non fossil energy in primary energy (%)	Forest coverage (%)	Ride frequency of public trolley-bus (time/capita/bus)	Energy consumption per unit residential building area (t/m^2)	COD emissions per unit of gross industrial output value (t/10000 yuan)	SO_2 emissions per unit of gross industrial output value (t/10000 yuan)
Beijing	0.78	22.60	7.72	0.045699	3.00	23.00	0.018298	85.58	0.000180	0.001326
Tianjin	0.52	20.36	10.39	0.015810	8.71	8.14	0.015217	40.62	0.000351	0.002590
Shijiazhuang	0.26	21.98	11.06	0.007639	5.00	25.70	0.011568	7.21	0.003151	0.009103
Tangshan	0.16	21.95	28.56	0.003837	3.50	28.70	0.021400	11.03	0.001572	0.007479
Qinhuangdao	0.33	25.05	7.94	0.012114	3.72	41.98	0.047775	14.08	0.001366	0.007339
Handan	0.18	21.42	11.73	0.005267	3.50	18.00	0.010365	5.91	0.000595	0.008722
Baoding	0.39	20.78	3.77	0.016253	4.50	19.27	0.008466	3.20	0.003009	0.008881
Taiyuan	0.21	20.36	17.25	0.013295	3.00	15.00	0.050179	51.41	0.000344	0.006327
Datong	0.22	21.45	7.09	0.018578	2.00	19.50	0.048985	16.43	0.006288	0.025201
Yangquan	0.19	20.68	11.93	0.014515	2.40	23.10	0.202661	69.64	0.000242	0.037276
Changzhi	0.14	16.25	12.95	0.008417	2.50	26.90	0.045591	5.72	0.000844	0.008272
Linfen	0.11	17.73	14.99	0.005628	3.00	31.40	0.026017	3.86	0.001162	0.009572
Huhhot	0.26	22.76	19.40	0.006013	4.50	29.90	0.093505	21.00	0.000484	0.014472
Baotou	0.18	20.46	36.76	0.003481	3.30	12.40	0.076944	22.63	0.000415	0.017481

(Contd.)

City	Economic indicator		Social indicator		Resource indicator		Facility indicator		Environmental indicator	
	Low carbon production index	Low carbon consumption index	Low carbon population index	Low carbon employment index	Clean energy index	Forest carbon sink index	Low carbon traffic index	Low carbon building index	Water environment index	Atmospheric environment index
	GDP output per unit of carbon dioxide emissions (10000 yuan/t) (Constant price)	Low carbon consumption expenditure of residents (%)	Per capita carbon emissions (t/capita)	Jobs per unit of carbon emissions (Jobs/t)	Proportion of non fossil energy in primary energy (%)	Forest coverage (%)	Ride frequency of public trolley-bus (time/capita/bus)	Energy consumption per unit residential building area (t/m^2)	COD emissions per unit of gross industrial output value (t/10000 yuan)	SO_2 emissions per unit of gross industrial output value (t/10000 yuan)
Chifeng	0.24	18.98	6.53	0.010646	3.70	32.56	0.037187	4.72	0.000193	0.066551
Shenyang	0.40	20.54	13.20	0.011129	4.20	24.00	0.031997	29.10	0.000450	0.004193
Dalian	0.46	19.65	14.59	0.011016	4.00	41.50	0.034986	25.99	0.000471	0.003484
An'shan	0.23	18.52	28.53	0.003958	3.30	46.40	0.044151	12.93	0.000638	0.010237
Fushun	0.24	21.68	22.11	0.005382	4.70	66.20	0.122955	39.17	0.000133	0.002582
Changchun	0.62	23.94	5.89	0.020000	5.00	15.20	0.019757	18.75	0.000980	0.002066
Jilin	0.25	18.99	10.80	0.007054	4.50	55.00	0.052292	36.92	0.003370	0.005614
Harbin	0.33	23.69	9.25	0.015698	4.70	44.80	0.021675	19.86	0.001191	0.004456
Qiqihar	0.27	20.78	4.49	0.016227	4.20	17.37	0.000159	10.60	0.002988	0.012826
Daqing	0.34	15.26	24.48	0.007718	3.00	10.60	0.030879	14.36	0.000264	0.001606
Mudanjiang	0.32	24.75	5.77	0.015570	4.00	62.30	0.072576	6.10	0.014870	0.035852
Shanghai	0.64	19.73	11.24	0.017845	11.10	11.60	0.011955	68.72	0.000253	0.002085
Nanjing	0.40	24.25	12.94	0.011786	4.00	25.00	0.027347	50.49	0.000630	0.003033
Wuxi	0.53	21.27	14.12	0.007936	3.00	26.20	0.024460	25.69	0.000710	0.002492
Xuzhou	0.34	19.46	7.00	0.010096	4.00	28.70	0.015503	6.81	0.001234	0.008170
Changzhou	0.45	20.46	10.81	0.007747	3.00	20.00	0.039676	27.28	0.001560	0.003643
Suzhou	0.48	19.91	15.10	0.008427	3.50	22.00	0.026654	23.09	0.000783	0.003221

(Contd.)

City	Economic indicator		Social indicator		Resource indicator		Facility indicator		Environmental indicator	
	Low carbon production index	Low carbon consumption index	Low carbon population index	Low carbon employment index	Clean energy index	Forest carbon sink index	Low carbon traffic index	Low carbon building index	Water environment index	Atmospheric environment index
	GDP output per unit of carbon dioxide emissions (10000 yuan/t) (Constant price)	Low carbon consumption expenditure of residents (%)	Per capita carbon emissions (t/capita)	Jobs per unit of carbon emissions (Jobs/t)	Proportion of non fossil energy in primary energy (%)	Forest coverage (%)	Ride frequency of public trolley-bus (time/capita/bus)	Energy consumption per unit residential building area (t/m^2)	COD emissions per unit of gross industrial output value (t/10000 yuan)	SO_2 emissions per unit of gross industrial output value (t/10000 yuan)
Nantong	0.58	22.61	6.16	0.014129	4.20	20.00	0.016465	7.44	0.001798	0.005421
Kianyungang	0.48	21.31	3.60	0.020794	4.30	19.70	0.043732	5.58	0.000810	0.006302
Yangzhou	0.58	24.13	6.09	0.014119	3.70	16.00	0.027939	9.38	0.001386	0.007016
Hangzhou	0.58	15.69	11.78	0.025378	4.20	64.44	0.026126	37.92	0.003387	0.003641
Ningbo	0.54	19.80	12.32	0.018830	3.50	50.20	0.025102	25.72	0.000467	0.003060
Wenzhou	0.66	18.56	4.76	0.029026	3.70	60.30	0.017559	15.37	0.002965	0.006963
Jiaxing	0.48	19.40	8.70	0.020262	2.50	17.00	0.022944	8.65	0.001167	0.006806
Huzhou	0.44	20.55	8.73	0.015013	2.70	51.30	0.036137	11.98	0.001212	0.008788
Shaoxing	0.47	21.38	5.50	0.019648	3.90	54.03	0.028684	13.84	0.002210	0.003362
Taizhou	0.81	21.32	4.22	0.026320	4.50	62.20	0.016763	9.98	0.000468	0.003819
Hefei	0.43	18.30	7.44	0.016525	3.50	16.10	0.046679	29.39	0.000113	0.002482
Wuhu	0.36	19.97	8.66	0.012217	2.50	21.23	0.073471	18.93	0.001721	0.005659
Ma'anshan	0.18	22.58	28.18	0.004242	2.50	8.44	0.180400	20.64	0.000771	0.007268
Fuzhou	0.61	16.22	5.78	0.024851	4.50	54.90	0.029230	29.87	0.000281	0.006380
Xiamen	0.73	18.74	8.94	0.035628	4.50	41.00	0.116268	75.29	0.000113	0.002247
Qunazhou	0.53	15.59	6.72	0.025925	4.50	56.70	0.021493	5.42	0.001798	0.003606
Nanchang	0.46	18.48	8.07	0.017665	4.00	16.10	0.040798	20.06	0.003193	0.003423

(Contd.)

City	Economic indicator		Social indicator		Resource indicator		Facility indicator		Environmental indicator	
	Low carbon production index	Low carbon consumption index	Low carbon population index	Low carbon employment index	Clean energy index	Forest carbon sink index	Low carbon traffic index	Low carbon building index	Water environment index	Atmospheric environment index
	GDP output per unit of carbon dioxide emissions (10000 yuan/t) (Constant price)	Low carbon consumption expenditure of residents (%)	Per capita carbon emissions (t/capita)	Jobs per unit of carbon emissions (Jobs/t)	Proportion of non fossil energy in primary energy (%)	Forest coverage (%)	Ride frequency of public trolley-bus (time/capita/bus)	Energy consumption per unit residential building area (t/m^2)	COD emissions per unit of gross industrial output value (t/10000 yuan)	SO_2 emissions per unit of gross industrial output value (t/10000 yuan)
Jiujiang	0.38	14.56	3.93	0.017278	4.50	50.00	0.041500	7.16	0.001275	0.015890
Ji'nan	0.40	20.18	11.96	0.015430	3.50	28.90	0.029148	24.89	0.000418	0.003931
Qingdao	0.54	16.30	10.03	0.014428	3.50	36.10	0.025039	42.14	0.000340	0.001241
Zibo	0.24	18.73	22.34	0.006043	3.20	33.40	0.016230	21.14	0.001242	0.008217
Zhaozhuang	0.23	17.51	12.69	0.007377	3.20	32.00	0.016301	7.91	0.003415	0.017852
Yantai	0.52	21.45	9.62	0.012680	3.20	39.00	0.028401	10.51	0.000459	0.003223
Weifang	0.36	19.48	7.90	0.010296	3.20	33.00	0.014378	5.66	0.000866	0.005020
Jining	0.30	21.38	9.02	0.008353	3.20	26.50	0.014642	3.56	0.000782	0.007465
Tai'an	0.33	19.28	8.22	0.011645	3.30	35.80	0.020000	4.90	0.000777	0.009699
Weihai	0.53	17.49	13.33	0.010426	3.20	39.20	0.058223	8.45	0.000428	0.006855
Sunshine	0.22	15.38	12.53	0.005558	3.00	35.20	0.047705	9.85	0.001675	0.006985
Zhengzhou	0.35	19.01	10.58	0.013069	2.80	24.00	0.024573	31.88	0.000999	0.011983
Kaifeng	0.36	17.64	3.96	0.016867	3.40	19.00	0.014424	7.56	0.008842	0.027100
Luoyang	0.31	21.55	9.56	0.008527	2.50	42.82	0.024532	6.27	0.000331	0.018088
Pingdingshan	0.21	20.90	9.21	0.010231	3.30	24.98	0.023200	3.69	0.001210	0.016702
Anyang	0.18	23.21	10.18	0.008105	2.00	18.49	0.019210	8.28	0.004654	0.011010
Jiaozuo	0.21	22.20	13.79	0.007134	3.20	22.00	0.033327	8.47	0.006970	0.012558

(Contd.)

City	Economic indicator		Social indicator		Resource indicator		Facility indicator		Environmental indicator	
	Low carbon production index	Low carbon consumption index	Low carbon population index	Low carbon employment index	Clean energy index	Forest carbon sink index	Low carbon traffic index	Low carbon building index	Water environment index	Atmospheric environment index
	GDP output per unit of carbon dioxide emissions (10000 yuan/t) (Constant price)	Low carbon consumption expenditure of residents (%)	Per capita carbon emissions (t/capita)	Jobs per unit of carbon emissions (Jobs/t)	Proportion of non fossil energy in primary energy (%)	Forest coverage (%)	Ride frequency of public trolley-bus (time/capita/bus)	Energy consumption per unit residential building area (t/m^2)	COD emissions per unit of gross industrial output value (t/10000 yuan)	SO_2 emissions per unit of gross industrial output value (t/10000 yuan)
Wuhan	0.40	18.06	10.77	0.017939	8.20	26.48	0.027448	41.12	0.000797	0.004189
Yichang	0.35	20.34	7.23	0.017473	10.50	55.30	0.040616	9.51	0.001436	0.004463
Jingzhou	0.26	19.25	4.00	0.016520	6.00	15.20	0.024932	3.63	0.007080	0.010972
Changsha	0.48	19.49	8.49	0.018159	4.58	53.28	0.054298	34.99	0.000394	0.004168
Zhuzhou	0.30	19.53	7.59	0.011466	3.60	61.10	0.046733	11.99	0.002518	0.012005
Xiangtan	0.23	20.17	9.16	0.009709	3.80	45.89	0.014404	5.24	0.003846	0.012822
Yueyang	0.31	19.46	6.27	0.013508	3.00	36.40	0.031466	6.95	0.005025	0.010324
Changde	0.43	16.50	3.90	0.014377	3.80	44.80	0.034699	4.85	0.007531	0.013814
Zhangjiajie	0.45	17.28	2.46	0.018916	3.00	67.16	0.078900	3.82	0.001027	0.063728
Guangzhou	0.67	24.21	12.16	0.018715	6.30	38.20	0.035403	101.12	0.000539	0.002110
Shaoguan	0.24	17.49	7.77	0.013108	5.50	66.80	0.057945	11.76	0.001727	0.012554
Shenzhen	0.82	17.21	11.01	0.022624	4.00	44.60	0.036962	403.31	0.000073	0.000559
Zhuhai	0.75	16.59	9.53	0.041363	4.00	28.60	0.213205	124.27	0.000336	0.001739
Shantou	0.69	24.77	2.83	0.021883	3.00	30.00	0.023691	40.43	0.001631	0.007229
Foshan	0.62	23.56	11.97	0.007731	3.00	18.20	0.021955	47.57	0.001066	0.005466
Zhanjiang	0.69	19.70	2.11	0.026858	4.00	28.10	0.014512	8.70	0.002883	0.010376
Zhongshan	0.65	15.07	8.81	0.012032	5.00	18.50	0.077836	76.41	0.001080	0.004230

(Contd.)

City	Economic indicator		Social indicator		Resource indicator		Facility indicator		Environmental indicator	
	Low carbon production index	Low carbon consumption index	Low carbon population index	Low carbon employment index	Clean energy index	Forest carbon sink index	Low carbon traffic index	Low carbon building index	Water environment index	Atmospheric environment index
	GDP output per unit of carbon dioxide emissions (10000 yuan/t) (Constant price)	Low carbon consumption expenditure of residents (%)	Per capita carbon emissions (t/capita)	Jobs per unit of carbon emissions (Jobs/t)	Proportion of non fossil energy in primary energy (%)	Forest coverage (%)	Ride frequency of public trolley-bus (time/capita/bus)	Energy consumption per unit residential building area (t/m^2)	COD emissions per unit of gross industrial output value (t/10000 yuan)	SO_2 emissions per unit of gross industrial output value (t/10000 yuan)
Nanning	0.53	23.66	3.45	0.028522	8.83	42.15	0.032195	18.75	0.022024	0.017895
Liuzhou	0.23	16.42	10.77	0.008852	6.00	61.12	-	14.59	0.003527	0.004976
Guilin	0.40	18.57	4.62	0.013795	4.20	66.50	0.058733	7.98	0.002419	0.018285
Beihai	0.41	14.32	5.21	0.012745	5.00	31.10	0.057375	9.06	0.017739	0.041456
Haikou	0.56	15.74	4.49	0.036839	2.00	38.38	1.567690	28.97	0.000239	0.000075
Sanya	0.56	20.20	4.35	0.023677	3.50	68.00	0.387616	27.95	0.000044	0.000361
Chongqing	0.35	20.00	5.27	0.016510	3.36	35.00	0.006642	20.29	0.003019	0.017644
Chengdu	0.54	20.19	5.72	0.022366	9.10	36.80	0.014108	40.78	0.002298	0.005152
Panzhihua	0.16	16.52	22.23	0.006663	4.50	58.97	0.151815	15.18	0.002121	0.039459
Luzhou	0.30	24.77	3.81	0.016147	6.00	46.80	0.043922	7.23	0.006186	0.034420
Mianyang	0.28	18.15	5.36	0.013457	4.50	46.70	0.038074	9.24	0.001026	0.005605
Yibin	0.31	19.23	4.56	0.016679	5.50	40.21	0.079754	6.38	0.008279	0.017477
Guiyang	0.24	20.42	9.50	0.018180	7.30	41.01	0.064142	50.78	0.000178	0.008406
Zunyi	0.22	21.31	3.91	0.010271	5.70	48.56	0.041159	5.23	0.000700	0.013607
Kunming	0.37	16.09	7.33	0.020425	12.50	45.05	0.031766	43.51	0.000242	0.006515
Qujing	0.23	17.58	5.35	0.010235	4.20	36.20	0.025243	2.57	0.000872	0.018069
Xi'an	0.52	23.99	4.97	0.032374	4.18	44.99	0.029760	32.87	0.002490	0.005055

(Contd.)

City	Economic indicator		Social indicator		Resource indicator		Facility indicator		Environmental indicator	
	Low carbon production index	Low carbon consumption index	Low carbon population index	Low carbon employment index	Clean energy index	Forest carbon sink index	Low carbon traffic index	Low carbon building index	Water environment index	Atmospheric environment index
	GDP output per unit of carbon dioxide emissions (10000 yuan/t) (Constant price)	Low carbon consumption expenditure of residents (%)	Per capita carbon emissions (t/capita)	Jobs per unit of carbon emissions (Jobs/t)	Proportion of non fossil energy in primary energy (%)	Forest coverage (%)	Ride frequency of public trolley-bus (time/capita/bus)	Energy consumption per unit residential building area (t/m^2)	COD emissions per unit of gross industrial output value (t/10000 yuan)	SO_2 emissions per unit of gross industrial output value (t/10000 yuan)
Tongchuan	0.22	21.95	6.87	0.017459	3.00	44.80	0.152398	53.42	0.000484	0.013234
Baoji	0.34	21.24	5.68	0.014672	5.00	53.78	0.067077	7.71	0.004330	0.012393
Xianyang	0.35	26.78	4.25	0.017937	2.50	26.00	0.036612	4.35	0.003558	0.017836
Yan'an	0.70	20.98	4.21	0.024233	1.60	36.60	0.102944	5.92	0.000661	0.002050
Lanzhou	0.24	21.31	11.03	0.013731	6.50	12.21	0.086222	35.46	0.000250	0.009625
Jinchang	0.23	21.73	17.72	0.008765	5.80	20.82	0.276822	0.70	0.001225	0.019770
Xining	0.14	18.13	11.79	0.010548	15.30	28.00	0.147966	29.15	0.005335	0.023488
Yinchuan	0.21	18.32	13.63	0.012788	5.00	14.00	0.086394	187.98	0.004366	0.007096
Shizuishan	0.06	16.51	44.04	0.002569	3.00	11.80	0.020371	14.57	0.002966	0.036496
Urumqi	0.17	18.25	22.66	0.008543	3.00	4.89	0.078232	67.51	0.001129	0.014064
Karamay	0.26	21.74	49.58	0.008039	1.90	4.00	0.254618	38.53	0.000167	0.005649

Appendix IV The Order of 110 Selected Chinese Cities for Low-carbon Development in 2008

Table A4-1 GDO of 110 cities (constant price in 2005)

City	Ranking	GDP(billion yuan)	City	Ranking	GDP(billion yuan)
Shanghai	1	13909.69	Zaozhuang	57	1090.34
Beijing	2	10561.84	Zhuhai	58	1063.75
Guangzhou	3	8412.65	Yueyang	59	1057.33
Shenzhen	4	8064.70	Yichang	60	1034.90
Suzhou	5	6739.57	Changde	61	1023.15
Tianjin	6	6619.61	Zhanjiang	62	1020.99
Chongqing	7	5230.28	Shantou	63	1003.97
Wuxi	8	4672.86	Huzhou	64	1000.72
Hangzhou	9	4668.53	Jiaozuo	65	999.63
Qingdao	10	4596.12	Pingdingshan	66	966.40
Foshan	11	4442.36	Anyang	67	948.55
Nanjing	12	4011.59	Urumqi	68	944.29
Chengdu	13	3999.72	Liuzhou	69	937.53
Dalian	14	3942.99	Guilin	70	901.76
Wuhan	15	3886.83	Guiyang	71	894.62
Ningbo	16	3807.40	Lanzhou	72	882.83
Shenyang	17	3792.16	Zhuzhou	73	878.40
Yantai	18	3538.87	Lianyungang	74	776.45
Tangshan	19	3363.81	Qinghuangdao	75	772.64
Ji'nan	20	3187.30	Sunshine	76	765.02
Harbin	21	3016.22	Mianyang	77	750.98
Shijiazhuang	22	2828.57	Xianyang	78	731.13
Zhengzhou	23	2805.55	Qujing	79	730.40
Quanzhou	24	2783.04	Wuhu	80	719.80
Changchun	25	2747.90	Baoji	81	715.81
Changsha	26	2671.13	Jiujiang	82	712.39
Nantong	27	2556.74	Linfen	83	707.82
Weifang	28	2535.79	Qigihar	84	695.38
Zibo	29	2459.55	Fushun	85	687.12
Wenzhou	30	2457.54	Jinzhou	86	686.37
Fuzhou	31	24391.00	Chifeng	87	674.55
Anshan	32	2340.09	Kaifeng	88	668.18
Daqing	33	2325.91	Ma'anshan	89	660.92
Shaoxing	34	2287.56	Zunyi	90	651.81
Jining	35	2187.35	Yan'an	91	640.69
Xi'an	36	2177.10	Yibin	92	635.44
Changzhou	37	2172.73	Xiangtan	93	620.38
Xuzhou	38	2072.51	Benxi	94	620.33
Taizhou	39	1983.45	Jingzhou	95	616.48
Weihai	40	1976.49	Changzhi	96	616.02
Luoyang	41	1899.11	Shaoguan	97	551.59
Handan	42	1884.28	Karamay	98	509.88

(Contd.)

City	Ranking	GDP(billion yuan)	City	Ranking	GDP(billion yuan)
Jiaxing	43	1819.33	Mudanjiang	99	504.69
Nanchang	44	1743.92	Luzhou	100	494.39
Baotou	45	1738.17	Datong	101	490.12
Kunming	46	1695.60	Yinchuan	102	476.66
Xiamen	47	1644.51	Haikou	103	468.61
Baoding	48	1636.30	Panzhihua	104	413.61
Hefei	49	1628.37	Xining	105	353.89
Yangzhou	50	1581.30	Beihai	106	337.68
Tai'an	51	1491.04	Yangquan	107	297.01
Zhongshan	52	1447.45	Jinchang	108	192.95
Hohhot	53	1364.56	Shizuishan	109	184.82
Taiyuan	54	1285.67	Zhangjiajie	110	182.92
Nanning	55	1283.75	Sanya	111	142.02
Jilin	56	1187.72	Tongchuan	112	124.35

Table A4-2 Energy consumption per unit of GDP in 110 cities

City	Ranking	GDP energy comsumption (t ce/10000 yuan)	City	Ranking	GDP energy comsumption (t ce/10000 yuan)
Taizhou	1	0.513	Xuzhou	56	1.164
Shenzhen	2	0.529	Chongqing	57	1.181
Beijing	3	0.541	Baoji	58	1.186
Xiamen	4	0.579	Tai'an	59	1.200
Zhuhai	5	0.581	Kunming	60	1.223
Shantou	6	0.608	Yichang	61	1.230
Yan'an	7	0.626	Qinghuangdao	62	1.231
Wenzhou	8	0.630	Harbin	63	1.241
Zhanjiang	9	0.641	Mudanjiang	64	1.257
Zhongshan	10	0.646	Luoyang	65	1.281
Guangzhou	11	0.651	Yueyang	66	1.292
Fuzhou	12	0.655	Jining	67	1.300
Changchun	13	0.659	Daqing	68	1.314
Nantong	14	0.692	Zhuzhou	69	1.315
Foshan	15	0.694	Yibin	70	1.376
Yangzhou	16	0.698	Mianyang	71	1.476
Hangzhou	17	0.700	Qigihar	72	1.479
Shanghai	18	0.727	Luzhou	73	1.488
Haikou	19	0.753	Shijiazhuang	74	1.537
Wuxi	20	0.754	Hohhot	75	1.552
Yantai	21	0.770	Jingzhou	76	1.600
Weihai	22	0.770	Chifeng	77	1.637

(Contd.)

City	Ranking	GDP energy comsumption (t ce/10000 yuan)	City	Ranking	GDP energy comsumption (t ce/10000 yuan)
Qingdao	23	0.780	Anshan	78	1.664
Sanya	24	0.780	Jilin	79	1.678
Quanzhou	25	0.795	Zibo	80	1.680
Jiaxing	26	0.817	Zaozhuang	81	1.680
Ningbo	27	0.820	Karamay	82	1.690
Xi'an	28	0.820	Xiangtan	83	1.718
Nanning	29	0.821	Qujing	84	1.730
Lianyungang	30	0.831	Shaoguan	85	1.737
Tianjin	31	0.836	Jinchang	86	1.741
Chengdu	32	0.840	Guiyang	87	1.742
Changsha	33	0.846	Liuzhou	88	1.756
Nanchang	34	0.855	Datong	89	1.760
Suzhou	35	0.856	Sunshine	90	1.780
Shaoxing	36	0.860	Tongchuan	91	1.798
Changzhou	37	0.886	Taiyuan	92	1.830
Huzhou	38	0.900	Zunyi	93	1.830
Zhangjiajie	39	0.901	Jiaozuo	94	1.857
Hefei	40	0.931	Lanzhou	95	1.870
Changde	41	0.937	Pingdingshan	96	1.884
Dalian	42	0.947	Yinchuan	97	2.020
Shenyang	43	0.999	Baotou	98	2.113
Baoding	44	1.006	Yangquan	99	2.130
Beihai	45	1.006	Ma'anshan	100	2.136
Ji'nan	46	1.040	Handan	101	2.144
Jiujiang	47	1.068	Anyang	102	2.251
Guilin	48	1.069	Urumqi	103	2.438
Wuhu	49	1.090	Tangshan	104	2.464
Nanjing	50	1.105	Panzhihua	105	2.499
Kaifeng	51	1.105	Changzhi	106	2.650
Weifang	52	1.110	Fushun	107	2.930
Wuhan	53	1.110	Linfen	108	3.450
Zhengzhou	54	1.115	Xining	109	3.556
Xianyang	55	1.143	Shizuishan	110	6.701

Table A4-3 CO_2 productivity in 110 cities

City	Ranking	Productivity level of CO_2 (10000 yuan GDP/t CO_2)	City	Ranking	Productivity level of CO_2 (10000 yuan GDP/t CO_2)
Shenzhen	1	0.8219	Zhengzhou	56	0.3527
Taizhou	2	0.8121	Xianyang	57	0.3525
Beijing	3	0.7793	Chongqing	58	0.3471
Zhuhai	4	0.7482	Xuzhou	59	0.3411
Xiamen	5	0.7297	Baoji	60	0.3393
Yan'an	6	0.7009	Daqing	61	0.3391
Shantou	7	0.6944	Tai'an	62	0.331
Zhanjiang	8	0.6933	Harbin	63	0.3287
Guangzhou	9	0.6695	Qinghuangdao	64	0.3268
Wenzhou	10	0.6627	Mudanjiang	65	0.323
Zhongshan	11	0.6525	Yibin	66	0.3125
Shanghai	12	0.6440	Luoyang	67	0.3093
Foshan	13	0.6190	Yueyang	68	0.3075
Changchun	14	0.6168	Zhuzhou	69	0.3022
Fuzhou	15	0.6133	Jining	70	0.3021
Nantong	16	0.5820	Luzhou	71	0.2989
Hangzhou	17	0.5797	Mianyang	72	0.2821
Yangzhou	18	0.5777	Qigihar	73	0.2712
Sanya	19	0.5633	Jingzhou	74	0.2633
Haikou	20	0.5551	Karamay	75	0.2614
Chengdu	21	0.5434	Hohhot	76	0.2597
Ningbo	22	0.5412	Shijiazhuang	77	0.2588
Qingdao	23	0.5393	Jilin	78	0.2532
Wuxi	24	0.5341	Zibo	79	0.2439
Nanning	25	0.5301	Lanzhou	80	0.2411
Quanzhou	26	0.5267	Shaoguan	81	0.2391
Weihai	27	0.5263	Chifeng	82	0.2388
Yantai	28	0.5228	Guiyang	83	0.2372
Xi'an	29	0.5196	Liuzhou	84	0.2348
Tianjin	30	0.5190	Qujing	85	0.2346
Jiaxing	31	0.4850	Zaozhuang	86	0.2345
Lianyungang	32	0.4847	Anshan	87	0.233
Changsha	33	0.4830	Xiangtan	88	0.2294
Suzhou	34	0.4765	Jinchang	89	0.2288
Shaoxing	35	0.4744	Sunshine	90	0.2214
Nanchang	36	0.4646	Zunyi	91	0.2206
Dalian	37	0.4623	Tongchuan	92	0.2173
Changzhou	38	0.4516	Datong	93	0.2164
Zhangjiajie	39	0.4499	Taiyuan	94	0.2128
Huzhou	40	0.4425	Jiaozuo	95	0.2118
Hefei	41	0.4290	Pingdingshan	96	0.2083

(Contd.)

City	Ranking	Productivity level of CO_2 (10000 yuan GDP/t CO_2)	City	Ranking	Productivity level of CO_2 (10000 yuan GDP/t CO_2)
Changde	42	0.4255	Yinchuan	97	0.2055
Beihai	43	0.4079	Yangquan	98	0.188
Nanjing	44	0.4019	Baotou	99	0.1838
Shenyang	45	0.4011	Ma'anshan	100	0.1818
Ji'nan	46	0.3990	Handan	101	0.1809
Guilin	47	0.3973	Anyang	102	0.1786
Wuhan	48	0.3967	Urumqi	103	0.1727
Baoding	49	0.3940	Panzhihua	104	0.159
Jiujiang	50	0.3783	Tangshan	105	0.1577
Kunming	51	0.3686	Changzhi	106	0.1442
Wuhu	52	0.3616	Fushun	107	0.1396
Weifang	53	0.3586	Xining	108	0.1361
Kaifeng	54	0.3579	Linfen	109	0.1118
Yichang	55	0.3539	Shizuishan	110	0.0577

Table A4-4 Carbon emission intensity from energy consumption in 110 cities

City	Ranking	Carbon emission intensityof energy consumption (t CO_2/t ce)	City	Ranking	Carbon emission intensityof energy consumption (t CO_2/t ce)
Xining	1	2.0662	Zhangjiajie	56	2.4667
Shanghai	2	2.1358	Weihai	57	2.4674
Chengdu	3	2.1906	Jiujiang	58	2.4753
Kunming	4	2.2184	Zunyi	59	2.4760
Lanzhou	5	2.2185	Yangzhou	60	2.4800
Daqing	6	2.2441	Hohhot	61	2.4810
Luzhou	7	2.2486	Xianyang	62	2.4817
Zhanjiang	8	2.2500	Lianyungang	63	2.4827
Nanjing	9	2.2525	Nantong	64	2.4829
Ningbo	10	2.2532	Wuxi	65	2.4836
Karamay	11	2.2636	Yantai	66	2.4810
Wuhan	12	2.2707	Baoji	67	2.4849
Sanya	13	2.2761	Qinghuangdao	68	2.4857
Yan'an	14	2.2790	Anyang	69	2.4871
Dalian	15	2.2850	Fuzhou	70	2.4893
Guangzhou	16	2.2942	Qigihar	71	2.4925
Nanning	17	2.2968	Shenyang	72	2.4950
Yichang	18	2.2975	Yangquan	73	2.4970
Shenzhen	19	2.3000	Changzhou	74	2.5000
Zhuhai	20	2.3005	Hefei	75	2.5040
Tianjin	21	2.3048	Changde	76	2.5081

(Contd.)

City	Ranking	Carbon emission intensityof energy consumption (t CO_2/t ce)	City	Ranking	Carbon emission intensityof energy consumption (t CO_2/t ce)
Yibin	22	2.3252	Jinchang	77	2.5093
Foshan	23	2.3278	Huzhou	78	2.5108
Xi'an	24	2.3470	Weifang	79	2.5124
Jilin	25	2.3544	Shijiazhuang	80	2.5143
Guilin	26	2.3547	Panzhihua	81	2.5164
Xiamen	27	2.3668	Zhuzhou	82	2.5165
Shantou	28	2.3686	Yueyang	83	2.5167
Beijing	29	2.3719	Nanchang	84	2.5176
Zhongshan	30	2.3724	Tai'an	85	2.5179
Jingzhou	31	2.3738	Xuzhou	86	2.5195
Urumqi	32	2.3741	Baoding	87	2.5229
Qingdao	33	2.3773	Jiaxing	88	2.5238
Quanzhou	34	2.3883	Luoyang	89	2.5240
Haikou	35	2.3925	Kaifeng	90	2.5285
Wenzhou	36	2.3952	Wuhu	91	2.5373
Taizhou	37	2.4004	Xiangtan	92	2.5377
Mianyang	38	2.4014	Sunshine	93	2.5380
Shaoguan	39	2.4078	Zaozhuang	94	2.5383
Yinchuan	40	2.4089	Jiaozuo	95	2.5426
Ji'nan	41	2.4100	Zhengzhou	96	2.5429
Guiyang	42	2.4202	Jining	97	2.5467
Liuzhou	43	2.4255	Pingdingshan	98	2.5479
Beihai	44	2.4365	Chifeng	99	2.5588
Chongqing	45	2.4392	Tongchuan	100	2.5593
Zibo	46	2.4403	Taiyuan	101	2.5680
Fushun	47	2.4449	Tangshan	102	2.5731
Changsha	48	2.4474	Ma'anshan	103	2.5744
Shaoxing	49	2.4499	Baotou	104	2.5753
Harbin	50	2.4516	Handan	105	2.5776
Suzhou	51	2.4527	Anshan	106	2.5791
Changchun	52	2.4597	Shizuishan	107	2.5862
Mudanjiang	53	2.4625	Linfen	108	2.5919
Hangzhou	54	2.4642	Changzhi	109	2.6172
Qujing	55	2.4643	Datong	110	2.6252

Table A4-5 Total energy consumption in 110 cities

City	Ranking	Total energy consumption (10000 t ce)	City	Ranking	Total energy consumption (10000 t ce)
Sanya	1	110.777	Changzhi	56	1632.422
Zhangjiajie	2	164.814	Liuzhou	57	1645.973
Tongchuan	3	223.582	Baoding	58	1646.116
Jinchang	4	336.015	Lanzhou	59	1650.890
Beihai	5	339.751	Nantong	60	1769.267
Haikou	6	352.864	Xi'an	61	1785.220
Yan'an	7	401.070	Tai'an	62	1789.246
Shantou	8	610.413	Changchun	63	1811.221
Zhuhai	9	618.039	Pingdingshan	64	1820.701
Yangquan	10	632.639	Zaozhuang	65	1831.765
Mudanjiang	11	634.501	Jiaozuo	66	1856.310
Lianyungang	12	645.228	Changzhou	67	1924.413
Zhanjiang	13	654.453	Shaoxing	68	1968.220
Luzhou	14	735.645	Jilin	69	1992.500
Kaifeng	15	738.338	Fushun	70	2013.255
Jiujiang	16	760.831	Kunming	71	2073.717
Wuhu	17	784.581	Hohhot	72	2117.655
Xianyang	18	835.686	Anyang	73	2135.178
Baoji	19	848.947	Quanzhou	74	2212.516
Karamay	20	861.690	Changsha	75	2259.777
Datong	21	862.612	Urumqi	76	2302.557
Yibin	22	874.363	Taiyuan	77	2352.787
Huzhou	23	900.652	Xuzhou	78	2411.845
Zhongshan	24	935.051	Luoyang	79	2432.756
Qinghuangdao	25	951.118	Linfen	80	2441.994
Xiamen	26	952.173	Yantai	81	2724.929
Shaoguan	27	958.107	Weifang	82	2814.726
Changde	28	958.691	Jining	83	2843.550
Yinchuan	29	962.855	Daqing	84	3056.716
Guilin	30	963.906	Foshan	85	3082.994
Jingzhou	31	986.361	Ningbo	86	3122.068
Taizhou	32	1017.512	Zhengzhou	87	3128.183
Qigihar	33	1028.535	Hangzhou	88	3267.974
Panzhihua	34	1033.612	Ji'nan	89	3314.787
Nanning	35	1054.366	Chengdu	90	3359.768
Xiangtan	36	1065.810	Wuxi	91	3522.842
Yangzhou	37	1103.804	Qingdao	92	3584.973
Chifeng	38	1103.901	Baotou	93	3671.894
Mianyang	39	1108.444	Dalian	94	3733.084
Zhuzhou	40	1155.095	Harbin	95	3742.824
Zunyi	41	1193.065	Shenyang	96	3789.548
Shizuishan	42	1238.484	Anshan	97	3893.627

(Contd.)

City	Ranking	Total energy consumption (10000 t ce)	City	Ranking	Total energy consumption (10000 t ce)
Xining	43	1258.582	Handan	98	4039.888
Qujing	44	1263.588	Zibo	99	4132.042
Yichang	45	1272.923	Shenzhen	100	4266.227
Sunshine	46	1361.741	Wuhan	101	4314.382
Yueyang	47	1366.071	Shijiazhuang	102	4347.519
Ma'anshan	48	1411.898	Nanjing	103	4431.889
Jiaxing	49	1486.395	Guangzhou	104	5476.635
Nanchang	50	1491.050	Tianjin	105	5533.990
Hefei	51	1516.011	Beijing	106	5713.953
Weihai	52	1521.901	Suzhou	107	5766.778
Wenzhou	53	1548.250	Chongqing	108	6176.956
Guiyang	54	1558.187	Tangshan	109	8288.421
Fuzhou	55	1595.523	Shanghai	110	10112.348

Table A4-6 Per-capita CO_2 emission in 110 cities

City	Ranking	CO_2 emissions per capita (t CO_2/capita)	City	Ranking	CO_2 emissions per capita (t CO_2/capita)
Zhanjiang	1	2.11	Jining	56	9.02
Zhangjiajie	2	2.46	Xiangtan	57	9.16
Shantou	3	2.83	Pingdingshan	58	9.21
Nanning	4	3.45	Harbin	59	9.25
Lianyungang	5	3.60	Guiyang	60	9.50
Baoding	6	3.77	Zhuhai	61	9.53
Luzhou	7	3.81	Luoyang	62	9.56
Changde	8	3.90	Yantai	63	9.62
Zunyi	9	3.91	Qingdao	64	10.03
Jiujiang	10	3.93	Anyang	65	10.18
Kaifeng	11	3.96	Tianjin	66	10.39
Jingzhou	12	4.00	Zhengzhou	67	10.58
Yan'an	13	4.21	Wuhan	68	10.77
Taizhou	14	4.22	Liuzhou	69	10.77
Xianyang	15	4.25	Jilin	70	10.80
Sanya	16	4.35	Changzhou	71	10.81
Qigihar	17	4.49	Shenzhen	72	11.01
Haikou	18	4.49	Shaoxing	73	11.02
Yibin	19	4.56	Lanzhou	74	11.03
Guilin	20	4.62	Shijiazhuang	75	11.06
Wenzhou	21	4.76	Shanghai	76	11.24
Xi'an	22	4.97	Handan	77	11.73
Beihai	23	5.21	Hangzhou	78	11.78
Chongqing	24	5.27	Xining	79	11.79

(Contd.)

City	Ranking	CO_2 emissions per capita (t CO_2/capita)	City	Ranking	CO_2 emissions per capita (t CO_2/capita)
Qujing	25	5.35	Yangquan	80	11.93
Mianyang	26	5.36	Ji'nan	81	11.96
Baoji	27	5.68	Foshan	82	11.97
Chengdu	28	5.72	Guangzhou	83	12.16
Mudanjiang	29	5.77	Ningbo	84	12.32
Fuzhou	30	5.78	Sunshine	85	12.53
Changchun	31	5.89	Zaozhuang	86	12.69
Yangzhou	32	6.09	Nanjing	87	12.94
Nantong	33	6.16	Changzhi	88	12.95
Yueyang	34	6.27	Shenyang	89	13.20
Chifeng	35	6.53	Weihai	90	13.33
Quanzhou	36	6.72	Yinchuan	91	13.63
Tongchuan	37	6.87	Jiaozuo	92	13.79
Xuzhou	38	7.00	Wuxi	93	14.12
Datong	39	7.09	Dalian	94	14.59
Yichang	40	7.23	Linfen	95	14.99
Kunming	41	7.33	Suzhou	96	15.10
Hefei	42	7.44	Taiyuan	97	17.25
Zhuzhou	43	7.59	Jinchang	98	17.72
Beijing	44	7.72	Hohhot	99	19.40
Shaoguan	45	7.77	Fushun	100	22.11
Weifang	46	7.90	Panzhihua	101	22.23
Qinghuangdao	47	7.94	Zibo	102	22.34
Nanchang	48	8.07	Urumqi	103	22.66
Tai'an	49	8.22	Daqing	104	24.48
Changsha	50	8.49	Ma'anshan	105	28.18
Wuhu	51	8.66	Anshan	106	28.53
Jiaxing	52	8.70	Tangshan	107	28.56
Huzhou	53	8.73	Baotou	108	36.76
Zhongshan	54	8.81	Shizuishan	109	44.04
Xiamen	55	8.94	Karamay	110	49.58

Table A4-7 Residents' low-carbon consumption coefficient in 110 cities (proportion of residents' low-carbon consumption expenditure)

City	Ranking	The proportion of residents' low-carbon consumption expenditure (%)	City	Ranking	The proportion of residents' low-carbon consumption expenditure (%)
Xianyang	1	26.78	Wuhu	56	19.97
Qinghuangdao	2	25.05	Suzhou	57	19.91
Shantou	3	24.77	Ningbo	58	19.80
Luzhou	4	24.77	Shanghai	59	19.73
Mudanjiang	5	24.75	Zhanjiang	60	19.70
Nanjing	6	24.25	Dalian	61	19.65
Guangzhou	7	24.21	Zhuzhou	62	19.53
Yangzhou	8	24.13	Changsha	63	19.49
Xi'an	9	23.99	Weifang	64	19.48
Changchun	10	23.94	Yueyang	65	19.46
Harbin	11	23.69	Xuzhou	66	19.46
Nanning	12	23.66	Jiaxing	67	19.40
Foshan	13	23.56	Tai'an	68	19.28
Anyang	14	23.21	Jingzhou	69	19.25
Hohhot	15	22.76	Yibin	70	19.23
Nantong	16	22.61	Zhengzhou	71	19.01
Beijing	17	22.60	Jilin	72	18.99
Ma'anshan	18	22.58	Chifeng	73	18.98
Jiaozuo	19	22.20	Xiamen	74	18.74
Shijiazhuang	20	21.98	Zibo	75	18.73
Tongchuan	21	21.95	Guilin	76	18.57
Tangshan	22	21.95	Wenzhou	77	18.56
Karamay	23	21.74	Anshan	78	18.52
Jinchang	24	21.73	Nanchang	79	18.48
Fushun	25	21.68	Yinchuan	80	18.32
Luoyang	26	21.55	Hefei	81	18.30
Datong	27	21.45	Urumqi	82	18.25
Yantai	28	21.45	Mianyang	83	18.15
Handan	29	21.42	Xining	84	18.13
Jining	30	21.38	Wuhan	85	18.06
Shaoxing	31	21.38	Linfen	86	17.73
Taizhou	32	21.32	Kaifeng	87	17.64
Lianyungang	33	21.31	Qujing	88	17.58
Zunyi	34	21.31	Zaozhuang	89	17.51
Lanzhou	35	21.31	Shaoguan	90	17.49
Wuxi	36	21.27	Weihai	91	17.49
Baoji	37	21.24	Zhangjiajie	92	17.28
Yan'an	38	20.98	Shenzhen	93	17.21

(Contd.)

City	Ranking	The proportion of residents' low-carbon consumption expenditure (%)	City	Ranking	The proportion of residents' low-carbon consumption expenditure (%)
Pingdingshan	39	20.90	Zhuhai	94	16.59
Baoding	40	20.78	Panzhihua	95	16.52
Qigihar	41	20.78	Shizuishan	96	16.51
Yangquan	42	20.68	Changde	97	16.50
Huzhou	43	20.55	Liuzhou	98	16.42
Shenyang	44	20.54	Qingdao	99	16.30
Changzhou	45	20.46	Changzhi	100	16.25
Baotou	46	20.46	Fuzhou	101	16.22
Guiyang	47	20.42	Kunming	102	16.09
Tianjin	48	20.36	Haikou	103	15.74
Taiyuan	49	20.36	Hangzhou	104	15.69
Yichang	50	20.34	Quanzhou	105	15.59
Sanya	51	20.20	Sunshine	106	15.38
Chengdu	52	20.19	Daqing	107	15.26
Ji'nan	53	20.18	Zhongshan	108	15.07
Xiangtan	54	20.17	Jiujiang	109	14.56
Chongqing	55	20.00	Beihai	110	14.32

Table A4-8 Job contribution from unit carbon emission in 110 cities

City	Ranking	Job contribution from unit carbon emission (job/t CO_2)	City	Ranking	Job contribution from unit carbon emission (job/t CO_2)
Beijing	1	0.04570	Lanzhou	56	0.01373
Zhuhai	2	0.04136	Yueyang	57	0.01351
Haikou	3	0.03684	Mianyang	58	0.01346
Xiamen	4	0.03563	Taiyuan	59	0.01330
Xi'an	5	0.03237	Shaoguan	60	0.01311
Wenzhou	6	0.02903	Zhengzhou	61	0.01307
Nanning	7	0.02852	Yinchuan	62	0.01279
Zhanjiang	8	0.02686	Beihai	63	0.01275
Taizhou	9	0.02630	Yantai	64	0.01268
Quanzhou	10	0.02593	Wuhu	65	0.01222
Hangzhou	11	0.02538	Qinghuangdao	66	0.01211
Fuzhou	12	0.02485	Zhongshan	67	0.01203
Yan'an	13	0.02423	Nanjing	68	0.01179
Sanya	14	0.02368	Tai'an	69	0.01165
Shenzhen	15	0.02262	Zhuzhou	70	0.01147

(Contd.)

City	Ranking	Job contribution from unit carbon emission (job/t CO_2)	City	Ranking	Job contribution from unit carbon emission (job/t CO_2)
Chengdu	16	0.02237	Shenyang	71	0.01113
Shantou	17	0.02188	Dalian	72	0.01102
Lianyungang	18	0.02079	Chifeng	73	0.01065
Kunming	19	0.02043	Xining	74	0.01055
Jiaxing	20	0.02026	Weihai	75	0.01043
Changchun	21	0.02000	Weifang	76	0.01030
Shaoxing	22	0.01965	Zunyi	77	0.01027
Zhangjiajie	23	0.01892	Qujing	78	0.01024
Ningbo	24	0.01880	Pingdingshan	79	0.01023
Guangzhou	25	0.01872	Xuzhou	80	0.01010
Datong	26	0.01858	Xiangtan	81	0.00971
Guiyang	27	0.01818	Liuzhou	82	0.00885
Changsha	28	0.01816	Jinchang	83	0.00877
Wuhan	29	0.01794	Urumqi	84	0.00854
Xianyang	30	0.01794	Luoyang	85	0.00853
Shanghai	31	0.01785	Suzhou	86	0.00843
Nanchang	32	0.01767	Changzhi	87	0.00842
Yichang	33	0.01747	Jining	88	0.00835
Tongchuan	34	0.01746	Anyang	89	0.00811
Jiujiang	35	0.01728	Karamay	90	0.00804
Kaifeng	36	0.01687	Wuxi	91	0.00794
Yibin	37	0.01668	Changzhou	92	0.00775
Hefei	38	0.01653	Foshan	93	0.00773
Jingzhou	39	0.01652	Daqing	94	0.00772
Chongqing	40	0.01651	Shijiazhuang	95	0.00764
Baoding	41	0.01625	Zaozhuang	96	0.00738
Qigihar	42	0.01623	Jiaozuo	97	0.00713
Luzhou	43	0.01615	Jilin	98	0.00705
Tianjin	44	0.01581	Panzhihua	99	0.00666
Harbin	45	0.01570	Zibo	100	0.00604
Mudanjiang	46	0.01551	Hohhot	101	0.00601
Ji'nan	47	0.01543	Linfen	102	0.00563
Huzhou	48	0.01501	Sunshine	103	0.00556
Baoji	49	0.01467	Fushun	104	0.00538
Yangquan	50	0.01452	Handan	105	0.00527
Qingdao	51	0.01443	Ma'anshan	106	0.00424
Changde	52	0.01438	Anshan	107	0.00396
Nantong	53	0.01413	Tangshan	108	0.00384
Yangzhou	54	0.01412	Baotou	109	0.00348
Guilin	55	0.01380	Shizuishan	110	0.00257

Table A4-9 Proportion of non-fossil energy consumption in 110 cities

City	Ranking	Proportion of non-fossil energy consumption (%)	City	Ranking	Proportion of non-fossil energy consumption (%)
Xining	1	15.30	Changde	56	3.80
Kunming	2	12.50	Qinghuangdao	57	3.72
Shanghai	3	11.10	Wenzhou	58	3.70
Yichang	4	10.50	Chifeng	59	3.70
Chengdu	5	9.10	Yangzhou	60	3.70
Nanning	6	8.83	Zhuzhou	61	3.60
Tianjin	7	8.71	Sanya	62	3.50
Wuhan	8	8.20	Ningbo	63	3.50
Guiyang	9	7.30	Qingdao	64	3.50
Lanzhou	10	6.50	Ji'nan	65	3.50
Guangzhou	11	6.30	Tangshan	66	3.50
Liuzhou	12	6.00	Suzhou	67	3.50
Luzhou	13	6.00	Handan	68	3.50
Jingzhou	14	6.00	Hefei	69	3.50
Jinchang	15	5.80	Kaifeng	70	3.40
Zunyi	16	5.70	Chongqing	71	3.36
Shaoguan	17	5.50	Anshan	72	3.30
Yibin	18	5.50	Tai'an	73	3.30
Baoji	19	5.00	Pingdingshan	74	3.30
Beihai	20	5.00	Baotou	75	3.30
Shijiazhuang	21	5.00	Weihai	76	3.20
Zhongshan	22	5.00	Yantai	77	3.20
Changchun	23	5.00	Zibo	78	3.20
Yinchuan	24	5.00	Weifang	79	3.20
Fushun	25	4.70	Zaozhuang	80	3.20
Harbin	26	4.70	Jining	81	3.20
Changsha	27	4.58	Jiaozuo	82	3.20
Taizhou	28	4.50	Zhangjiajie	83	3.00
Panzhihua	29	4.50	Tongchuan	84	3.00
Quanzhou	30	4.50	Yueyang	85	3.00
Jilin	31	4.50	Sunshine	86	3.00
Fuzhou	32	4.50	Linfen	87	3.00
Jiujiang	33	4.50	Shantou	88	3.00
Mianyang	34	4.50	Wuxi	89	3.00
Xiamen	35	4.50	Beijing	90	3.00
Hohhot	36	4.50	Changzhou	91	3.00
Baoding	37	4.50	Foshan	92	3.00
Lianyungang	38	4.30	Taiyuan	93	3.00
Guilin	39	4.20	Daqing	94	3.00
Hangzhou	40	4.20	Urumqi	95	3.00

(Contd.)

City	Ranking	Proportion of non-fossil energy consumption (%)	City	Ranking	Proportion of non-fossil energy consumption (%)
Qujing	41	4.20	Zhengzhou	96	2.80
Shenyang	42	4.20	Huzhou	97	2.70
Nantong	43	4.20	Luoyang	98	2.50
Qigihar	44	4.20	Changzhi	99	2.50
Xi'an	45	4.18	Xianyang	100	2.50
Mudanjiang	46	4.00	Wuhu	101	2.50
Shenzhen	47	4.00	Jiaxing	102	2.50
Dalian	48	4.00	Ma'anshan	103	2.50
Xuzhou	49	4.00	Yangquan	104	2.40
Zhuhai	50	4.00	Haikou	105	2.00
Zhanjiang	51	4.00	Datong	106	2.00
Nanjing	52	4.00	Anyang	107	2.00
Nanchang	53	4.00	Shizuishan	108	2.00
Shaoxing	54	3.90	Karamay	109	1.90
Xiangtan	55	3.80	Yan'an	110	1.60

Table A4-10 Urban forest coverage rate in 110 cities

City	Ranking	Urban forest coverage rate (%)	City	Ranking	Urban forest coverage rate (%)
Sanya	1	68.00	Zaozhuang	56	32.00
Zhangjiajie	2	67.16	Linfen	57	31.40
Shaoguan	3	66.80	Beihai	58	31.10
Guilin	4	66.50	Shantou	59	30.00
Fushun	5	66.20	Hohhot	60	29.90
Hangzhou	6	64.44	Ji'nan	61	28.90
Mudanjiang	7	62.30	Xuzhou	62	28.70
Taizhou	8	62.20	Tangshan	63	28.70
Liuzhou	9	61.12	Zhuhai	64	28.60
Zhuzhou	10	61.10	Zhanjiang	65	28.10
Wenzhou	11	60.03	Xining	66	28.00
Panzhihua	12	58.97	Changzhi	67	26.90
Quanzhou	13	56.70	Jining	68	26.50
Yichang	14	55.30	Wuhan	69	26.48
Jilin	15	55.00	Wuxi	70	26.20
Fuzhou	16	54.90	Xianyang	71	26.00
Shaoxing	17	54.30	Shijiazhuang	72	25.70
Baoji	18	53.78	Nanjing	73	25.00
Changsha	19	53.28	Pingdingshan	74	24.98
Huzhou	20	51.30	Shenyang	75	24.00

(Contd.)

City	Ranking	Urban forest coverage rate (%)	City	Ranking	Urban forest coverage rate (%)
Ningbo	21	50.20	Zhengzhou	76	24.00
Jiujiang	22	50.00	Yangquan	77	23.10
Zunyi	23	48.56	Beijing	78	23.00
Luzhou	24	46.80	Suzhou	79	22.00
Mianyang	25	46.70	Jiaozuo	80	22.00
Anshan	26	46.40	Wuhu	81	21.23
Xiangtan	27	45.89	Jinchang	82	20.82
Kunming	28	45.05	Nantong	83	20.00
Xi'an	29	44.99	Changzhou	84	20.00
Harbin	30	44.80	Lianyungang	85	19.70
Changde	31	44.80	Datong	86	19.50
Tongchuan	32	44.80	Baoding	87	19.27
Shenzhen	33	44.60	Kaifeng	88	19.00
Luoyang	34	42.82	Zhongshan	89	18.50
Nanning	35	42.15	Anyang	90	18.49
Qinghuangdao	36	41.98	Foshan	91	18.20
Dalian	37	41.50	Handan	92	18.00
Guiyang	38	41.01	Qigihar	93	17.37
Xiamen	39	41.00	Jiaxing	94	17.00
Yibin	40	40.21	Nanchang	95	16.10
Weihai	41	39.20	Hefei	96	16.10
Yantai	42	39.00	Yangzhou	97	16.00
Haikou	43	38.38	Jingzhou	98	15.20
Guangzhou	44	38.20	Changchun	99	15.20
Chengdu	45	36.80	Taiyuan	100	15.00
Yan'an	46	36.60	Yinchuan	101	14.00
Yueyang	47	36.40	Baotou	102	12.40
Qujing	48	36.20	Lanzhou	103	12.21
Qingdao	49	36.10	Shizuishan	104	11.80
Tai'an	50	35.80	Shanghai	105	11.60
Sunshine	51	35.20	Daqing	106	10.60
Chongqing	52	35.00	Ma'anshan	107	8.44
Zibo	53	33.40	Tianjin	108	8.14
Weifang	54	33.00	Urumqi	109	4.89
Chifeng	55	32.56	Karamay	110	4.00

Table A4-11 Per-capital riding number per public electric vehicle in 110 cities

City	Ranking	Public electric vehicle transportation (time/(capita•car))	City	Ranking	Public electric vehicle transportation (time/(capita•car))
Haikou	1	1.56769	Nanning	56	0.03220
Sanya	2	0.38762	Shenyang	57	0.03200
Jinchang	3	0.27682	Kunming	58	0.03177
Karamay	4	0.25462	Yueyang	59	0.03147
Zhuhai	5	0.21321	Daqing	60	0.03088
Yangquan	6	0.20266	Xi'an	61	0.02976
Ma'anshan	7	0.18040	Fuzhou	62	0.02923
Tongchuan	8	0.15240	Ji'nan	63	0.02915
Panzhihua	9	0.15182	Shaoxing	64	0.02868
Xining	10	0.14797	Yantai	65	0.02840
Fushun	11	0.12296	Yangzhou	66	0.02794
Xiamen	12	0.11627	Wuhan	67	0.02745
Yan'an	13	0.10294	Nanjing	68	0.02735
Hohhot	14	0.09351	Suzhou	69	0.02665
Yinchuan	15	0.08639	Hangzhou	70	0.02613
Lanzhou	16	0.08622	Linfen	71	0.02602
Yibin	17	0.07975	Qujing	72	0.02524
Urumqi	18	0.07823	Ningbo	73	0.02510
Zhangjiajie	19	0.07809	Qingdao	74	0.02504
Zhongshan	20	0.07784	Jingzhou	75	0.02493
Baotou	21	0.07694	Zhengzhou	76	0.02457
Wuhu	22	0.07347	Luoyang	77	0.02453
Mudanjiang	23	0.07258	Wuxi	78	0.02446
Baoji	24	0.06708	Shantou	79	0.02369
Guiyang	25	0.06414	Pingdingshan	80	0.02320
Guilin	26	0.05873	Jiaxing	81	0.02294
Weihai	27	0.05822	Foshan	82	0.02196
Shaoguan	28	0.05795	Harbin	83	0.02168
Beihai	29	0.05738	Quanzhou	84	0.02149
Changsha	30	0.05430	Tangshan	85	0.02140
Jilin	31	0.05229	Shizuishan	86	0.02037
Taiyuan	32	0.05018	Tai'an	87	0.02012
Datong	33	0.04899	Changchun	88	0.01976
Qinghuangdao	34	0.04778	Anyang	89	0.01921
Sunshine	35	0.04771	Beijing	90	0.01830
Zhuzhou	36	0.04673	Wenzhou	91	0.01756
Hefei	37	0.04668	Taizhou	92	0.01676
Changzhi	38	0.04559	Nantong	93	0.01647
Anshan	39	0.04415	Zaozhuang	94	0.01630
Luzhou	40	0.04392	Zibo	95	0.01623
Lianyungang	41	0.04373	Xuzhou	96	0.01550
Jiujiang	42	0.04150	Tianjin	97	0.01522

(Contd.)

City	Ranking	Public electric vehicle transportation (time/(capita•car))	City	Ranking	Public electric vehicle transportation (time/(capita•car))
Zunyi	43	0.04116	Jining	98	0.01464
Nanchang	44	0.04080	Zhanjiang	99	0.01451
Yichang	45	0.04062	Kaifeng	100	0.01442
Changzhou	46	0.03968	Xiangtan	101	0.01440
Mianyang	47	0.03807	Weifang	102	0.01438
Chifeng	48	0.03719	Chengdu	103	0.01411
Shenzhen	49	0.03696	Shanghai	104	0.01196
Xianyang	50	0.03661	Shijiazhuang	105	0.01157
Huzhou	51	0.03614	Handan	106	0.01037
Guangzhou	52	0.03540	Baoding	107	0.00847
Dalian	53	0.03499	Chongqing	108	0.00664
Changde	54	0.03470	Qigihar	109	0.00016
Jiaozuo	55	0.03333			

Table A4-12 Energy consumption level (t/m^2) per living area in 110 cities

City	Ranking	Energy consumption per residential building area (t/m^2)		City	Ranking	Energy consumption per residential building area (t/m^2)	
		Absolute value	Relative value			Absolute value	Relative value
Jinchang	1	0.70	1.000000	Liuzhou	56	14.59	0.965480
Qujing	2	2.57	0.995339	Panzhihua	57	15.18	0.964037
Baoding	3	3.20	0.995768	Wenzhou	58	15.37	0.963543
Jining	4	3.56	0.992887	Datong	59	16.43	0.960924
Jingzhou	5	3.63	0.992720	Nanning	60	18.75	0.955170
Pingdingshan	6	3.69	0.992558	Changchun	61	18.75	0.955161
Zhangjiajie	7	3.82	0.992252	Wuhu	62	18.93	0.954706
Linfen	8	3.86	0.992134	Harbin	63	19.86	0.952397
Xianyang	9	4.35	0.990913	Nanchang	64	20.06	0.951912
Chifeng	10	4.72	0.989994	Chongqing	65	20.29	0.951340
Changde	11	4.85	0.989694	Ma'anshan	66	20.64	0.950455
Tai'an	12	4.90	0.989561	Hohhot	67	21.00	0.949572
Zunyi	13	5.23	0.988742	Zibo	68	21.14	0.949233
Xiangtan	14	5.24	0.988718	Baotou	69	22.63	0.945530
Quanzhou	15	5.42	0.988255	Suzhou	70	23.09	0.944385
Lianyungang	16	5.58	0.987869	Ji'nan	71	24.89	0.939901
Weifang	17	5.66	0.987678	Wuxi	72	25.69	0.937925
Changzhi	18	5.72	0.987511	Ningbo	73	25.72	0.937858
Handan	19	5.91	0.987051	Dalian	74	25.99	0.937165
Yan'an	20	5.92	0.987035	Changzhou	75	27.28	0.933983
Mudanjiang	21	6.10	0.986581	Sanya	76	27.95	0.932315

(Contd.)

City	Ranking	Energy consumption per residential building area (t/m²)		City	Ranking	Energy consumption per residential building area (t/m²)	
		Absolute value	Relative value			Absolute value	Relative value
Luoyang	22	6.27	0.986161	Haikou	77	28.97	0.929763
Yibin	23	6.38	0.985895	Shenyang	78	29.10	0.929444
Xuzhou	24	6.81	0.984803	Xining	79	29.15	0.929328
Yueyang	25	6.95	0.984479	Hefei	80	29.39	0.928736
Jiujiang	26	7.16	0.983938	Fuzhou	81	29.87	0.927529
Shijiazhuang	27	7.21	0.983811	Zhengzhou	82	31.88	0.922554
Luzhou	28	7.23	0.983764	Xi'an	83	32.87	0.920093
Nantong	29	7.44	0.983248	Changsha	84	34.99	0.914819
Kaifeng	30	7.56	0.982944	Lanzhou	85	35.46	0.913661
Baoji	31	7.71	0.982586	Jilin	86	36.92	0.910034
Zaozhuang	32	7.91	0.982077	Hangzhou	87	37.92	0.907550
Guilin	33	7.98	0.981915	Karamay	88	38.53	0.906023
Anyang	34	8.28	0.981168	Fushun	89	39.17	0.904452
Weihai	35	8.45	0.980737	Shantou	90	40.43	0.901323
Jiaozuo	36	8.47	0.980681	Tianjin	91	40.62	0.900840
Jiaxing	37	8.65	0.980237	Chengdu	92	40.78	0.900438
Zhanjiang	38	8.70	0.980128	Wuhan	93	41.12	0.899586
Beihai	39	9.06	0.979233	Qingdao	94	42.14	0.897070
Mianyang	40	9.24	0.978774	Kunming	95	43.51	0.893655
Yangzhou	41	9.38	0.978423	Foshan	96	47.57	0.883580
Yichang	42	9.51	0.978097	Nanjing	97	50.49	0.876322
Sunshine	43	9.85	0.977267	Guiyang	98	50.78	0.875595
Taizhou	44	9.98	0.976951	Taiyuan	99	51.41	0.874042
Yantai	45	10.51	0.975628	Tongchuan	100	53.42	0.869041
Qigihar	46	10.60	0.975395	Urumqi	101	67.51	0.834057
Tangshan	47	11.03	0.974344	Shanghai	102	68.72	0.831047
Shaoguan	48	11.76	0.972524	Yangquan	103	69.64	0.828766
Huzhou	49	11.98	0.971975	Xiamen	104	75.29	0.814737
Zhuzhou	50	11.99	0.971954	Zhongshan	105	76.41	0.811956
Anshan	51	12.93	0.969617	Beijing	106	85.58	0.789173
Shaoxing	52	13.84	0.967350	Guangzhou	107	101.12	0.750575
Qinghuangdao	53	14.08	0.966758	Zhuhai	108	124.27	0.693087
Daqing	54	14.36	0.966070	Yinchuan	109	187.98	0.534824
Shizuishan	55	14.57	0.965539	Shenzhen	110	403.31	0.000000

Table A4-13 Water pollutant discharge intensity (t/10,000 yuan) in 110 cities

City	Ranking	Water pollution discharge intensity ($t/10^4$ yuan)		City	Ranking	Water pollution discharge intensity ($t/10^4$ yuan)	
		Absolute value	Relative value			Absolute value	Relative value
Sanya	1	0.000044	1.000000	Harbin	56	0.001191	0.947816
Shenzhen	2	0.000073	0.998681	Pingdingshan	57	0.001210	0.946952
Beijing	3	0.000108	0.997088	Huzhou	58	0.001212	0.946861
Hefei	4	0.000113	0.996861	Jinchang	59	0.001225	0.946269
Xiamen	5	0.000113	0.996861	Xuzhou	60	0.001234	0.945860
Fushun	6	0.000133	0.995951	Zibo	61	0.001242	0.945496
Karamay	7	0.000167	0.994404	Jiujiang	62	0.001275	0.943995
Guiyang	8	0.000178	0.993904	Qinghuangdao	63	0.001366	0.939854
Haikou	9	0.000239	0.991128	Yangzhou	64	0.001386	0.938944
Yangquan	10	0.000242	0.990992	Yichang	65	0.001436	0.936670
Kunming	11	0.000242	0.990992	Changzhou	66	0.001560	0.931028
Lanzhou	12	0.000250	0.990628	Tangshan	67	0.001572	0.930482
Shanghai	13	0.000253	0.990491	Shantou	68	0.001631	0.927798
Daqing	14	0.000264	0.989991	Sunshine	69	0.001675	0.925796
Fuzhou	15	0.000281	0.989217	Wuhu	70	0.001721	0.923703
Qingdao	16	0.000304	0.988171	Shaoguan	71	0.001727	0.923430
Luoyang	17	0.000331	0.986943	Nantong	72	0.001798	0.920200
Zhuhai	18	0.000336	0.986715	Quanzhou	73	0.001798	0.920200
Taiyuan	19	0.000344	0.986351	Chifeng	74	0.001930	0.914195
Tianjin	20	0.000351	0.986033	Panzhihua	75	0.002121	0.905505
Changsha	21	0.000394	0.984076	Shaoxing	76	0.002201	0.901865
Shenyang	22	0.000405	0.983576	Chengdu	77	0.002298	0.897452
Baotou	23	0.000415	0.983121	Guilin	78	0.002419	0.891947
Ji'nan	24	0.000418	0.982985	Xi'an	79	0.002490	0.887170
Weihai	25	0.000428	0.982530	Zhuzhou	80	0.002518	0.887443
Yantai	26	0.000459	0.981119	Zhanjiang	81	0.002883	0.870837
Ningbo	27	0.000467	0.980755	Wenzhou	82	0.002965	0.867106
Taizhou	28	0.000468	0.980710	Qigihar	83	0.002988	0.866060
Dalian	29	0.000471	0.980573	Shizuishan	84	0.002996	0.865696
Hohhot	30	0.000484	0.979982	Baoding	85	0.003009	0.865105
Tongchuan	31	0.000484	0.979982	Chongqing	86	0.003019	0.864650
Guangzhou	32	0.000539	0.977480	Shijiazhuang	87	0.003151	0.858644
Handan	33	0.000595	0.974932	Nanchang	88	0.003193	0.856733
Nanjing	34	0.000603	0.974568	Jilin	89	0.003370	0.848681
Anshan	35	0.000638	0.972975	Hangzhou	90	0.003387	0.847907
Yan'an	36	0.000661	0.971929	Zaozhuang	91	0.003415	0.846633
Zunyi	37	0.000700	0.970155	Liuzhou	92	0.003527	0.841538
Wuxi	38	0.000710	0.969700	Xianyang	93	0.003558	0.840127

(Contd.)

City	Ranking	Water pollution discharge intensity ($t/10^4$ yuan)		City	Ranking	Water pollution discharge intensity ($t/10^4$ yuan)	
		Absolute value	Relative value			Absolute value	Relative value
Ma'anshan	39	0.000771	0.966924	Xiangtan	94	0.003846	0.827025
Tai'an	40	0.000777	0.966652	Baoji	95	0.004330	0.805005
Jining	41	0.000782	0.966424	Yinchuan	96	0.004366	0.803367
Suzhou	42	0.000783	0.966379	Anyang	97	0.004654	0.790264
Wuhan	43	0.000797	0.965742	Yueyang	98	0.005025	0.773385
Lianyungang	44	0.000801	0.965560	Xining	99	0.005335	0.759281
Changzhi	45	0.000844	0.963603	Luzhou	100	0.006186	0.720564
Weifang	46	0.000866	0.962602	Datong	101	0.006288	0.715924
Qujing	47	0.000872	0.962329	Jiaozuo	102	0.006907	0.687762
Changchun	48	0.000980	0.957416	Jingzhou	103	0.007080	0.679891
Zhengzhou	49	0.000999	0.956551	Changde	104	0.007531	0.659372
Zhongshan	50	0.001008	0.956142	Yibin	105	0.008279	0.625341
Mianyang	51	0.001026	0.955323	Kaifeng	106	0.008842	0.599727
Foshan	52	0.001066	0.953503	Zhangjiajie	107	0.010272	0.534668
Urumqi	53	0.001129	0.950637	Mudanjiang	108	0.014870	0.325478
Linfen	54	0.001162	0.949136	Beihai	109	0.017739	0.194950
Jiaxing	55	0.001167	0.948908	Nanning	110	0.022024	0.000000

Table A4-14 Air pollutant emission intensity (t/10,000 yuan) in top 110 cities

City	Ranking	Air pollution emission intensity ($t/10^4$ yuan)		City	Ranking	Air pollution emission intensity ($t/10^4$ yuan)	
		Absolute value	Relative value			Absolute value	Relative value
Haikou	1	0.000075	1.000000	Ma'anshan	56	0.007268	0.891796
Sanya	2	0.000361	0.995698	Qinghuangdao	57	0.007339	0.890727
Shenzhen	3	0.000559	0.992719	Jining	58	0.007465	0.888832
Qingdao	4	0.001241	0.982460	Tangshan	59	0.007479	0.888621
Beijing	5	0.001326	0.981181	Xuzhou	60	0.008170	0.878227
Daqing	6	0.001606	0.976969	Zibo	61	0.008217	0.877520
Zhuhai	7	0.001739	0.974968	Changzhi	62	0.008272	0.876692
Yan'an	8	0.002050	0.970290	Guiyang	63	0.008406	0.874677
Changchun	9	0.002066	0.970049	Handan	64	0.008722	0.869923
Shanghai	10	0.002085	0.969764	Huzhou	65	0.008788	0.868930
Guangzhou	11	0.002110	0.969387	Baoding	66	0.008881	0.867531
Xiamen	12	0.002247	0.967327	Shijiazhuang	67	0.009103	0.864192
Hefei	13	0.002482	0.963791	Linfen	68	0.009572	0.857136
Wuxi	14	0.002492	0.963641	Lanzhou	69	0.009625	0.856339

(Contd.)

City	Ranking	Air pollution emission intensity ($t/10^4$ yuan)		City	Ranking	Air pollution emission intensity ($t/10^4$ yuan)	
		Absolute value	Relative value			Absolute value	Relative value
Fushun	15	0.002582	0.962287	Tai'an	70	0.009699	0.855226
Tianjin	16	0.002590	0.962167	Anshan	71	0.010237	0.847133
Nanjing	17	0.003033	0.955503	Yueyang	72	0.010324	0.845824
Ningbo	18	0.003060	0.955097	Zhanjiang	73	0.010376	0.845042
Suzhou	19	0.003221	0.952675	Jingzhou	74	0.010972	0.836076
Yantai	20	0.003223	0.952645	Anyang	75	0.011010	0.835505
Shaoxing	21	0.003362	0.950554	Zhengzhou	76	0.011983	0.820868
Nanchang	22	0.003423	0.949636	Zhuzhou	77	0.012005	0.820537
Dalian	23	0.003484	0.948718	Baoji	78	0.012393	0.814700
Quanzhou	24	0.003606	0.946883	Shaoguan	79	0.012554	0.812278
Hangzhou	25	0.003641	0.946357	Jiaozuo	80	0.012558	0.812218
Changzhou	26	0.003643	0.946326	Xiangtan	81	0.012822	0.808247
Taizhou	27	0.003819	0.943679	Qigihar	82	0.012826	0.808186
Ji'nan	28	0.003931	0.941994	Tongchuan	83	0.013234	0.802049
Changsha	29	0.004168	0.938429	Zunyi	84	0.013607	0.796438
Wuhan	30	0.004189	0.938113	Changde	85	0.013814	0.793324
Shenyang	31	0.004193	0.938053	Urumqi	86	0.014064	0.789563
Zhongshan	32	0.004230	0.937496	Hohhot	87	0.014472	0.783426
Harbin	33	0.004456	0.934097	Jiujiang	88	0.015890	0.762095
Yichang	34	0.004463	0.933991	Pingdingshan	89	0.016702	0.749880
Liuzhou	35	0.004976	0.926274	Yibin	90	0.017477	0.738221
Weifang	36	0.005020	0.925612	Baotou	91	0.017481	0.738161
Xi'an	37	0.005055	0.925086	Chongqing	92	0.017644	0.735709
Chengdu	38	0.005152	0.923627	Xianyang	93	0.017836	0.732821
Nantong	39	0.005421	0.919580	Zaozhuang	94	0.017852	0.732580
Foshan	40	0.005466	0.918903	Nanning	95	0.017895	0.731933
Mianyang	41	0.005605	0.916812	Qujing	96	0.018069	0.729316
Jilin	42	0.005614	0.916677	Luoyang	97	0.018088	0.729030
Karamay	43	0.005649	0.916150	Guilin	98	0.018285	0.726067
Wuhu	44	0.005659	0.916000	Jinchang	99	0.019770	0.703728
Lianyungang	45	0.006302	0.906327	Xining	100	0.023488	0.647798
Taiyuan	46	0.006327	0.905951	Datong	101	0.025201	0.622029
Fuzhou	47	0.006380	0.905154	Kaifeng	102	0.027100	0.593462
Kunming	48	0.006515	0.903132	Luzhou	103	0.034420	0.483347
Jiaxing	49	0.006806	0.898745	Mudanjiang	104	0.035852	0.461806
Weihai	50	0.006855	0.898008	Shizuishan	105	0.036496	0.452118
Wenzhou	51	0.006963	0.896384	Yangquan	106	0.037276	0.440384
Sunshine	52	0.006985	0.896053	Panzhihua	107	0.039459	0.407546
Yangzhou	53	0.007016	0.895586	Beihai	108	0.041456	0.377505
Yinchuan	54	0.007096	0.894383	Zhangjiajie	109	0.063728	0.042466
Shantou	55	0.007229	0.892382	Chifeng	110	0.066551	0.000000

Appendix V Dimensional Data of Low-carbon Development Indicators for 110 Selected Chinese Cities in 2009

Table A5-1 GDP in 110 cities (constant price in 2005)

A5-1

City	GDP(billion yuan)	City	GDP(billion yuan)	City	GDP(billion yuan)	City	GDP(billion yuan)
Beijing	9592.95	Xuzhou	1826.00	Tai'an	1312.53	Guilin	791.02
Tianjin	5682.07	Changzhou	1950.39	Weihai	1752.21	Beihai	291.11
Shijiazhuang	2545.97	Suzhou	6071.69	Sunshine	670.49	Haikou	422.93
Tangshan	3022.29	Nantong	2242.76	Zhengzhou	2504.95	Sanya	121.18
Qinghuangdao	702.40	Lianyungang	684.10	Kaifeng	591.31	Chongqing	4552.03
Handan	1694.49	Yangzhou	1394.22	Luoyang	1710.91	Chengdu	3487.12
Baoding	1474.14	Hangzhou	4244.12	Pingdingshan	859.02	Panzhihua	371.95
Taiyuan	1253.09	Ningbo	3505.89	Anyang	862.32	Luzhou	429.15
Datong	479.10	Wenzhou	2265.01	Jiaozuo	890.93	Mianyang	656.45
Yangquan	277.32	Jiaxing	1664.53	Wuhan	3418.50	Yibin	553.04
Changzhi	560.01	Huzhou	922.33	Yichang	899.13	Guiyang	789.60
Linfen	679.29	Shaoxing	2092.92	Jingzhou	543.15	Zunyi	578.87
Hohhot	1177.36	Taizhou	1793.36	Changsha	2328.80	Kunming	1503.19
Baotou	1448.48	Hefei	1388.21	Zhuzhou	769.18	Qujing	646.94
Chifeng	578.52	Wuhu	623.74	Xiangtan	545.63	Xi'an	1901.40
Shenyang	3323.54	Ma'anshan	589.58	Yueyang	927.48	Tongchuan	107.94
Dalian	3428.69	Fuzhou	2155.68	Changde	912.71	Baoji	622.44
Anshan	1974.76	Xiamen	1522.70	Zhangjiajie	160.88	Xianyang	640.22
Fushun	604.86	Quanzhou	2473.81	Guangzhou	7544.98	Yan'an	571.02
Changchun	2389.48	Nanchang	1541.93	Shaoguan	500.99	Lanzhou	796.78
Jilin	1015.14	Jiujiang	626.00	Shenzhen	7285.19	Jinchang	168.96
Harbin	2669.22	Ji'nan	2840.73	Zhuhai	997.89	Xining	312.35
Qigihar	610.52	Qingdao	4096.36	Shantou	906.93	Yinchuan	421.82
Daqing	1920.66	Zibo	2172.75	Foshan	3913.97	Shizuishan	163.27
Mudanjiang	442.71	Zaozhuang	968.33	Zhanjiang	923.14	Urumqi	862.37
Shanghai	12855.54	Yantai	3117.95	Zhongshan	1313.47	Karamay	516.07
Nanjing	3597.84	Weifang	2246.05	Nanning	1116.30		
Wuxi	4190.91	Jining	1940.86	Liuzhou	806.13		

Table A5-2 Energy consumption per unit of GDP in 110 cities

A5-2

City	GDP energy comsumption (t ce/10000 yuan)	City	GDP energy comsumption (t ce/10000 yuan)	City	GDP energy comsumption (t ce/10000 yuan)	City	GDP energy comsumption (t ce/10000 yuan)
Beijing	0.662	Xuzhou	1.227	Tai'an	1.281	Guilin	1.096
Tianjin	0.947	Changzhou	0.934	Weihai	0.812	Beihai	1.021
Shijiazhuang	1.622	Suzhou	0.906	Sunshine	1.342	Haikou	0.799
Tangshan	2.600	Nantong	0.730	Zhengzhou	1.187	Sanya	0.801
Qinghuangdao	1.299	Lianyungang	0.833	Kaifeng	1.117	Chongqing	1.267
Handan	2.262	Yangzhou	0.739	Luoyang	1.364	Chengdu	0.905
Baoding	1.060	Hangzhou	0.750	Pingdingshan	2.009	Panzhihua	3.299
Taiyuan	2.230	Ningbo	0.840	Anyang	2.419	Luzhou	1.596
Datong	2.100	Wenzhou	0.650	Jiaozuo	1.991	Mianyang	1.591
Yangquan	2.340	Jiaxing	0.860	Wuhan	1.190	Yibin	1.364
Changzhi	2.930	Huzhou	1.020	Yichang	1.160	Guiyang	1.820
Linfen	3.760	Shaoxing	0.900	Jingzhou	1.690	Zunyi	1.900
Hohhot	1.677	Taizhou	0.560	Changsha	0.886	Kunming	1.282
Baotou	2.387	Hefei	0.980	Zhuzhou	1.390	Qujing	1.833
Chifeng	1.821	Wuhu	1.400	Xiangtan	1.816	Xi'an	0.873
Shenyang	1.046	Ma'anshan	2.249	Yueyang	1.371	Tongchuan	1.930
Dalian	1.031	Fuzhou	0.676	Changde	0.997	Baoji	1.253
Anshan	1.757	Xiamen	0.600	Zhangjiajie	0.947	Xianyang	1.207
Fushun	3.084	Quanzhou	0.814	Guangzhou	0.680	Yan'an	0.642
Changchun	0.698	Nanchang	0.906	Shaoguan	1.819	Lanzhou	1.987
Jilin	1.815	Jiujiang	1.068	Shenzhen	0.544	Jinchang	1.806
Harbin	1.316	Ji'nan	1.101	Zhuhai	0.603	Xining	3.705
Qigihar	1.594	Qingdao	0.821	Shantou	0.632	Yinchuan	2.154
Daqing	1.382	Zibo	1.791	Foshan	0.802	Shizuishan	7.206
Mudanjiang	1.335	Zaozhuang	1.785	Zhanjiang	0.659	Urumqi	2.503
Shanghai	0.801	Yantai	0.815	Zhongshan	0.673	Karamay	1.770
Nanjing	1.178	Weifang	1.187	Nanning	0.838		
Wuxi	0.799	Jining	1.380	Liuzhou	1.951		

Table A5-3 CO_2 productivity in 110 cities

A5-3

City	Productivity level of CO_2(10000 yuan GDP/t CO_2)	City	Productivity level of CO_2(10000 yuan GDP/t CO_2)	City	Productivity level of CO_2 (10000 yuan GDP/t CO_2)	City	Productivity level of CO_2(10000 yuan GDP/t CO_2)
Beijing	0.630	Xuzhou	0.321	Tai'an	0.307	Guilin	0.385
Tianjin	0.456	Changzhou	0.425	Weihai	0.494	Beihai	0.401
Shijiazhuang	0.242	Suzhou	0.446	Sunshine	0.292	Haikou	0.532
Tangshan	0.148	Nantong	0.542	Zhengzhou	0.329	Sanya	0.544
Qinghuangdao	0.307	Lianyungang	0.481	Kaifeng	0.352	Chongqing	0.320
Handan	0.170	Yangzhou	0.543	Luoyang	0.287	Chengdu	0.504
Baoding	0.370	Hangzhou	0.538	Pingdingshan	0.194	Panzhihua	0.119
Taiyuan	0.173	Ningbo	0.522	Anyang	0.165	Luzhou	0.272
Datong	0.179	Wenzhou	0.633	Jiaozuo	0.197	Mianyang	0.259
Yangquan	0.170	Jiaxing	0.456	Wuhan	0.366	Yibin	0.313
Changzhi	0.130	Huzhou	0.388	Yichang	0.370	Guiyang	0.225
Linfen	0.102	Shaoxing	0.451	Jingzhou	0.248	Zunyi	0.209
Hohhot	0.239	Taizhou	0.736	Changsha	0.457	Kunming	0.347
Baotou	0.162	Hefei	0.404	Zhuzhou	0.284	Qujing	0.221
Chifeng	0.213	Wuhu	0.279	Xiangtan	0.216	Xi'an	0.485
Shenyang	0.381	Ma'anshan	0.171	Yueyang	0.288	Tongchuan	0.202
Dalian	0.419	Fuzhou	0.594	Changde	0.397	Baoji	0.320
Anshan	0.219	Xiamen	0.698	Zhangjiajie	0.424	Xianyang	0.332
Fushun	0.131	Quanzhou	0.509	Guangzhou	0.635	Yan'an	0.677
Changchun	0.575	Nanchang	0.437	Shaoguan	0.227	Lanzhou	0.225
Jilin	0.232	Jiujiang	0.374	Shenzhen	0.795	Jinchang	0.219
Harbin	0.307	Ji'nan	0.373	Zhuhai	0.714	Xining	0.129
Qigihar	0.251	Qingdao	0.507	Shantou	0.665	Yinchuan	0.191
Daqing	0.319	Zibo	0.228	Foshan	0.533	Shizuishan	0.053
Mudanjiang	0.302	Zaozhuang	0.219	Zhanjiang	0.678	Urumqi	0.165
Shanghai	0.582	Yantai	0.491	Zhongshan	0.623	Karamay	0.248
Nanjing	0.372	Weifang	0.334	Nanning	0.512		
Wuxi	0.500	Jining	0.282	Liuzhou	0.210		

Table A5-4 Carbon emission intensity from energy consumption in 110 cities

A5-4

City	Carbon emission intensityof energy consumption ($t CO_2/t ce$)	City	Carbon emission intensity of energy consumption ($t CO_2/t ce$)	City	Carbon emission intensityof energy consumption ($t CO_2/t ce$)	City	Carbon emission intensity of energy consumption ($t CO_2/t ce$)
Beijing	2.399	Xuzhou	2.541	Tai'an	2.539	Guilin	2.370
Tianjin	2.316	Changzhou	2.518	Weihai	2.493	Beihai	2.444
Shijiazhuang	2.544	Suzhou	2.473	Sunshine	2.549	Haikou	2.412
Tangshan	2.598	Nantong	2.527	Zhengzhou	2.562	Sanya	2.295
Qinghuangdao	2.507	Lianyungang	2.498	Kaifeng	2.541	Chongqing	2.468
Handan	2.603	Yangzhou	2.491	Luoyang	2.552	Chengdu	2.193
Baoding	2.551	Hangzhou	2.478	Pingdingshan	2.569	Panzhihua	2.538
Taiyuan	2.593	Ningbo	2.280	Anyang	2.503	Luzhou	2.305
Datong	2.657	Wenzhou	2.430	Jiaozuo	2.554	Mianyang	2.426
Yangquan	2.512	Jiaxing	2.549	Wuhan	2.296	Yibin	2.345
Changzhi	2.627	Huzhou	2.529	Yichang	2.328	Guiyang	2.447
Linfen	2.609	Shaoxing	2.465	Jingzhou	2.385	Zunyi	2.518
Hohhot	2.498	Taizhou	2.426	Changsha	2.469	Kunming	2.249
Baotou	2.591	Hefei	2.523	Zhuzhou	2.531	Qujing	2.474
Chifeng	2.576	Wuhu	2.559	Xiangtan	2.549	Xi'an	2.364
Shenyang	2.511	Ma'anshan	2.605	Yueyang	2.533	Tongchuan	2.570
Dalian	2.317	Fuzhou	2.492	Changde	2.526	Baoji	2.497
Anshan	2.595	Xiamen	2.389	Zhangjiajie	2.493	Xianyang	2.499
Fushun	2.477	Quanzhou	2.413	Guangzhou	2.317	Yan'an	2.300
Changchun	2.495	Nanchang	2.529	Shaoguan	2.427	Lanzhou	2.234
Jilin	2.372	Jiujiang	2.501	Shenzhen	2.311	Jinchang	2.530
Harbin	2.477	Ji'nan	2.434	Zhuhai	2.323	Xining	2.092
Qigihar	2.505	Qingdao	2.404	Shantou	2.380	Yinchuan	2.430
Daqing	2.271	Zibo	2.451	Foshan	2.339	Shizuishan	2.614
Mudanjiang	2.481	Zaozhuang	2.554	Zhanjiang	2.240	Urumqi	2.417
Shanghai	2.147	Yantai	2.500	Zhongshan	2.384	Karamay	2.278
Nanjing	2.280	Weifang	2.522	Nanning	2.329		
Wuxi	2.503	Jining	2.566	Liuzhou	2.437		

Table A5-5 Total energy consumption in 110 cities

A5-5

City	Total energy consumption (10000 t ce)	City	Total energy consumption (10000 t ce)	City	Total energy consumption (10000 t ce)	City	Total energy consumption (10000 t ce)
Beijing	6350.53	Xuzhou	2240.50	Tai'an	1681.64	Guilin	867.21
Tianjin	5380.92	Changzhou	1821.66	Weihai	1423.55	Beihai	297.35
Shijiazhuang	4129.57	Suzhou	5500.95	Sunshine	900.04	Haikou	329.47
Tangshan	7857.95	Nantong	1637.21	Zhengzhou	2973.38	Sanya	97.06
Qinghuangdao	912.42	Lianyungang	569.85	Kaifeng	660.49	Chongqing	5767.42
Handan	3832.94	Yangzhou	1029.59	Luoyang	2333.68	Chengdu	3155.84
Baoding	1562.59	Hangzhou	3183.09	Pingdingshan	1725.78	Panzhihua	1227.07
Taiyuan	2794.40	Ningbo	2944.95	Anyang	2085.94	Luzhou	684.93
Datong	1006.11	Wenzhou	1472.26	Jiaozuo	1773.85	Mianyang	1044.41
Yangquan	648.94	Jiaxing	1431.50	Wuhan	4068.01	Yibin	754.34
Changzhi	1640.84	Huzhou	940.77	Yichang	1042.99	Guiyang	1437.07
Linfen	2554.15	Shaoxing	1883.63	Jingzhou	917.92	Zunyi	1099.85
Hohhot	1974.68	Taizhou	1004.28	Changsha	2063.32	Kunming	1927.09
Baotou	3457.16	Hefei	1360.44	Zhuzhou	1069.16	Qujing	1185.85
Chifeng	1053.44	Wuhu	873.24	Xiangtan	990.86	Xi'an	1659.92
Shenyang	3477.75	Ma'anshan	1325.79	Yueyang	1271.58	Tongchuan	208.33
Dalian	3534.33	Fuzhou	1457.24	Changde	909.97	Baoji	779.92
Anshan	3469.65	Xiamen	913.62	Zhangjiajie	152.36	Xianyang	772.75
Fushun	1865.51	Quanzhou	2013.68	Guangzhou	5130.59	Yan'an	366.60
Changchun	1666.64	Nanchang	1396.55	Shaoguan	911.30	Lanzhou	1583.03
Jilin	1842.91	Jiujiang	668.77	Shenzhen	3963.14	Jinchang	305.19
Harbin	3513.23	Ji'nan	3128.26	Zhuhai	601.73	Xining	1157.13
Qigihar	973.10	Qingdao	3363.55	Shantou	573.18	Yinchuan	908.44
Daqing	2654.15	Zibo	3891.72	Foshan	3139.00	Shizuishan	1176.42
Mudanjiang	590.98	Zaozhuang	1728.82	Zhanjiang	608.35	Urumqi	2158.32
Shanghai	10297.29	Yantai	2539.45	Zhongshan	883.97	Karamay	913.17
Nanjing	4238.25	Weifang	2664.96	Nanning	935.55		
Wuxi	3348.54	Jining	2678.72	Liuzhou	1572.54		

Table A5-6 Per-capita CO_2 emission in 110 cities

A5-6

City	CO_2 emissions per capita (t CO_2/capita)	City	CO_2 emissions per capita (t CO_2/capita)	City	CO_2 emissions per capita (t CO_2/capita)	City	CO_2 emissions per capita (t CO_2/capita)
Beijing	8.99	Xuzhou	6.55	Tai'an	7.83	Guilin	3.58
Tianjin	10.60	Changzhou	10.41	Weihai	12.65	Beihai	4.61
Shijiazhuang	10.67	Suzhou	11.52	Sunshine	8.06	Haikou	4.33
Tangshan	27.47	Nantong	5.09	Zhengzhou	10.25	Sanya	3.94
Qinghuangdao	7.74	Lianyungang	2.88	Kaifeng	3.58	Chongqing	5.01
Handan	10.75	Yangzhou	5.24	Luoyang	9.28	Chengdu	5.45
Baoding	3.49	Hangzhou	8.44	Pingdingshan	8.85	Panzhihua	28.01
Taiyuan	20.88	Ningbo	9.34	Anyang	10.01	Luzhou	3.67
Datong	8.41	Wenzhou	3.94	Jiaozuo	12.54	Mianyang	4.69
Yangquan	12.35	Jiaxing	6.51	Wuhan	10.43	Yibin	3.33
Changzhi	13.13	Huzhou	8.30	Yichang	6.01	Guiyang	8.93
Linfen	15.87	Shaoxing	9.15	Jingzhou	3.39	Zunyi	3.68
Hohhot	18.46	Taizhou	3.51	Changsha	7.74	Kunming	6.95
Baotou	35.38	Hefei	6.47	Zhuzhou	7.29	Qujing	5.07
Chifeng	6.24	Wuhu	9.03	Xiangtan	8.59	Xi'an	4.69
Shenyang	11.26	Ma'anshan	26.02	Yueyang	5.91	Tongchuan	6.38
Dalian	13.36	Fuzhou	5.32	Changde	3.72	Baoji	5.18
Anshan	24.28	Xiamen	8.77	Zhangjiajie	2.32	Xianyang	13.04
Fushun	20.52	Quanzhou	6.24	Guangzhou	11.68	Yan'an	3.93
Changchun	5.38	Nanchang	7.65	Shaoguan	6.47	Lanzhou	10.35
Jilin	9.90	Jiujiang	3.52	Shenzhen	10.45	Jinchang	15.87
Harbin	8.64	Ji'nan	11.49	Zhuhai	9.44	Xining	11.11
Qigihar	4.28	Qingdao	9.56	Shantou	2.69	Yinchuan	12.58
Daqing	21.75	Zibo	21.17	Foshan	12.33	Shizuishan	41.31
Mudanjiang	5.23	Zaozhuang	12.05	Zhanjiang	1.97	Urumqi	19.11
Shanghai	11.71	Yantai	9.04	Zhongshan	8.39	Karamay	41.63
Nanjing	12.73	Weifang	7.56	Nanning	3.11		
Wuxi	13.73	Jining	8.63	Liuzhou	10.33		

Table A5-7 Residents' low-carbon consumption coefficient in 110 cities

A5-7

City	The proportion of residents' low-carbon consumption expenditure (%)	City	The proportion of residents' low-carbon consumption expenditure (%)	City	The proportion of residents' low-carbon consumption expenditure (%)	City	The proportion of residents' low-carbon consumption expenditure (%)
Beijing	29.76	Xuzhou	17.45	Tai'an	19.69	Guilin	17.03
Tianjin	21.08	Changzhou	22.07	Weihai	18.32	Beihai	11.47
Shijiazhuang	21.86	Suzhou	20.12	Sunshine	18.62	Haikou	16.79
Tangshan	22.03	Nantong	26.66	Zhengzhou	20.16	Sanya	18.13
Qinghuangdao	25.13	Lianyungang	20.34	Kaifeng	16.28	Chongqing	20.04
Handan	17.80	Yangzhou	24.60	Luoyang	22.56	Chengdu	17.96
Baoding	21.22	Hangzhou	17.43	Pingdingshan	20.47	Panzhihua	16.38
Taiyuan	24.03	Ningbo	19.85	Anyang	23.22	Luzhou	24.75
Datong	22.02	Wenzhou	20.65	Jiaozuo	18.91	Mianyang	6.57
Yangquan	21.66	Jiaxing	13.39	Wuhan	17.69	Yibin	7.77
Changzhi	17.76	Huzhou	19.87	Yichang	18.60	Guiyang	17.75
Linfen	15.63	Shaoxing	23.28	Jingzhou	19.53	Zunyi	19.27
Hohhot	23.15	Taizhou	21.06	Changsha	20.19	Kunming	16.49
Baotou	22.31	Hefei	18.43	Zhuzhou	22.48	Qujing	16.32
Chifeng	19.62	Wuhu	20.19	Xiangtan	19.99	Xi'an	23.57
Shenyang	18.86	Ma'anshan	21.19	Yueyang	20.94	Tongchuan	19.45
Dalian	18.12	Fuzhou	16.26	Changde	15.80	Baoji	21.35
Anshan	21.22	Xiamen	15.77	Zhangjiajie	19.65	Xianyang	26.14
Fushun	21.21	Quanzhou	15.43	Guangzhou	22.80	Yan'an	21.36
Changchun	22.46	Nanchang	19.06	Shaoguan	24.48	Lanzhou	21.09
Jilin	28.06	Jiujiang	13.10	Shenzhen	17.10	Jinchang	25.63
Harbin	21.36	Ji'nan	21.28	Zhuhai	24.69	Xining	18.57
Qigihar	20.33	Qingdao	17.34	Shantou	24.79	Yinchuan	18.52
Daqing	19.44	Zibo	26.21	Foshan	24.71	Shizuishan	15.17
Mudanjiang	25.53	Zaozhuang	19.01	Zhanjiang	18.80	Urumqi	18.93
Shanghai	18.71	Yantai	19.67	Zhongshan	15.63	Karamay	20.61
Nanjing	24.50	Weifang	21.36	Nanning	17.61		
Wuxi	20.14	Jining	19.72	Liuzhou	15.80		

Table A5-8 Job contribution from unit carbon emission in 110 cities

A5-8

City	Job contribution from unit carbon emission (job/10000 t CO_2)	City	Job contribution from unit carbon emission (job/10000 t CO_2)	City	Job contribution from unit carbon emission (job/10000 t CO_2)	City	Job contribution from unit carbon emission (job/10000 t CO_2)
Beijing	365.80	Xuzhou	107.60	Tai'an	117.90	Guilin	149.40
Tianjin	160.80	Changzhou	79.60	Weihai	106.50	Beihai	138.70
Shijiazhuang	81.50	Suzhou	90.10	Sunshine	82.00	Haikou	364.20
Tangshan	38.00	Nantong	146.20	Zhengzhou	131.30	Sanya	257.60
Qinghuangdao	123.30	Lianyungang	233.20	Kaifeng	162.90	Chongqing	165.60
Handan	56.90	Yangzhou	143.80	Luoyang	87.90	Chengdu	223.00
Baoding	163.90	Hangzhou	220.10	Pingdingshan	108.10	Panzhihua	56.90
Taiyuan	110.90	Ningbo	161.00	Anyang	82.60	Luzhou	160.20
Datong	174.90	Wenzhou	287.00	Jiaozuo	74.30	Mianyang	131.30
Yangquan	132.60	Jiaxing	201.30	Wuhan	192.60	Yibin	179.20
Changzhi	82.20	Huzhou	130.00	Yichang	155.50	Guiyang	193.40
Linfen	54.10	Shaoxing	168.90	Jingzhou	168.20	Zunyi	109.10
Hohhot	62.70	Taizhou	224.10	Changsha	164.10	Kunming	224.00
Baotou	37.10	Hefei	142.40	Zhuzhou	115.70	Qujing	100.50
Chifeng	117.20	Wuhu	107.10	Xiangtan	113.60	Xi'an	321.40
Shenyang	117.80	Ma'anshan	44.10	Yueyang	123.10	Tongchuan	169.90
Dalian	114.20	Fuzhou	262.30	Changde	140.70	Baoji	157.40
Anshan	55.40	Xiamen	382.50	Zhangjiajie	194.30	Xianyang	187.50
Fushun	63.20	Quanzhou	254.50	Guangzhou	205.00	Yan'an	258.80
Changchun	233.30	Nanchang	181.10	Shaoguan	135.10	Lanzhou	146.10
Jilin	77.40	Jiujiang	195.70	Shenzhen	219.30	Jinchang	97.30
Harbin	170.30	Ji'nan	158.20	Zhuhai	416.20	Xining	92.70
Qigihar	181.70	Qingdao	160.00	Shantou	229.40	Yinchuan	130.30
Daqing	90.60	Zibo	59.90	Foshan	77.10	Shizuishan	25.70
Mudanjiang	174.00	Zaozhuang	77.90	Zhanjiang	297.50	Urumqi	89.90
Shanghai	168.40	Yantai	138.10	Zhongshan	125.30	Karamay	79.20
Nanjing	110.90	Weifang	103.80	Nanning	327.60		
Wuxi	77.20	Jining	88.90	Liuzhou	90.90		

Table A5-9 Proportion of non-fossil energy consumption in 110 cities

A5-9

City	Proportion of non-fossil energy consumption (%)	City	Proportion of non-fossil energy consumption (%)	City	Proportion of non-fossil energy consumption (%)	City	Proportion of non-fossil energy consumption (%)
Beijing	4.00	Xuzhou	3.50	Tai'an	2.00	Guilin	2.15
Tianjin	4.71	Changzhou	2.60	Weihai	2.20	Beihai	5.50
Shijiazhuang	4.00	Suzhou	3.20	Sunshine	4.00	Haikou	1.50
Tangshan	3.00	Nantong	2.90	Zhengzhou	2.30	Sanya	4.70
Qinghuangdao	3.00	Lianyungang	8.00	Kaifeng	3.20	Chongqing	11.73
Handan	3.00	Yangzhou	3.60	Luoyang	2.00	Chengdu	16.20
Baoding	3.70	Hangzhou	8.10	Pingdingshan	4.60	Panzhihua	4.40
Taiyuan	3.00	Ningbo	1.30	Anyang	1.80	Luzhou	3.50
Datong	1.50	Wenzhou	3.50	Jiaozuo	7.50	Mianyang	4.00
Yangquan	3.00	Jiaxing	2.00	Wuhan	16.90	Yibin	5.00
Changzhi	2.50	Huzhou	2.30	Yichang	20.36	Guiyang	9.00
Linfen	2.50	Shaoxing	3.70	Jingzhou	6.00	Zunyi	4.70
Hohhot	5.00	Taizhou	4.00	Changsha	4.28	Kunming	19.00
Baotou	4.00	Hefei	4.00	Zhuzhou	3.40	Qujing	4.00
Chifeng	5.00	Wuhu	3.00	Xiangtan	3.80	Xi'an	3.60
Shenyang	4.60	Ma'anshan	2.00	Yueyang	3.00	Tongchuan	2.80
Dalian	2.98	Fuzhou	4.90	Changde	3.60	Baoji	4.90
Anshan	8.00	Xiamen	8.00	Zhangjiajie	2.70	Xianyang	0.70
Fushun	10.00	Quanzhou	4.00	Guangzhou	13.00	Yan'an	1.40
Changchun	10.00	Nanchang	5.60	Shaoguan	9.00	Lanzhou	6.00
Jilin	4.00	Jiujiang	3.40	Shenzhen	4.50	Jinchang	5.60
Harbin	5.00	Ji'nan	4.00	Zhuhai	8.00	Xining	20.00
Qigihar	5.00	Qingdao	4.00	Shantou	3.00	Yinchuan	4.50
Daqing	1.70	Zibo	2.50	Foshan	5.00	Shizuishan	1.50
Mudanjiang	4.70	Zaozhuang	2.00	Zhanjiang	6.50	Urumqi	0.40
Shanghai	11.10	Yantai	4.00	Zhongshan	10.00	Karamay	1.70
Nanjing	3.80	Weifang	4.00	Nanning	18.00		
Wuxi	2.50	Jining	3.00	Liuzhou	8.00		

Table A5-10 Forest coverage rate in 110 cities

A5-10

City	Urban forest coverage rate (%)	City	Urban forest coverage rate (%)	City	Urban forest coverage rate (%)	City	Urban forest coverage rate (%)
Beijing	21.50	Xuzhou	27.70	Tai'an	34.40	Guilin	58.00
Tianjin	8.30	Changzhou	20.00	Weihai	38.20	Beihai	28.89
Shijiazhuang	26.92	Suzhou	20.30	Sunshine	34.20	Haikou	37.00
Tangshan	24.35	Nantong	16.80	Zhengzhou	23.30	Sanya	64.40
Qinghuangdao	41.98	Lianyungang	18.05	Kaifeng	15.60	Chongqing	34.00
Handan	11.60	Yangzhou	17.60	Luoyang	32.82	Chengdu	36.80
Baoding	20.00	Hangzhou	64.35	Pingdingshan	23.00	Panzhihua	58.97
Taiyuan	15.00	Ningbo	50.50	Anyang	15.00	Luzhou	38.40
Datong	8.00	Wenzhou	59.60	Jiaozuo	13.50	Mianyang	45.70
Yangquan	25.25	Jiaxing	16.80	Wuhan	25.12	Yibin	39.01
Changzhi	26.90	Huzhou	50.90	Yichang	55.30	Guiyang	39.93
Linfen	29.50	Shaoxing	54.03	Jingzhou	15.00	Zunyi	40.00
Hohhot	20.00	Taizhou	62.20	Changsha	52.25	Kunming	45.00
Baotou	10.00	Hefei	15.80	Zhuzhou	60.63	Qujing	36.20
Chifeng	24.00	Wuhu	20.00	Xiangtan	45.01	Xi'an	44.99
Shenyang	24.00	Ma'anshan	8.30	Yueyang	36.00	Tongchuan	43.90
Dalian	41.50	Fuzhou	54.90	Changde	43.80	Baoji	50.00
Anshan	46.40	Xiamen	43.00	Zhangjiajie	67.50	Xianyang	23.50
Fushun	66.20	Quanzhou	58.70	Guangzhou	38.30	Yan'an	37.00
Changchun	15.00	Nanchang	16.10	Shaoguan	66.10	Lanzhou	12.21
Jilin	54.00	Jiujiang	52.00	Shenzhen	45.00	Jinchang	20.82
Harbin	44.00	Ji'nan	27.80	Zhuhai	30.00	Xining	27.00
Qigihar	13.37	Qingdao	35.37	Shantou	30.00	Yinchuan	12.00
Daqing	10.40	Zibo	32.40	Foshan	18.00	Shizuishan	11.00
Mudanjiang	62.30	Zaozhuang	31.00	Zhanjiang	27.70	Urumqi	2.94
Shanghai	11.60	Yantai	38.00	Zhongshan	17.00	Karamay	2.50
Nanjing	23.00	Weifang	30.60	Nanning	42.15		
Wuxi	22.00	Jining	26.00	Liuzhou	55.00		

Table A5-11 Per-capital riding number per public electric vehicle in 110 cities

A5-11

City	Public electric vehicle transportation(time/(10000 capita•car))	City	Public electric vehicle transportation(time/(10000 capita•car))	City	Public electric vehicle transportation(time/(10000 capita•car))	City	Public electric vehicle transportation(time/(10000 capita•car))
Beijing	152	Xuzhou	328	Tai'an	700	Guilin	1288
Tianjin	157	Changzhou	695	Weihai	908	Beihai	1931
Shijiazhuang	354	Suzhou	473	Sunshine	919	Haikou	23735
Tangshan	309	Nantong	472	Zhengzhou	374	Sanya	1536
Qinghuangdao	1189	Lianyungang	981	Kaifeng	851	Chongqing	126
Handan	316	Yangzhou	623	Luoyang	875	Chengdu	518
Baoding	546	Hangzhou	348	Pingdingshan	412	Panzhihua	2362
Taiyuan	666	Ningbo	748	Anyang	810	Luzhou	1279
Datong	1208	Wenzhou	318	Jiaozuo	1215	Mianyang	1548
Yangquan	3002	Jiaxing	708	Wuhan	377	Yibin	2469
Changzhi	1530	Huzhou	606	Yichang	705	Guiyang	948
Linfen	758	Shaoxing	543	Jingzhou	950	Zunyi	1202
Hohhot	1067	Taizhou	242	Changsha	836	Kunming	439
Baotou	793	Hefei	1030	Zhuzhou	1098	Qujing	930
Chifeng	1155	Wuhu	1373	Xiangtan	809	Xi'an	637
Shenyang	364	Ma'anshan	2870	Yueyang	797	Tongchuan	1700
Dalian	497	Fuzhou	531	Changde	820	Baoji	2305
Anshan	941	Xiamen	2039	Zhangjiajie	2184	Xianyang	1807
Fushun	1920	Quanzhou	382	Guangzhou	364	Yan'an	8283
Changchun	500	Nanchang	529	Shaoguan	1492	Lanzhou	1281
Jilin	967	Jiujiang	807	Shenzhen	802	Jinchang	2873
Harbin	395	Ji'nan	420	Zhuhai	2329	Xining	2681
Qigihar	297	Qingdao	378	Shantou	393	Yinchuan	1514
Daqing	663	Zibo	356	Foshan	141	Shizuishan	1858
Mudanjiang	981	Zaozhuang	483	Zhanjiang	359	Urumqi	1128
Shanghai	96	Yantai	539	Zhongshan	1374	Karamay	2917
Nanjing	667	Weifang	562	Nanning	725		
Wuxi	237	Jining	480	Liuzhou	1280		

Table A5-12 Energy consumption level (standard coal) per living area in 110 cities

A5-12

City	Energy consumption per residential building area (kg/m^2)	City	Energy consumption per residential building area (kg/m^2)	City	Energy consumption per residential building area (kg/m^2)	City	Energy consumption per residential building area (kg/m^2)
Beijing	8.731	Xuzhou	1.163	Tai'an	1.306	Guilin	1.697
Tianjin	3.811	Changzhou	4.424	Weihai	2.152	Beihai	3.039
Shijiazhuang	2.420	Suzhou	2.620	Sunshine	3.131	Haikou	3.901
Tangshan	3.493	Nantong	1.591	Zhengzhou	4.606	Sanya	6.896
Qinghuangdao	3.987	Lianyungang	1.567	Kaifeng	4.709	Chongqing	4.631
Handan	1.092	Yangzhou	1.925	Luoyang	1.697	Chengdu	8.495
Baoding	1.270	Hangzhou	6.548	Pingdingshan	0.715	Panzhihua	6.959
Taiyuan	3.756	Ningbo	6.843	Anyang	3.004	Luzhou	2.752
Datong	3.520	Wenzhou	2.556	Jiaozuo	3.784	Mianyang	3.007
Yangquan	10.417	Jiaxing	1.692	Wuhan	5.141	Yibin	2.120
Changzhi	1.149	Huzhou	1.641	Yichang	2.318	Guiyang	11.560
Linfen	4.732	Shaoxing	1.623	Jingzhou	2.503	Zunyi	2.012
Hohhot	3.560	Taizhou	0.826	Changsha	5.027	Kunming	9.121
Baotou	5.733	Hefei	8.113	Zhuzhou	2.686	Qujing	0.000
Chifeng	0.779	Wuhu	5.011	Xiangtan	2.377	Xi'an	6.499
Shenyang	4.566	Ma'anshan	3.202	Yueyang	1.579	Tongchuan	4.180
Dalian	7.514	Fuzhou	3.762	Changde	1.586	Baoji	2.881
Anshan	3.058	Xiamen	8.935	Zhangjiajie	0.458	Xianyang	1.706
Fushun	10.243	Quanzhou	0.922	Guangzhou	8.770	Yan'an	2.327
Changchun	3.467	Nanchang	2.382	Shaoguan	3.825	Lanzhou	4.034
Jilin	2.738	Jiujiang	2.252	Shenzhen	23.085	Jinchang	0.000
Harbin	2.678	Ji'nan	4.905	Zhuhai	5.366	Xining	4.923
Qigihar	2.770	Qingdao	3.791	Shantou	5.811	Yinchuan	11.732
Daqing	4.913	Zibo	7.404	Foshan	5.863	Shizuishan	3.129
Mudanjiang	1.531	Zaozhuang	2.774	Zhanjiang	2.732	Urumqi	10.914
Shanghai	7.307	Yantai	1.959	Zhongshan	12.435	Karamay	7.307
Nanjing	4.978	Weifang	1.506	Nanning	4.531		
Wuxi	5.153	Jining	1.251	Liuzhou	3.459		

Table A5-13 Industrial wastewater COD discharge intensity in 110 cities

A5-13

City	Water pollution discharge intensity (kg COD/10^4 yuan)	City	Water pollution discharge intensity (kg COD/10^4 yuan)	City	Water pollution discharge intensity (kg COD/10^4 yuan)	City	Water pollution discharge intensity (kg COD/10^4 yuan)
Beijing	0.125	Xuzhou	1.262	Tai'an	0.713	Guilin	2.177
Tianjin	0.418	Changzhou	1.828	Weihai	0.460	Beihai	23.153
Shijiazhuang	3.642	Suzhou	1.013	Sunshine	1.707	Haikou	0.275
Tangshan	1.817	Nantong	1.884	Zhengzhou	1.141	Sanya	0.123
Qinghuangdao	1.123	Lianyungang	0.800	Kaifeng	12.284	Chongqing	3.258
Handan	0.806	Yangzhou	1.801	Luoyang	0.423	Chengdu	3.728
Baoding	3.025	Hangzhou	3.125	Pingdingshan	1.361	Panzhihua	1.147
Taiyuan	0.290	Ningbo	0.419	Anyang	4.608	Luzhou	7.869
Datong	6.089	Wenzhou	3.422	Jiaozuo	6.734	Mianyang	0.941
Yangquan	0.302	Jiaxing	1.260	Wuhan	0.876	Yibin	6.801
Changzhi	0.940	Huzhou	1.329	Yichang	1.565	Guiyang	0.272
Linfen	1.059	Shaoxing	2.260	Jingzhou	10.151	Zunyi	0.725
Hohhot	0.843	Taizhou	0.550	Changsha	0.503	Kunming	0.317
Baotou	0.410	Hefei	0.308	Zhuzhou	2.867	Qujing	0.813
Chifeng	2.491	Wuhu	1.711	Xiangtan	5.273	Xi'an	3.236
Shenyang	0.295	Ma'anshan	0.668	Yueyang	5.692	Tongchuan	0.534
Dalian	0.403	Fuzhou	0.502	Changde	12.052	Baoji	4.211
Anshan	0.511	Xiamen	0.146	Zhangjiajie	6.081	Xianyang	3.704
Fushun	0.452	Quanzhou	2.335	Guangzhou	0.445	Yan'an	1.446
Changchun	0.841	Nanchang	3.221	Shaoguan	1.567	Lanzhou	0.306
Jilin	2.303	Jiujiang	1.180	Shenzhen	0.147	Jinchang	1.582
Harbin	1.223	Ji'nan	0.352	Zhuhai	0.364	Xining	3.556
Qigihar	4.110	Qingdao	0.277	Shantou	1.577	Yinchuan	3.803
Daqing	0.449	Zibo	1.100	Foshan	0.999	Shizuishan	3.341
Mudanjiang	10.058	Zaozhuang	3.614	Zhanjiang	2.180	Urumqi	1.048
Shanghai	0.204	Yantai	0.656	Zhongshan	1.128	Karamay	0.181
Nanjing	0.613	Weifang	0.864	Nanning	22.612		
Wuxi	0.787	Jining	0.661	Liuzhou	4.699		

Table A5-14 Industrial waste gas SO_2 emission intensity in 110 cities

A5-14

City	Air pollution emission intensity (kg $SO_2/10^4$ yuan)	City	Air pollution emission intensity (kg $SO_2/10^4$ yuan)	City	Air pollution emission intensity (kg $SO_2/10^4$ yuan)	City	Air pollution emission intensity (kg $SO_2/10^4$ yuan)
Beijing	1.466	Xuzhou	13.11	Tai'an	11.469	Guilin	13.518
Tianjin	3.148	Changzhou	4.279	Weihai	8.665	Beihai	59.8
Shijiazhuang	11.671	Suzhou	4.012	Sunshine	6.479	Haikou	0.15
Tangshan	7.784	Nantong	6.674	Zhengzhou	14.45	Sanya	0.28
Qinghuangdao	8.627	Lianyungang	8.069	Kaifeng	14.647	Chongqing	20.188
Handan	11.075	Yangzhou	8.664	Luoyang	22.411	Chengdu	6.374
Baoding	9.179	Hangzhou	3.207	Pingdingshan	16.353	Panzhihua	21.753
Taiyuan	6.145	Ningbo	3.442	Anyang	11.792	Luzhou	26.133
Datong	22.119	Wenzhou	6.675	Jiaozuo	15.839	Mianyang	6.603
Yangquan	46.119	Jiaxing	8.624	Wuhan	4.24	Yibin	20.002
Changzhi	10.869	Huzhou	8.507	Yichang	4.937	Guiyang	11.108
Linfen	9.121	Shaoxing	3.689	Jingzhou	14.901	Zunyi	24.057
Hohhot	15.102	Taizhou	6.334	Changsha	6.128	Kunming	7.124
Baotou	15.392	Hefei	2.906	Zhuzhou	12.985	Qujing	13.526
Chifeng	86.530	Wuhu	7.186	Xiangtan	15.627	Xi'an	7.433
Shenyang	2.977	Ma'anshan	5.815	Yueyang	11.887	Tongchuan	14.661
Dalian	4.246	Fuzhou	7.805	Changde	15.378	Baoji	12.715
Anshan	8.118	Xiamen	2.207	Zhangjiajie	35.072	Xianyang	21.82
Fushun	6.491	Quanzhou	4.854	Guangzhou	2.314	Yan'an	3.868
Changchun	2.478	Nanchang	3.849	Shaoguan	14.582	Lanzhou	9.977
Jilin	4.727	Jiujiang	19.178	Shenzhen	1.095	Jinchang	20.657
Harbin	3.999	Ji'nan	4.119	Zhuhai	1.867	Xining	19.339
Qigihar	15.094	Qingdao	2.403	Shantou	9.954	Yinchuan	4.806
Daqing	1.827	Zibo	8.726	Foshan	6.064	Shizuishan	49.86
Mudanjiang	27.838	Zaozhuang	20.886	Zhanjiang	7.494	Urumqi	16.468
Shanghai	2.195	Yantai	4.851	Zhongshan	5.164	Karamay	3.109
Nanjing	3.258	Weifang	5.437	Nanning	17.258		
Wuxi	2.618	Jining	8.259	Liuzhou	5.653		

Appendix VI Dimensional and Dimensionless Data of Low-carbon Development Indicators for 110 Selected Chinese Cities in 2009

Table A6-1 Rankings of 110 cities in terms of total GDP (constant price in 2005)

City	Ranking	GDP(billion yuan)	Relative value	City	Ranking	GDP(billion yuan)	Relative value
Shanghai	1	12855.54	1.000	jilin	56	1015.14	0.071
Beijing	2	9592.95	0.744	Zhuhai	57	997.89	0.070
Guangzhou	3	7544.98	0.583	Zaozhuang	58	968.33	0.067
Shenzhen	4	7285.19	0.563	Yueyang	59	927.48	0.064
Suzhou	5	6071.69	0.468	Zhanjiang	60	923.14	0.064
Tianjin	6	5682.06	0.437	Huzhou	61	922.33	0.064
Chongqing	7	4552.02	0.349	Changde	62	912.71	0.063
Hangzhou	8	4244.12	0.324	Shantou	63	906.93	0.063
Wuxi	9	4190.91	0.320	Yichang	64	899.13	0.062
Qingdao	10	4096.36	0.313	Jiaozuo	65	890.93	0.061
Foshan	11	3913.97	0.299	Urumqi	66	862.37	0.059
Nanjing	12	3597.84	0.274	Anyang	67	862.31	0.059
Ningbo	13	3505.89	0.267	Pingdingshan	68	859.02	0.059
Chengdu	14	3487.12	0.265	Liuzhou	69	806.13	0.055
Dalian	15	3428.69	0.260	Lanzhou	70	796.78	0.054
Wuhan	16	3418.50	0.260	Guilin	71	791.02	0.054
Shenyang	17	3323.54	0.252	Guiyang	72	789.60	0.053
Yantai	18	3117.95	0.236	Zhuzhou	73	769.18	0.052
Tangshan	19	3022.29	0.229	Qinghuangdao	74	702.40	0.047
Ji'nan	20	2840.73	0.214	Lianyungang	75	684.09	0.045
Harbin	21	2669.22	0.201	Linfen	76	679.29	0.045
Shijiazhuang	22	2545.97	0.191	Sunshine	77	670.48	0.044
Zhengzhou	23	2504.95	0.188	Mianyang	78	656.45	0.043
Quanzhou	24	2473.81	0.186	Qujing	79	646.94	0.042
Changchun	25	2389.48	0.179	Xianyang	80	640.22	0.042
Changsha	26	2328.80	0.174	Jiujiang	81	626.00	0.041
Wenzhou	27	2265.01	0.169	Wuhu	82	623.74	0.040
Weifang	28	2246.05	0.168	Baoji	83	622.44	0.040
Nantong	29	2242.76	0.167	Qigihar	84	610.52	0.039
Zibo	30	2172.75	0.162	Fushun	85	604.86	0.039
Fuzhou	31	2155.67	0.161	Kaifeng	86	591.31	0.038
Shaoxing	32	2092.92	0.156	Ma'anshan	87	589.58	0.038
Anshan	33	1974.76	0.146	Zunyi	88	578.87	0.037
Changzhou	34	1950.39	0.145	Chifeng	89	578.52	0.037
Jining	35	1940.86	0.144	Yan'an	90	571.02	0.036
Daqing	36	1920.66	0.142	Changzhi	91	560.01	0.035
Xi'an	37	1901.40	0.141	Yibin	92	553.04	0.035
Xuzhou	38	1826.00	0.135	Xiangtan	93	545.63	0.034
Taizhou	39	1793.36	0.132	Jingzhou	94	543.15	0.034
Weihai	40	1752.21	0.129	Karamay	95	516.07	0.032

(Contd.)

City	Ranking	GDP(billion yuan)	Relative value	City	Ranking	GDP(billion yuan)	Relative value
Luoyang	41	1710.91	0.126	Shaoguan	96	500.99	0.031
Handan	42	1694.49	0.124	Datong	97	479.10	0.029
Jiaxing	43	1664.53	0.122	Mudanjiang	98	442.71	0.026
Nanchang	44	1541.93	0.112	Luzhou	99	429.15	0.025
Xiamen	45	1522.70	0.111	Haikou	100	422.93	0.025
Kunming	46	1503.19	0.109	Yinchuan	101	421.82	0.025
Baoding	47	1474.14	0.107	Panzhihua	102	371.95	0.021
Baotou	48	1448.48	0.105	Xining	103	312.34	0.016
Yangzhou	49	1393.22	0.101	Beihai	104	291.11	0.014
Hefei	50	1388.21	0.100	Yangquan	105	277.32	0.013
Zhongshan	51	1313.47	0.095	Jinchang	106	168.96	0.005
Tai'an	52	1312.53	0.094	Shizuishan	107	163.27	0.004
Taiyuan	53	1253.09	0.090	Zhangjiajie	108	160.88	0.004
Hohhot	54	1177.36	0.084	Sanya	109	121.18	0.001
Nanning	55	1116.30	0.079	Tongchuan	110	107.94	0.000

Table A6-2 Rankings of 110 cities in terms of energy consumption per unit of GDP

City	Ranking	GDP energy comsumption (t ce/10000 yuan)	Relative value	City	Ranking	GDP energy comsumption (t ce/10000 yuan)	Relative value
Shenzhen	1	0.544	1.000	Xuzhou	56	1.227	0.897
Taizhou	2	0.560	0.998	Baoji	57	1.253	0.894
Xiamen	3	0.600	0.992	Chongqing	58	1.267	0.891
Zhuhai	4	0.603	0.991	Tai'an	59	1.281	0.889
Shantou	5	0.632	0.987	Kunming	60	1.282	0.889
Yan'an	6	0.642	0.985	Qinghuangdao	61	1.299	0.887
Wenzhou	7	0.650	0.984	Harbin	62	1.316	0.884
Zhanjiang	8	0.659	0.983	Mudanjiang	63	1.335	0.881
Beijing	9	0.662	0.982	Sunshine	64	1.342	0.880
Zhongshan	10	0.673	0.981	Luoyang	65	1.364	0.877
Fuzhou	11	0.676	0.980	Yibin	66	1.364	0.877
Guangzhou	12	0.680	0.980	Yueyang	67	1.371	0.876
Changchun	13	0.697	0.977	Jining	68	1.380	0.874
Nantong	14	0.730	0.972	Daqing	69	1.382	0.874
Yangzhou	15	0.739	0.971	Zhuzhou	70	1.390	0.873
Hangzhou	16	0.750	0.969	Wuhu	71	1.400	0.872
Haikou	17	0.779	0.965	Mianyang	72	1.591	0.843
Wuxi	18	0.799	0.962	Qigihar	73	1.594	0.842
Shanghai	19	0.801	0.961	Luzhou	74	1.596	0.842
Sanya	20	0.801	0.961	Shijiazhuang	75	1.622	0.838
Foshan	21	0.802	0.961	Hohhot	76	1.677	0.830
Harm	22	0.812	0.960	Jingzhou	77	1.690	0.828
Quanzhou	23	0.814	0.959	Anshan	78	1.757	0.818

(Contd.)

City	Ranking	GDP energy comsumption (t ce/10000 yuan)	Relative value	City	Ranking	GDP energy comsumption (t ce/10000 yuan)	Relative value
Yantai	24	0.814	0.959	Karamay	79	1.769	0.816
Qingdao	25	0.821	0.958	Zaozhuang	80	1.785	0.814
Lianyungang	26	0.833	0.957	Zibo	81	1.791	0.813
Nanning	27	0.838	0.956	Jinchang	82	1.806	0.811
Ningbo	28	0.840	0.956	Jilin	83	1.815	0.809
Jiaxing	29	0.860	0.953	Xiangtan	84	1.816	0.809
Xi'an	30	0.873	0.951	Shaoguan	85	1.819	0.809
Changsha	31	0.886	0.949	Guiyang	86	1.820	0.808
Shaoxing	32	0.900	0.947	Chifeng	87	1.821	0.808
Chengdu	33	0.905	0.946	Qujing	88	1.833	0.807
Nanchang	34	0.906	0.946	Zunyi	89	1.900	0.796
Suzhou	35	0.906	0.946	Tongchuan	90	1.930	0.972
Changzhou	36	0.934	0.941	Liuzhou	91	1.951	0.789
Tianjin	37	0.947	0.940	Lanzhou	92	1.987	0.783
Zhangjiajie	38	0.947	0.940	Jiaozuo	93	1.991	0.783
Hefei	39	0.980	0.935	Pingdingshan	94	2.009	0.780
Changde	40	0.997	0.932	Datong	95	2.100	0.766
Huzhou	41	1.020	0.929	Yinchuan	96	2.154	0.758
Beihai	42	1.021	0.928	Taiyuan	97	2.230	0.747
Dalian	43	1.031	0.927	Ma'anshan	98	2.249	0.744
Shenyang	44	1.046	0.925	Handan	99	2.262	0.742
Baoding	45	1.060	0.923	Yangquan	100	2.340	0.730
Jiujiang	46	1.068	0.921	Baotou	101	2.387	0.723
Guilin	47	1.096	0.917	Anyang	102	2.419	0.719
Ji'nan	48	1.101	0.916	Urumqi	103	2.503	0.706
Kaifeng	49	1.117	0.914	Tangshan	104	2.600	0.691
Yichang	50	1.160	0.908	Changzhi	105	2.930	0.642
Nanjing	51	1.178	0.905	Fushun	106	3.084	0.619
Weifang	52	1.187	0.904	Panzhihua	107	3.299	0.586
Zhengzhou	53	1.187	0.903	Xining	108	3.705	0.526
Wuhan	54	1.190	0.903	Linfen	109	3.760	0.517
Xianyang	55	1.207	0.900	Shizuishan	110	7.206	0.000

Table A6-3 Rankings of 110 cities in terms of CO_2 productivity

City	Ranking	Productivity level of CO_2 (10000 yuan GDP/t CO_2)	Relative value	City	Ranking	Productivity level of CO_2 (10000 yuan GDP/t CO_2)	Relative value
Shenzhen	1	0.795	1.000	Zhengzhou	56	0.329	0.371
Taizhou	2	0.736	0.920	Xuzhou	57	0.321	0.361
Zhuhai	3	0.714	0.890	Chongqing	58	0.320	0.359
Xiamen	4	0.698	0.868	Baoji	59	0.320	0.359

(Contd.)

City	Ranking	Productivity level of CO_2 (10000 yuan GDP/t CO_2)	Relative value	City	Ranking	Productivity level of CO_2 (10000 yuan GDP/t CO_2)	Relative value
Zhanjiang	5	0.699	0.841	Daqing	60	0.319	0.358
Yan'an	6	0.677	0.841	Yibin	61	0.313	0.350
Shantou	7	0.665	0.842	Tai'an	62	0.307	0.343
Guangzhou	8	0.635	0.784	Qinghuangdao	63	0.307	0.342
Wenzhou	9	0.633	0.781	Harbin	64	0.308	0.342
Beijing	10	0.630	0.777	Mudanjiang	65	0.302	0.335
Zhongshan	11	0.623	0.768	Sunshine	66	0.292	0.322
Fuzhou	12	0.594	0.728	Yueyang	67	0.288	0.316
Shanghai	13	0.581	0.712	Luoyang	68	0.287	0.316
Changchun	14	0.575	0.703	Zhuzhou	69	0.284	0.311
Sanya	15	0.544	0.661	Jining	70	0.282	0.309
Yangzhou	16	0.543	0.660	Wuhu	71	0.279	0.305
Nantong	17	0.542	0.659	Luzhou	72	0.272	0.295
Hangzhou	18	0.538	0.654	Mianyang	73	0.259	0.277
Foshan	19	0.533	0.647	Qigihar	74	0.251	0.266
Haikou	20	0.532	0.646	Jingzhou	75	0.248	0.263
Ningbo	21	0.522	0.632	Karamay	76	0.248	0.263
Nanning	22	0.512	0.619	Shijiazhuang	77	0.242	0.255
Quanzhou	23	0.509	0.614	Hohhot	78	0.239	0.250
Qingdao	24	0.507	0.611	Jilin	79	0.232	0.241
Chengdu	25	0.504	0.607	Zibo	80	0.228	0.235
Wuxi	26	0.500	0.602	Shaoguan	81	0.227	0.234
Weihai	27	0.494	0.594	Lanzhou	82	0.225	0.231
Yantai	28	0.491	0.590	Guiyang	83	0.225	0.231
Xi'an	29	0.485	0.581	Qujing	84	0.221	0.226
Lianyungang	30	0.481	0.576	Anshan	85	0.219	0.224
Changsha	31	0.457	0.544	Zaozhuang	86	0.219	0.224
Jiaxing	32	0.456	0.543	Jinchang	87	0.219	0.224
Tianjin	33	0.456	0.543	Xiangtan	88	0.216	0.220
Shaoxing	34	0.451	0.536	Chifeng	89	0.213	0.216
Suzhou	35	0.446	0.530	Liuzhou	90	0.210	0.212
Nanchang	36	0.437	0.517	Zunyi	91	0.209	0.210
Changzhou	37	0.425	0.501	Tongchuan	92	0.202	0.200
Zhangjiajie	38	0.424	0.499	Jiaozuo	93	0.197	0.193
Dalian	39	0.419	0.493	Pingdingshan	94	0.194	0.190
Hefei	40	0.404	0.473	Yinchuan	95	0.191	0.186
Beihai	41	0.401	0.468	Datong	96	0.179	0.170
Changde	42	0.397	0.463	Taiyuan	97	0.173	0.161
Huzhou	43	0.388	0.451	Ma'anshan	98	0.171	0.158
Guilin	44	0.385	0.447	Yangquan	99	0.170	0.158
Shenyang	45	0.381	0.441	Handan	100	0.170	0.157
Jiujiang	46	0.374	0.433	Urumqi	101	0.165	0.151
Ji'nan	47	0.373	0.431	Anyang	102	0.165	0.151

(Contd.)

City	Ranking	Productivity level of CO_2 (10000 yuan GDP/t CO_2)	Relative value	City	Ranking	Productivity level of CO_2 (10000 yuan GDP/t CO_2)	Relative value
Nanjing	48	0.372	0.430	Baotou	103	0.162	0.146
Yichang	49	0.370	0.427	Tangshan	104	0.148	0.128
Baoding	50	0.370	0.427	Fushun	105	0.131	0.105
Wuhan	51	0.366	0.421	Changzhi	106	0.130	0.103
Kaifeng	52	0.352	0.403	Xining	107	0.129	0.102
Kunming	53	0.347	0.396	Panzhihua	108	0.119	0.089
Weifang	54	0.334	0.379	Linfen	109	0.102	0.066
Xianyang	55	0.332	0.375	Shizuishan	110	0.053	0.000

Table A6-4 Rankings of 110 cities in terms of carbon emission intensity from energy consumption

City	Ranking	Carbon emission intensity of energy consumption (t CO_2/t ce)	Relative value	City	Ranking	Carbon emission intensity of energy consumption (t CO_2/t ce)	Relative value
Xining	1	2.092	1.000	Fuzhou	56	2.492	0.291
Shanghai	2	2.147	0.902	Zhangjiajie	57	2.493	0.29
Chengdu	3	2.193	0.820	Weihai	58	2.493	0.29
Lanzhou	4	2.234	0.749	Changchun	59	2.495	0.287
Zhanjiang	5	2.240	0.738	Baoji	60	2.497	0.283
Kunming	6	2.249	0.722	Hohhot	61	2.498	0.281
Daqing	7	2.271	0.682	Lianyungang	62	2.498	0.281
Karamay	8	2.278	0.670	Xianyang	63	2.499	0.279
Ningbo	9	2.280	0.667	Yantai	64	2.499	0.278
Nanjing	10	2.280	0.667	Jiujiang	65	2.501	0.276
Sanya	11	2.295	0.639	Anyang	66	2.503	0.273
Wuhan	12	2.296	0.638	Wuxi	67	2.503	0.272
Yan'an	13	2.300	0.631	Qigihar	68	2.504	0.269
Luzhou	14	2.305	0.622	Qinghuangdao	69	2.507	0.265
Shenzhen	15	2.311	0.612	Shenyang	70	2.511	0.257
Tianjin	16	2.316	0.603	Yangquan	71	2.512	0.257
Dalian	17	2.317	0.602	Zunyi	72	2.517	0.246
Guangzhou	18	2.317	0.601	Changzhou	73	2.518	0.246
Zhuhai	19	2.323	0.591	Weifang	74	2.522	0.238
Yichang	20	2.328	0.582	Hefei	75	2.523	0.236
Nanning	21	2.329	0.580	Changde	76	2.526	0.232
Foshan	22	2.339	0.562	Nantong	77	2.527	0.229
Yibin	23	2.345	0.553	Nanchang	78	2.529	0.227
Xi'an	24	2.364	0.519	Huzhou	79	2.529	0.226
Guilin	25	2.370	0.507	Jinchang	80	2.530	0.225
Jilin	26	2.372	0.504	Zhuzhou	81	2.531	0.223

(Contd.)

City	Ranking	Carbon emission intensity of energy consumption (t CO_2/t ce)	Relative value	City	Ranking	Carbon emission intensity of energy consumption (t CO_2/t ce)	Relative value
Shantou	27	2.380	0.490	Yueyang	82	2.533	0.218
Zhongshan	28	2.383	0.484	Panzhihua	83	2.538	0.210
Jingzhou	29	2.385	0.482	Tai'an	84	2.539	0.208
Xiamen	30	2.389	0.473	Xuzhou	85	2.541	0.205
Beijing	31	2.399	0.456	Kaifeng	86	2.541	0.204
Qingdao	32	2.404	0.447	Shijiazhuang	87	2.544	0.200
Haikou	33	2.412	0.434	Jiaxing	88	2.549	0.191
Quanzhou	34	2.413	0.431	Xiangtan	89	2.549	0.191
Urumqi	35	2.417	0.424	Sunshine	90	2.549	0.190
Taizhou	36	2.426	0.409	Baoding	91	2.551	0.187
Mianyang	37	2.426	0.408	Luoyang	92	2.552	0.185
Shaoguan	38	2.427	0.406	Zaozhuang	93	2.554	0.182
Wenzhou	39	2.430	0.402	Jiaozuo	94	2.554	0.181
Yinchuan	40	2.430	0.402	Wuhu	95	2.559	0.173
Ji'nan	41	2.433	0.395	Zhengzhou	96	2.562	0.167
Liuzhou	42	2.437	0.390	Jining	97	2.566	0.161
Beihai	43	2.444	0.376	Pingdingshan	98	2.568	0.156
Guiyang	44	2.447	0.371	Tongchuan	99	2.570	0.153
Zibo	45	2.451	0.365	Chifeng	100	2.576	0.144
Shaoxing	46	2.465	0.339	Baotou	101	2.591	0.116
Chongqing	47	2.468	0.334	Taiyuan	102	2.593	0.112
Changsha	48	2.469	0.332	Anshan	103	2.595	0.110
Suzhou	49	2.473	0.325	Tangshan	104	2.598	0.103
Qujing	50	2.474	0.323	Handan	105	2.603	0.095
Fushun	51	2.477	0.318	Ma'anshan	106	2.605	0.091
Harbin	52	2.477	0.318	Linfen	107	2.609	0.084
Hangzhou	53	2.478	0.317	Shizuishan	108	2.614	0.076
Mudanjiang	54	2.481	0.312	Changzhi	109	2.627	0.052
Yangzhou	55	2.491	0.294	Datong	110	2.657	0.000

Table A6-5 Rankings of 110 cities in terms of total energy consumption

City	Ranking	Total energy consumption (10000 t ce)	Relative value	City	Ranking	Total energy consumption (10000 t ce)	Relative value
Sanya	1	97.064	1.000	Baoding	56	1562.591	0.856
Zhangjiajie	2	162.356	0.989	Liuzhou	57	1572.535	0.855
Tongchuan	3	208.330	0.980	Lanzhou	58	1583.031	0.854
Beihai	4	297.349	0.980	Nantong	59	1637.213	0.849
Jinchang	5	305.190	0.977	Changzhi	60	1640.842	0.849
Haikou	6	329.466	0.974	Xi'an	61	1659.918	0.847
Yan'an	7	366.596	0.954	Changchun	62	1666.640	0.846

(Contd.)

City	Ranking	Total energy consumption (10000 t ce)	Relative value	City	Ranking	Total energy consumption (10000 t ce)	Relative value
Lianyungang	8	569.851	0.953	Tai'an	63	1681.642	0.845
Shantou	9	573.178	0.952	Pingdingshan	64	1725.779	0.840
Mudanjiang	10	590.978	0.951	Zaozhuang	65	1728.815	0.840
Zhuhai	11	601.728	0.950	Jiaozuo	66	1773.850	0.836
Zhanjiang	12	608.346	0.946	Changzhou	67	1821.660	0.831
Yangquan	13	648.938	0.945	Jilin	68	1842.910	0.829
Kaifeng	14	660.492	0.944	Fushun	69	1865.507	0.827
Jiujiang	15	668.770	0.942	Shaoxing	70	1883.630	0.825
Luzhou	16	684.930	0.936	Kunming	71	1927.089	0.821
Yibin	17	754.341	0.934	Hohhot	72	1974.681	0.816
Xianyang	18	772.748	0.933	Quanzhou	73	2013.683	0.812
Baoji	19	779.918	0.924	Changsha	74	2063.315	0.807
Guilin	20	867.212	0.924	Anyang	75	2085.940	0.805
Wuhu	21	873.240	0.923	Urumqi	76	2158.323	0.798
Zhongshan	22	883.968	0.921	Xuzhou	77	2240.504	0.790
Sunshine	23	900.042	0.920	Luoyang	78	2333.678	0.781
Yinchuan	24	908.440	0.920	Yantai	79	2539.447	0.761
Changde	25	909.973	0.920	Linfen	80	2554.146	0.759
Shaoguan	26	911.296	0.920	Daqing	81	2654.154	0.749
Qinghuangdao	27	912.416	0.920	Weifang	82	2664.958	0.748
Karamay	28	913.169	0.920	Jining	83	2678.720	0.747
Xiamen	29	913.618	0.920	Taiyuan	84	2794.400	0.736
Jingzhou	30	917.924	0.920	Ningbo	85	2994.951	0.721
Nanning	31	935.551	0.918	Zhengzhou	86	2973.377	0.718
Huzhou	32	940.773	0.917	Ji'nan	87	3128.257	0.703
Qigihar	33	973.103	0.914	Foshan	88	3139.003	0.702
Xiangtan	34	990.859	0.912	Chengdu	89	3155.842	0.700
Taizhou	35	1004.281	0.911	Hangzhou	90	3183.092	0.697
Datong	36	1006.112	0.911	Wuxi	91	3348.535	0.681
Yangzhou	37	1029.587	0.909	Qingdao	92	3363.554	0.680
Yichang	38	1042.989	0.907	Baotou	93	3457.160	0.671
Mianyang	39	1044.412	0.907	Anshan	94	3469.652	0.669
Chifeng	40	1053.436	0.906	Shenyang	95	3477.750	0.669
Zhuzhou	41	1069.155	0.905	Harbin	96	3513.226	0.665
Zunyi	42	1099.850	0.902	Dalian	97	2534.326	0.663
Xining	43	1157.125	0.896	Handan	98	3832.943	0.634
Shizuishan	44	1176.416	0.894	Zibo	99	3891.717	0.628
Qujing	45	1185.845	0.893	Shenzhen	100	3963.141	0.621
Panzhihua	46	1227.069	0.889	Wuhan	101	4068.011	0.611
Yueyang	47	1271.579	0.885	Shijiazhuang	102	4129.566	0.605
Ma'anshan	48	1325.788	0.880	Nanjing	103	4238.252	0.594
Hefei	49	1360.444	0.876	Guangzhou	104	5130.585	0.507
Nanchang	50	1396.554	0.873	Tianjin	105	5380.915	0.482
Weihai	51	1423.547	0.870	Suzhou	106	5500.948	0.470

(Contd.)

City	Ranking	Total energy consumption (10000 t ce)	Relative value	City	Ranking	Total energy consumption (10000 t ce)	Relative value
Jiaxing	52	1431.497	0.869	Chongqing	107	5767.415	0.444
Guiyang	53	1437.070	0.869	Beijing	108	6350.531	0.387
Fuzhou	54	1457.236	0.867	Tangshan	109	7857.950	0.239
Wenzhou	55	1472.259	0.865	Shanghai	110	10297.288	0.000

Table A6-6 Rankings of 110 cities in terms of per-capita CO_2 emission

City	Ranking	CO_2 emissions per capita (t CO_2/capita)	Relative value	City	Ranking	CO_2 emissions per capita (t CO_2/capita)	Relative value
Zhanjiang	1	1.974	1.000	Harbin	56	8.641	0.832
Zhangjiajie	2	2.317	0.991	Xiamen	57	8.766	0.829
Shantou	3	2.693	0.982	Pingdingshan	58	8.846	0.827
Lianyungang	4	2.878	0.977	Guiyang	59	8.928	0.825
Nanning	5	3.106	0.971	Beijing	60	8.989	0.823
Yibin	6	3.332	0.966	Wuhu	61	9.034	0.822
Jingzhou	7	3.390	0.964	Yantai	62	9.043	0.822
Baoding	8	3.491	0.962	Shaoxing	63	9.149	0.819
Taizhou	9	3.511	0.961	Luoyang	64	9.276	0.816
Jiujiang	10	3.517	0.961	Ningbo	65	9.337	0.814
Kaifeng	11	3.576	0.960	Zhuhai	66	9.437	0.812
Guilin	12	3.584	0.959	Qingdao	67	9.562	0.809
Luzhou	13	3.668	0.957	Jilin	68	9.903	0.800
Zunyi	14	3.684	0.957	Anyang	69	10.012	0.797
Changde	15	3.716	0.956	Zhengzhou	70	10.245	0.791
Yan'an	16	3.932	0.951	Liuzhou	71	10.333	0.789
Wenzhou	17	3.937	0.950	Lanzhou	72	10.349	0.789
Sanya	18	3.943	0.950	Changzhou	73	10.407	0.787
Qigihar	19	4.282	0.942	Wuhan	74	10.426	0.787
Haikou	20	4.330	0.941	Shenzhen	75	10.446	0.786
Beihai	21	4.608	0.934	Tianjin	76	10.598	0.783
Xi'an	22	4.685	0.932	Shijiazhuang	77	10.668	0.781
Mianyang	23	4.687	0.932	Handan	78	10.750	0.779
Chongqing	24	5.014	0.923	Xining	79	11.113	0.770
Qujing	25	5.074	0.922	Shenyang	80	11.255	0.766
Nantong	26	5.087	0.921	Ji'nan	81	11.487	0.760
Baoji	27	5.176	0.919	Suzhou	82	11.520	0.759
Mudanjiang	28	5.234	0.918	Guangzhou	83	11.675	0.755
Yangzhou	29	5.241	0.918	Shanghai	84	11.707	0.755
Fuzhou	30	5.318	0.916	Zaozhuang	85	12.045	0.746
Changchun	31	5.384	0.914	Foshan	86	12.334	0.739
Chengdu	32	5.448	0.912	Yangquan	87	12.351	0.738
Yueyang	33	5.907	0.901	Jiaozuo	88	12.542	0.733

(Contd.)

City	Ranking	CO_2 emissions per capita (t CO_2/capita)	Relative value	City	Ranking	CO_2 emissions per capita (t CO_2/capita)	Relative value
Yichang	34	6.011	0.898	Yinchuan	89	12.583	0.732
Chifeng	35	6.236	0.893	Weihai	90	12.647	0.731
Quanzhou	36	6.238	0.892	Nanjing	91	12.732	0.729
Tongchuan	37	6.380	0.889	Xianyang	92	13.038	0.721
Hefei	38	6.465	0.887	Changzhi	93	13.131	0.719
Shaoguan	39	6.467	0.887	Dalian	94	13.357	0.713
Jiaxing	40	6.505	0.886	Wuxi	95	13.725	0.704
Xuzhou	41	6.550	0.885	Jinchang	96	15.872	0.650
Kunming	42	6.946	0.875	Linfen	97	15.874	0.649
Zhuzhou	43	7.294	0.866	Hohhot	98	18.459	0.584
Weifang	44	7.557	0.859	Urumqi	99	19.109	0.568
Nanchang	45	7.652	0.857	Fushun	100	20.519	0.532
Qinghuangdao	46	7.735	0.855	Taiyuan	101	20.876	0.523
Changsha	47	7.735	0.855	Zibo	102	21.170	0.516
Tai'an	48	7.826	0.852	Daqing	103	21.746	0.501
Sunshine	49	8.063	0.846	Anshan	104	24.275	0.438
Huzhou	50	8.300	0.840	Ma'anshan	105	26.018	0.394
Zhongshan	51	8.391	0.838	Tangshan	106	27.470	0.357
Datong	52	8.409	0.838	Panzhihua	107	28.014	0.343
Hangzhou	53	8.442	0.837	Baotou	108	35.377	0.158
Xiangtan	54	8.591	0.833	Shizuishan	109	41.311	0.008
Jining	55	8.626	0.832	Karamay	110	41.628	0.000

Table A6-7 Rankings of 110 cities in terms of residents' low-carbon consumption coefficient

City	Ranking	The proportion of residents' low-carbon consumption expenditure (%)	Relative value	City	Ranking	The proportion of residents' low-carbon consumption expenditure (%)	Relative value
Beijing	1	26.759	1.000	Huzhou	56	19.869	0.574
Jilin	2	28.058	0.927	Ningbo	57	19.849	0.573
Nantong	3	26.662	0.866	Jining	58	19.718	0.567
Zibo	4	26.206	0.847	Tai'an	59	19.687	0.566
Xianyang	5	26.136	0.844	Yantai	60	19.671	0.565
Jinchang	6	25.628	0.822	Zhangjiajie	61	19.650	0.564
Mudanjiang	7	25.525	0.817	Chifeng	62	19.624	0.563
Qinghuangdao	8	25.133	0.801	Jingzhou	63	19.534	0.559
Shantou	9	24.789	0.786	Tongchuan	64	19.450	0.555
Luzhou	10	24.750	0.784	Daqing	65	19.442	0.555
Foshan	11	24.713	0.782	Zunyi	66	19.267	0.548
Zhuhai	12	24.687	0.781	Nanchang	67	19.057	0.539
Yangzhou	13	24.600	0.778	Zaozhuang	68	19.015	0.537

(Contd.)

City	Ranking	The proportion of residents' low-carbon consumption expenditure (%)	Relative value	City	Ranking	The proportion of residents' low-carbon consumption expenditure (%)	Relative value
Nanjing	14	24.500	0.773	Urumqi	69	18.930	0.533
Shaoguan	15	24.484	0.773	Jiaozuo	70	18.912	0.532
Taiyuan	16	24.030	0.753	Shenyang	71	18.862	0.530
Xi'an	17	23.575	0.733	Zhanjiang	72	18.800	0.527
Shaoxing	18	23.281	0.721	Shanghai	73	18.713	0.524
Anyang	19	23.222	0.718	Sunshine	74	18.619	0.520
Hohhot	20	23.150	0.715	Yichang	75	18.599	0.519
Guangzhou	21	22.800	0.700	Xining	76	18.570	0.517
Luoyang	22	22.564	0.690	Yinchuan	77	18.523	0.515
Zhuzhou	23	22.480	0.686	Hefei	78	18.429	0.511
Changchun	24	22.462	0.685	Weihai	79	18.324	0.507
Baotou	25	22.311	0.679	Sanya	80	18.129	0.499
Changzhou	26	22.074	0.669	Dalian	81	18.116	0.498
Tangshan	27	22.028	0.667	Chengdu	82	17.964	0.491
Datong	28	22.022	0.666	Handan	83	17.799	0.484
Shijiazhuang	29	21.865	0.660	Changzhi	84	17.764	0.483
Yangquan	30	21.657	0.651	Guiyang	85	17.748	0.482
Harbin	31	21.359	0.638	Wuhan	86	17.688	0.479
Weifang	32	21.356	0.638	Nanning	87	17.610	0.476
Yan'an	33	21.355	0.638	Xuzhou	88	17.451	0.469
Baoji	34	21.349	0.637	Hangzhou	89	17.432	0.468
Ji'nan	35	21.284	0.635	Qingdao	90	17.344	0.465
Baoding	36	21.218	0.632	Shenzhen	91	17.100	0.454
Anshan	37	21.215	0.632	Guilin	92	17.033	0.451
Fushun	38	21.208	0.631	Haikou	93	17.791	0.441
Ma'anshan	39	21.190	0.630	Kunming	94	16.490	0.428
Lanzhou	40	21.087	0.626	Panzhihua	95	16.380	0.423
Tianjin	41	21.085	0.626	Qujing	96	16.318	0.420
Taizhou	42	21.064	0.625	Kaifeng	97	16.279	0.419
Yueyang	43	20.937	0.620	Fuzhou	98	16.257	0.418
Wenzhou	44	20.646	0.607	Liuzhou	99	15.802	0.398
Karamay	45	20.606	0.605	Changde	100	15.801	0.398
Pingdingshan	46	20.472	0.600	Xiamen	101	15.772	0.397
Lianyungang	47	20.340	0.594	Linfen	102	15.634	0.391
Qigihar	48	20.328	0.593	Zhongshan	103	15.633	0.391
Wuhu	49	20.187	0.587	Quanzhou	104	15.428	0.382
Changsha	50	20.187	0.587	Shizuishan	105	15.169	0.371
Zhengzhou	51	20.156	0.586	Jiaxing	106	13.394	0.294
Wuxi	52	20.143	0.585	Jiujiang	107	13.100	0.282
Suzhou	53	20.116	0.584	Beihai	108	11.473	0.211
Chongqing	54	20.042	0.581	Yibin	109	7.773	0.052
Xiangtan	55	19.995	0.579	Mianyang	110	6.569	0.000

Table A6-8 Rankings of 110 cities in terms of job contribution per unit carbon emission

City	Ranking	Job contribution from unit carbon emission (job/t CO_2)	Relative value	City	Ranking	Job contribution from unit carbon emission (job/t CO_2)	Relative value
Zhuhai	1	0.042	1.000	Yantai	56	0.014	0.288
Xiamen	2	0.038	0.914	Shaoguan	57	0.014	0.280
Beijing	3	0.037	0.871	Yangquan	58	0.013	0.274
Haikou	4	0.036	0.867	Mianyang	59	0.013	0.270
Nanning	5	0.033	0.773	Zhengzhou	60	0.013	0.270
Xi'an	6	0.032	0.757	Yinchuan	61	0.013	0.268
Zhanjiang	7	0.030	0.696	Huzhou	62	0.013	0.267
Wenzhou	8	0.029	0.669	Zhongshan	63	0.013	0.255
Fuzhou	9	0.026	0.606	Qinghuangdao	64	0.012	0.250
Yan'an	10	0.026	0.597	Yueyang	65	0.012	0.249
Sanya	11	0.026	0.594	Tai'an	66	0.012	0.236
Quanzhou	12	0.025	0.586	Shenyang	67	0.012	0.236
Changchun	13	0.023	0.532	Chifeng	68	0.012	0.234
Lianyungang	14	0.023	0.531	Zhuzhou	69	0.012	0.230
Shantou	15	0.023	0.522	Dalian	70	0.011	0.227
Taizhou	16	0.022	0.508	Xiangtan	71	0.011	0.225
Kunming	17	0.022	0.508	Taiyuan	72	0.011	0.218
Chengdu	18	0.022	0.505	Nanjing	73	0.011	0.218
Hangzhou	19	0.022	0.498	Zunyi	74	0.011	0.213
Shenzhen	20	0.022	0.496	Pingdingshan	75	0.011	0.211
Guangzhou	21	0.021	0.459	Xuzhou	76	0.011	0.210
Jiaxing	22	0.020	0.450	Wuhu	77	0.011	0.208
Jiujiang	23	0.020	0.435	Weihai	78	0.011	0.207
Zhangjiajie	24	0.019	0.432	Weifang	79	0.010	0.200
Guiyang	25	0.019	0.429	Qujing	80	0.010	0.192
Wuhan	26	0.019	0.427	Jinchang	81	0.010	0.183
Xianyang	27	0.019	0.414	Xining	82	0.009	0.179
Qigihar	28	0.018	0.399	Liuzhou	83	0.009	0.167
Nanchang	29	0.018	0.398	Daqing	84	0.009	0.166
Yibin	30	0.018	0.393	Suzhou	85	0.009	0.165
Datong	31	0.017	0.382	Urumqi	86	0.009	0.164
Mudanjiang	32	0.017	0.380	Jining	87	0.009	0.162
Harbin	33	0.017	0.370	Luoyang	88	0.009	0.159
Tongchuan	34	0.017	0.369	Anyang	89	0.008	0.146
Shaoxing	35	0.017	0.367	Changzhi	90	0.008	0.145
Shanghai	36	0.017	0.365	Sunshine	91	0.008	0.144
Jingzhou	37	0.017	0.365	Shijiazhuang	92	0.008	0.143
Chongqing	38	0.017	0.358	Changzhou	93	0.008	0.138
Changsha	39	0.016	0.354	Karamay	94	0.008	0.137
Baoding	40	0.016	0.354	Zaozhuang	95	0.008	0.134
Kaifeng	41	0.016	0.351	Jilin	96	0.008	0.132

(Contd.)

City	Ranking	Job contribution from unit carbon emission (job/t CO_2)	Relative value	City	Ranking	Job contribution from unit carbon emission (job/t CO_2)	Relative value
Ningbo	42	0.016	0.346	Wuxi	97	0.008	0.132
Tianjin	43	0.016	0.346	Foshan	98	0.008	0.132
Luzhou	44	0.016	0.344	Jiaozuo	99	0.007	0.124
Qingdao	45	0.016	0.344	Fushun	100	0.006	0.096
Ji'nan	46	0.016	0.339	Hohhot	101	0.006	0.095
Baoji	47	0.016	0.337	Zibo	102	0.006	0.087
Yichang	48	0.016	0.332	Handan	103	0.006	0.080
Guilin	49	0.015	0.317	Panzhihua	104	0.006	0.080
Nantong	50	0.015	0.308	Anshan	105	0.006	0.076
Lanzhou	51	0.015	0.308	Linfen	106	0.005	0.073
Yangzhou	52	0.014	0.302	Ma'anshan	107	0.004	0.047
Hefei	53	0.014	0.299	Tangshan	108	0.004	0.031
Changde	54	0.014	0.295	Baotou	109	0.004	0.029
Beihai	55	0.014	0.289	Shizuishan	110	0.003	0.000

Table A6-9 Rankings of 110 cities in terms of proportion of non-fossil energy consumption

City	Ranking	Proportion of non-fossil energy consumption (%)	Relative value	City	Ranking	Proportion of non-fossil energy consumption (%)	Relative value
Yichang	1	20.36	1.000	Jilin	56	4.00	0.180
Xining	2	20.00	0.982	Baotou	57	4.00	0.180
Kunming	3	19.00	0.932	Shijiazhuang	58	4.00	0.180
Nanning	4	18.00	0.882	Beijing	59	4.00	0.180
Wuhan	5	16.90	0.827	Xiangtan	60	3.80	0.170
Chengdu	6	16.20	0.792	Nanjing	61	3.80	0.170
Guangzhou	7	13.00	0.631	Shaoxing	62	3.70	0.165
Chongqing	8	11.73	0.568	Baoding	63	3.70	0.165
Shanghai	9	11.10	0.536	Xi'an	64	3.60	0.160
Zhongshan	10	10.00	0.481	Changde	65	3.60	0.160
Changchun	11	10.00	0.481	Yangzhou	66	3.60	0.160
Fushun	12	10.00	0.481	Luzhou	67	3.50	0.155
Guiyang	13	9.00	0.431	Wenzhou	68	3.50	0.155
Shaoguan	14	9.00	0.431	Xuzhou	69	3.50	0.155
Hangzhou	15	8.10	0.386	Zhuzhou	70	3.40	0.150
Liuzhou	16	8.00	0.381	Jiujiang	71	3.40	0.150
Zhuhai	17	8.00	0.381	Kaifeng	72	3.20	0.140
Xiamen	18	8.00	0.381	Suzhou	73	3.20	0.140
Lianyungang	19	8.00	0.381	Shantou	74	3.00	0.130
Anshan	20	8.00	0.381	Yueyang	75	3.00	0.130
Jiaozuo	21	7.50	0.356	Jining	76	3.00	0.130

(Contd.)

City	Ranking	Proportion of non-fossil energy consumption (%)	Relative value	City	Ranking	Proportion of non-fossil energy consumption (%)	Relative value
Zhanjiang	22	6.50	0.306	Wuhu	77	3.00	0.130
Lanzhou	23	6.00	0.281	Yangquan	78	3.00	0.130
Jingzhou	24	6.00	0.281	Taiyuan	79	3.00	0.130
Jinchang	25	5.60	0.261	Handan	80	3.00	0.130
Nanchang	26	5.60	0.261	Qinghuangdao	81	3.00	0.130
Beihai	27	5.50	0.256	Tangshan	82	3.00	0.130
Yibin	28	5.00	0.230	Dalian	83	2.98	0.129
Foshan	29	5.00	0.230	Nantong	84	2.90	0.125
Qigihar	30	5.00	0.230	Tongchuan	85	2.80	0.120
Harbin	31	5.00	0.230	Zhangjiajie	86	2.70	0.115
Chifeng	32	5.00	0.230	Changzhou	87	2.60	0.110
Hohhot	33	5.00	0.230	Zibo	88	2.50	0.105
Baoji	34	4.90	0.225	Wuxi	89	2.50	0.105
Fuzhou	35	4.90	0.225	Linfen	90	2.50	0.105
Tianjin	36	4.71	0.216	Changzhi	91	2.50	0.105
Zunyi	37	4.70	0.215	Zhengzhou	92	2.30	0.095
Sanya	38	4.70	0.215	Huzhou	93	2.20	0.095
Mudanjiang	39	4.70	0.215	Weihai	94	2.20	0.090
Pingdingshan	40	4.60	0.210	Guilin	95	2.15	0.088
Shenyang	41	4.60	0.210	Luoyang	96	2.00	0.080
Yinchuan	42	4.50	0.205	Tai'an	97	2.00	0.080
Shenzhen	43	4.50	0.205	Zaozhuang	98	2.00	0.080
Panzhihua	44	4.40	0.200	Ma'anshan	99	2.00	0.080
Changsha	45	4.28	0.194	Jiaxing	100	2.00	0.080
Qujing	46	4.00	0.180	Anyang	101	1.80	0.070
Mianyang	47	4.00	0.180	Karamay	102	1.70	0.065
Sunshine	48	4.00	0.180	Daqing	103	1.70	0.065
Weifang	49	4.00	0.180	Shizuishan	104	1.50	0.055
Yantai	50	4.00	0.180	Haikou	105	1.50	0.055
Qingdao	51	4.00	0.180	Datong	106	1.50	0.055
Ji'nan	52	4.00	0.180	Yan'an	107	1.40	0.050
Quanzhou	53	4.00	0.180	Ningbo	108	1.30	0.045
Hefei	54	4.00	0.180	Xianyang	109	0.70	0.015
Taizhou	55	4.00	0.180	Urumqi	110	0.40	0.000

Table A6-10 Rankings of 110 cities in terms of forest coverage rate

City	Ranking	Urban forest coverage rate (%)	Relative value	City	Ranking	Urban forest coverage rate (%)	Relative value
Zhangjiajie	1	67.5	1.000	Shantou	56	30.0	0.423
Fushun	2	66.2	0.980	Zhuhai	57	30.0	0.423

(Contd.)

City	Ranking	Urban forest coverage rate (%)	Relative value	City	Ranking	Urban forest coverage rate (%)	Relative value
Shaoguan	3	66.1	0.978	Linfen	58	29.5	0.415
Sanya	4	64.4	0.952	Beihai	59	28.9	0.406
Hangzhou	5	64.4	0.952	Ji’nan	60	27.8	0.389
Mudanjiang	6	62.3	0.920	Zhanjiang	61	27.7	0.388
Taizhou	7	62.2	0.918	Xuzhou	62	27.7	0.388
Zhuzhou	8	60.6	0.894	Xining	63	27.0	0.377
Wenzhou	9	59.6	0.878	Shijiazhuang	64	26.9	0.376
Panzhihua	10	59.0	0.869	Changzhi	65	26.9	0.375
Quanzhou	11	58.7	0.865	Jining	66	26.0	0.362
Guilin	12	58.0	0.854	Yangquan	67	25.3	0.350
Yichang	13	55.3	0.812	Wuhan	68	25.1	0.348
Liuzhou	14	55.0	0.808	Tangshan	69	24.4	0.336
Fuzhou	15	54.9	0.806	Shenyang	70	24.0	0.331
Shaoxing	16	54.0	0.793	Chifeng	71	24.0	0.331
Jilin	17	54.0	0.792	Xianyang	72	23.5	0.323
Changsha	18	52.3	0.765	Zhengzhou	73	23.3	0.320
Jiujiang	19	52.0	0.762	Pingdingshan	74	23.0	0.315
Huzhou	20	50.9	0.745	Nanjing	75	23.0	0.315
Ningbo	21	50.5	0.738	Wuxi	76	22.0	0.300
Baoji	22	50.0	0.731	Beijing	77	21.5	0.292
Anshan	23	46.4	0.675	Jinchang	78	20.8	0.282
Mianyang	24	45.7	0.665	Suzhou	79	20.3	0.274
Xiangtan	25	45.0	0.654	Wuhu	80	20.0	0.269
Kunming	26	45.0	0.654	Changzhou	81	20.0	0.269
Shenzhen	27	45.0	0.654	Hohhot	82	20.0	0.269
Xi’an	28	45.0	0.654	Baoding	83	20.0	0.269
Harbin	29	44.6	0.648	Lianyungang	84	18.1	0.239
Tongchuan	30	43.9	0.637	Foshan	85	18.0	0.238
Changde	31	43.8	0.635	Yangzhou	86	17.6	0.232
Xiamen	32	43.0	0.623	Zhongshan	87	17.0	0.223
Nanning	33	42.2	0.610	Jiaxing	88	16.8	0.220
Qinghuangdao	34	42.0	0.607	Nantong	89	16.8	0.220
Dalian	35	42.5	0.600	Nanchang	90	16.1	0.209
Zunyi	36	40.0	0.577	Hefei	91	15.8	0.205
Guiyang	37	39.9	0.576	Kaifeng	92	15.6	0.202
Yibin	38	39.0	0.562	Jingzhou	93	15.0	0.192
Luzhou	39	38.4	0.552	Anyang	94	15.0	0.192
Guangzhou	40	38.3	0.551	Changchun	95	15.0	0.192
Weihai	41	38.2	0.549	Taiyuan	96	15.0	0.192
Yantai	42	38.0	0.546	Jiaozuo	97	13.5	0.169
Yan’an	43	37.0	0.531	Qigihar	98	13.4	0.167
Haikou	44	37.0	0.531	Lanzhou	99	12.2	0.149
Chengdu	45	36.8	0.528	Yinchuan	100	12.0	0.146
Qujing	46	36.2	0.518	Shanghai	101	11.6	0.140

(Contd.)

City	Ranking	Urban forest coverage rate (%)	Relative value	City	Ranking	Urban forest coverage rate (%)	Relative value
Yueyang	47	36.0	0.515	Handan	102	11.6	0.140
Qingdao	48	35.4	0.506	Shizuishan	103	11.0	0.131
Tai'an	49	34.4	0.491	Daqing	104	10.4	0.122
Sunshine	50	34.2	0.488	Baotou	105	10.0	0.115
Chongqing	51	34.0	0.485	Ma'anshan	106	8.3	0.089
Luoyang	52	32.8	0.466	Tianjin	107	8.3	0.089
Zibo	53	32.4	0.460	Datong	108	8.0	0.085
Zaozhuang	54	31.0	0.438	Urumqi	109	2.9	0.007
Weifang	55	30.6	0.432	Karamay	110	2.5	0.000

Table A6-11 Rankings of 110 cities in terms of per-capital riding number per public electric vehicle

City	Ranking	Public electric vehicle transportation (time/(capita•car))	Relative value	City	Ranking	Public electric vehicle transportation (time/(capita•car))	Relative value
Haikou	1	2.373	1.000	Yucyang	56	0.080	0.030
Yan'an	2	0.828	0.346	Baotou	57	0.079	0.029
Yangquan	3	0.300	0.123	Linfen	58	0.076	0.028
Karamay	4	0.292	0.119	Ningbo	59	0.075	0.028
Jinchang	5	0.287	0.117	Nanning	60	0.072	0.027
Ma'anshan	6	0.287	0.117	Jiaxing	61	0.071	0.026
Xining	7	0.268	0.109	Yichang	62	0.071	0.026
Yibin	8	0.247	0.100	Tai'an	63	0.070	0.026
Panzhihua	9	0.236	0.096	Changzhou	64	0.070	0.025
Zhuhai	10	0.233	0.094	Nanjing	65	0.067	0.024
Baoji	11	0.231	0.093	Taiyuan	66	0.067	0.024
Zhangjiajie	12	0.218	0.088	Daqing	67	0.066	0.024
Xiamen	13	0.204	0.082	Xi'an	68	0.064	0.023
Beihai	14	0.193	0.078	Yangzhou	69	0.062	0.022
Fushun	15	0.192	0.077	Huzhou	70	0.061	0.022
Shizuishan	16	0.186	0.075	Weifang	71	0.056	0.020
Xianyang	17	0.181	0.072	Baoding	72	0.055	0.019
Tongchuan	18	0.170	0.068	Shaoxing	73	0.054	0.019
Mianyang	19	0.155	0.061	Yantai	74	0.054	0.019
Sanya	20	0.154	0.061	Fuzhou	75	0.053	0.018
Changzhi	21	0.153	0.061	Nanchang	76	0.053	0.018
Yinchuan	22	0.151	0.060	Chengdu	77	0.052	0.018
Shaoguan	23	0.149	0.059	Changchun	78	0.050	0.017

(Contd.)

City	Ranking	Public electric vehicle transportation (time/(capita•car))	Relative value	City	Ranking	Public electric vehicle transportation (time/(capita•car))	Relative value
Zhongshan	24	0.137	0.054	Dalian	79	0.050	0.017
Wuhu	25	0.137	0.054	Zaozhuang	80	0.048	0.016
Guilin	26	0.129	0.050	Jining	81	0.048	0.016
Lanzhou	27	0.128	0.050	Suzhou	82	0.047	0.016
Liuzhou	28	0.128	0.050	Nantong	83	0.047	0.016
Luzhou	29	0.128	0.050	Kunming	84	0.044	0.014
Jiaozuo	30	0.122	0.047	Ji'nan	85	0.042	0.014
Datong	31	0.121	0.047	Pingdingshan	86	0.041	0.013
Zunyi	32	0.120	0.047	Harbin	87	0.040	0.013
Qinghuangdao	33	0.119	0.046	Shantou	88	0.039	0.013
Chifeng	34	0.115	0.045	Quanzhou	89	0.038	0.012
Urumqi	35	0.113	0.044	Qingdao	90	0.038	0.012
Zhuzhou	36	0.110	0.042	Wuhan	91	0.038	0.012
Hohhot	37	0.107	0.041	Zhengzhou	92	0.037	0.012
Hefei	38	0.103	0.040	Shenyang	93	0.036	0.011
Lianyungang	39	0.098	0.037	Guangzhou	94	0.036	0.011
Mudanjiang	40	0.098	0.037	Zhanjiang	95	0.036	0.011
Jilin	41	0.097	0.037	Zibo	96	0.036	0.011
Jingzhou	42	0.095	0.036	Shijiazhuang	97	0.035	0.011
Guiyang	43	0.095	0.036	Hangzhou	98	0.035	0.011
Anshan	44	0.094	0.036	Xuzhou	99	0.033	0.010
Qujing	45	0.093	0.035	Wenzhou	100	0.032	0.009
Sunshine	46	0.092	0.035	Handan	101	0.032	0.009
Weihai	47	0.091	0.034	Tangshan	102	0.031	0.009
Luoyang	48	0.087	0.033	Qigihar	103	0.030	0.008
Kaifeng	49	0.085	0.032	Taizhou	104	0.024	0.006
Changsha	50	0.084	0.031	Wuxi	105	0.024	0.006
Changde	51	0.082	0.031	Tianjin	106	0.016	0.003
Anyang	52	0.081	0.030	Beijing	107	0.015	0.002
Xiangtan	53	0.081	0.030	Foshan	108	0.014	0.002
Jiujiang	54	0.081	0.030	Chongqing	109	0.013	0.001
Shenzhen	55	0.080	0.030	Shanghai	110	0.010	0.000

Table A6-12 Rankings of 110 cities in terms of energy consumption level (standard coal) per living area

City	Ranking	Energy consumption per residential building area (kg ce/m^2)	Relative value	City	Ranking	Energy consumption per residential building area (kg ce/m^2)	Relative value
Zhangjiajie	1	0.458	1.000	Tangshan	55	3.493	0.866
Pingdingshan	2	0.715	0.989	Datong	56	3.520	0.865
Chifeng	3	0.779	0.986	Hohhot	57	3.560	0.863
Taizhou	4	0.826	0.984	Taiyuan	58	3.756	0.854
Quanzhou	5	0.922	0.979	Fuzhou	59	3.762	0.854
Handan	6	1.092	0.972	Jiaozuo	60	3.784	0.853
Changzhi	7	1.149	0.969	Qingdao	61	3.791	0.853
Xuzhou	8	1.163	0.969	Tianjin	62	3.811	0.852
Jining	9	1.251	0.965	Shaoguan	63	3.825	0.851
Baoding	10	1.270	0.964	Haikou	64	3.901	0.848
Tai'an	11	1.306	0.963	Qinghuangdao	65	3.987	0.844
Weifang	12	1.506	0.954	Lanzhou	66	4.034	0.842
Mudanjiang	13	1.531	0.953	Tongchuan	67	4.180	0.835
Lianyungang	14	1.567	0.951	Changzhou	68	4.434	0.824
Yueyang	15	1.579	0.950	Nanning	69	4.531	0.820
Changde	16	1.586	0.950	Shenyang	70	4.566	0.818
Nantong	17	1.591	0.950	Zhengzhou	71	4.606	0.817
Shaoxing	18	1.623	0.949	Chongqing	72	4.631	0.816
Huzhou	19	1.641	0.948	Kaifeng	73	4.709	0.812
Jiaxing	20	1.692	0.945	Linfen	74	4.732	0.811
Guilin	21	1.697	0.945	Ji'nan	75	4.905	0.803
Luoyang	22	1.697	0.945	Daqing	76	4.913	0.803
Xianyang	23	1.706	0.945	Xining	77	4.923	0.803
Yangzhou	24	1.925	0.935	Nanjing	78	4.978	0.800
Yantai	25	1.959	0.934	Wuhu	79	5.011	0.799
Zunyi	26	2.012	0.931	Changsha	80	5.027	0.798
Yibin	27	2.120	0.927	Wuhan	81	5.141	0.793
Weihai	28	2.152	0.925	Wuxi	82	5.153	0.793
Jiujiang	29	2.252	0.921	Zhuhai	83	5.366	0.783
Yichang	30	2.318	0.918	Baotou	84	5.733	0.767
Yan'an	31	2.327	0.917	Shantou	85	5.811	0.763
Xiangtan	32	2.377	0.915	Foshan	86	5.863	0.761
Nanchang	33	2.382	0.915	Xi'an	87	6.499	0.733

(Contd.)

City	Ranking	Energy consumption per residential building area (kg ce/m^2)	Relative value	City	Ranking	Energy consumption per residential building area (kg ce/m^2)	Relative value
Shijiazhuang	34	2.420	0.913	Hangzhou	88	6.548	0.731
Jingzhou	35	2.503	0.910	Ningbo	89	6.843	0.718
Wenzhou	36	2.556	0.907	Sanya	90	6.896	0.715
Suzhou	37	2.620	0.904	Panzhihua	91	6.959	0.713
Zhuzhou	38	2.686	0.902	Karamay	92	7.307	0.697
Harbin	39	2.687	0.901	Shanghai	93	7.307	0.697
Zhanjiang	40	2.732	0.899	Zibo	94	7.404	0.693
Jilin	41	2.738	0.899	Dalian	95	7.514	0.688
Luzhou	42	2.752	0.899	Hefei	96	8.113	0.662
Qigihar	43	2.770	0.898	Chengdu	97	8.495	0.645
Zaozhuang	44	2.774	0.898	Beijing	98	8.731	0.634
Baoji	45	2.881	0.893	Guangzhou	99	8.770	0.633
Anyang	46	3.004	0.887	Xiamen	100	8.935	0.625
Mianyang	47	3.007	0.887	Kunming	101	9.121	0.617
Beihai	48	3.039	0.886	Fushun	102	10.243	0.568
Anshan	49	3.058	0.885	Yangquan	103	10.417	0.560
Shizuishan	50	3.129	0.882	Urumqi	104	10.914	0.538
Sunshine	51	3.131	0.882	Guiyang	105	11.560	0.509
Ma'anshan	52	3.202	0.879	Yinchuan	106	11.732	0.502
Liuzhou	53	3.459	0.867	Zhongshan	107	12.435	0.471
Changchun	54	3.467	0.867	Shenzhen	108	23.085	0.000

Table A6-13 Rankings of 110 cities in terms of industrial wastewater COD discharge intensity

City	Ranking	Water pollution discharge intensity (t COD/10^4 yuan)	Relative value	City	Ranking	Water pollution discharge intensity (t COD/10^4 yuan)	Relative value
Sanya	1	0.000	1.000	Panzhihua	56	0.001	0.956
Beijing	2	0.000	1.000	Jiujiang	57	0.001	0.954
Xiamen	3	0.000	0.999	Harbin	58	0.001	0.952
Shenzhen	4	0.000	0.999	Jiaxing	59	0.001	0.951
Karamay	5	0.000	0.997	Xuzhou	60	0.001	0.890
Shanghai	6	0.000	0.996	Huzhou	61	0.001	0.948
Guiyang	7	0.000	0.994	Pingdingshan	62	0.001	0.946
Haikou	8	0.000	0.993	Yan'an	63	0.001	0.943
Qingdao	9	0.000	0.993	Yichang	64	0.002	0.937
Taiyuan	10	0.000	0.993	Shaoguan	65	0.002	0.937
Shenyang	11	0.000	0.993	Shantou	66	0.002	0.937

(Contd.)

City	Ranking	Water pollution discharge intensity (t COD/10^4 yuan)	Relative value	City	Ranking	Water pollution discharge intensity (t COD/10^4 yuan)	Relative value
Yangquan	12	0.000	0.992	Jinchang	67	0.002	0.937
Lanzhou	13	0.000	0.992	Sunshine	68	0.002	0.931
Hefei	14	0.000	0.992	Wuhu	69	0.002	0.931
Kunming	15	0.000	0.992	Yangzhou	70	0.002	0.927
Ji'nan	16	0.000	0.990	Tangshan	71	0.002	0.926
Zhuhai	17	0.000	0.990	Changzhou	72	0.002	0.926
Dalian	18	0.000	0.988	Nantong	73	0.002	0.924
Baotou	19	0.000	0.988	Guilin	74	0.002	0.911
Tianjin	20	0.000	0.987	Zhanjiang	75	0.002	0.911
Ningbo	21	0.000	0.987	Shaoxing	76	0.002	0.907
Luoyang	22	0.000	0.987	Jilin	77	0.002	0.905
Guangzhou	23	0.000	0.986	Quanzhou	78	0.002	0.904
Daqing	24	0.000	0.986	Chifeng	79	0.002	0.897
Fushun	25	0.000	0.986	Zhuzhou	80	0.003	0.881
Weihai	26	0.000	0.985	Baoding	81	0.003	0.874
Fuzhou	27	0.001	0.984	Hangzhou	82	0.003	0.870
Changsha	28	0.001	0.983	Nanchang	83	0.003	0.865
Anshan	29	0.001	0.983	Xi'an	84	0.003	0.865
Tongchuan	30	0.001	0.982	Chongqing	85	0.003	0.864
Taizhou	31	0.001	0.981	Shizuishan	86	0.003	0.860
Nanjing	32	0.001	0.979	Wenzhou	87	0.003	0.857
Yantai	33	0.001	0.977	Xining	88	0.004	0.851
Jining	34	0.001	0.977	Zaozhuang	89	0.004	0.848
Ma'anshan	35	0.001	0.976	Shijiazhuang	90	0.004	0.847
Tai'an	36	0.001	0.974	Xianyang	91	0.004	0.845
Zunyi	37	0.001	0.974	Chengdu	92	0.004	0.843
Wuxi	38	0.001	0.971	Yinchuan	93	0.004	0.840
Lianyungang	39	0.001	0.971	Qigihar	94	0.004	0.827
Handan	40	0.001	0.970	Baoji	95	0.004	0.822
Qujing	41	0.001	0.970	Anyang	96	0.005	0.805
Changchun	42	0.001	0.969	Liuzhou	97	0.005	0.801
Hohhot	43	0.001	0.969	Xiangtan	98	0.005	0.776
Weifang	44	0.001	0.968	Yueyang	99	0.006	0.758
Wuhan	45	0.001	0.967	Zhangjiajie	100	0.006	0.741
Changzhi	46	0.001	0.965	Datong	101	0.006	0.741
Mianyang	47	0.001	0.964	Jiaozuo	102	0.007	0.713
Foshan	48	0.001	0.962	Yibin	103	0.007	0.710
Suzhou	49	0.001	0.961	Luzhou	104	0.008	0.664
Urumqi	50	0.001	0.960	Mudanjiang	105	0.010	0.569
Linfen	51	0.001	0.959	Jingzhou	106	0.010	0.565
Zibo	52	0.001	0.958	Changde	107	0.012	0.482
Qinghuangdao	53	0.001	0.957	Kaifeng	108	0.012	0.472
Zhongshan	54	0.001	0.956	Nanning	109	0.023	0.023
Zhengzhou	55	0.001	0.956	Beihai	110	0.023	0.000

Table A6-14 Rankings of 110 cities in terms of industrial waste gas SO_2 emission intensity

City	Ranking	Air pollution emission intensity (t $SO_2/10^4$ yuan)	Relative value	City	Ranking	Air pollution emission intensity (t $SO_2/10^4$ yuan)	Relative value
Haikou	1	0.000	1.000	Jining	56	0.008	0.906
Sanya	2	0.000	0.998	Huzhou	57	0.009	0.903
Shenzhen	3	0.001	0.989	Jiaxing	58	0.009	0.902
Beijing	4	0.001	0.985	Qinghuangdao	59	0.009	0.902
Daqing	5	0.002	0.981	Yangzhou	60	0.009	0.901
Zhuhai	6	0.002	0.980	Weihai	61	0.009	0.901
Shanghai	7	0.002	0.976	Zibo	62	0.009	0.901
Xiamen	8	0.002	0.976	Linfen	63	0.009	0.896
Guangzhou	9	0.002	0.975	Baoding	64	0.009	0.895
Qingdao	10	0.002	0.974	Shantou	65	0.010	0.887
Changchun	11	0.002	0.973	Lanzhou	66	0.010	0.886
Wuxi	12	0.003	0.971	Changzhi	67	0.011	0.876
Hefei	13	0.003	0.968	Handan	68	0.011	0.874
Shenyang	14	0.003	0.967	Guiyang	69	0.011	0.873
Karamay	15	0.003	0.966	Tai'an	70	0.011	0.869
Tianjin	16	0.003	0.965	Shijiazhuang	71	0.012	0.867
Hangzhou	17	0.003	0.965	Anyang	72	0.012	0.865
Nanjing	18	0.003	0.964	Yueyang	73	0.012	0.864
Ningbo	19	0.003	0.962	Baoji	74	0.013	0.855
Shaoxing	20	0.004	0.959	Zhuzhou	75	0.013	0.851
Nanchang	21	0.004	0.957	Xuzhou	76	0.013	0.850
Yan'an	22	0.004	0.957	Guilin	77	0.014	0.845
Harbin	23	0.004	0.955	Qujing	78	0.014	0.845
Suzhou	24	0.004	0.955	Zhengzhou	79	0.014	0.834
Ji'nan	25	0.004	0.954	Shaoguan	80	0.015	0.833
Wuhan	26	0.004	0.953	Kaifeng	81	0.015	0.832
Dalian	27	0.004	0.953	Tongchuan	82	0.015	0.832
Changzhou	28	0.004	0.952	Jingzhou	83	0.015	0.829
Jilin	29	0.005	0.947	Qigihar	84	0.015	0.827
Yinchuan	30	0.005	0.946	Hohhot	85	0.015	0.827
Yantai	31	0.005	0.946	Changde	86	0.015	0.824
Quanzhou	32	0.005	0.946	Baotou	87	0.015	0.824
Yichang	33	0.005	0.945	Xiangtan	88	0.016	0.821
Zhongshan	34	0.005	0.942	Jiaozuo	89	0.016	0.818

(Contd.)

City	Ranking	Air pollution emission intensity (t $SO_2/10^4$ yuan)	Relative value	City	Ranking	Air pollution emission intensity (t $SO_2/10^4$ yuan)	Relative value
Weifang	35	0.005	0.939	Pingdingshan	90	0.016	0.812
Liuzhou	36	0.006	0.936	Urumqi	91	0.016	0.811
Ma'anshan	37	0.006	0.934	Nanning	92	0.017	0.802
Foshan	38	0.006	0.932	Jiujiang	93	0.019	0.780
Changsha	39	0.006	0.931	Xining	94	0.019	0.778
Taiyuan	40	0.006	0.931	Yibin	95	0.020	0.770
Taizhou	41	0.006	0.928	Chongqing	96	0.020	0.768
Chengdu	42	0.006	0.928	Jinchang	97	0.021	0.763
Sunshine	43	0.006	0.927	Zaozhuang	98	0.021	0.760
Fushun	44	0.006	0.927	Panzhihua	99	0.022	0.750
Mianyang	45	0.007	0.925	Xianyang	100	0.022	0.749
Nantong	46	0.007	0.924	Datong	101	0.022	0.746
Wenzhou	47	0.007	0.924	Luoyang	102	0.022	0.742
Kunming	48	0.007	0.919	Zunyi	103	0.024	0.723
Wuhu	49	0.007	0.919	Luzhou	104	0.026	0.699
Xi'an	50	0.007	0.916	Mudanjiang	105	0.028	0.679
Zhanjiang	51	0.007	0.915	Zhangjiajie	106	0.035	0.596
Tangshan	52	0.008	0.912	Yangquan	107	0.046	0.468
Fuzhou	53	0.008	0.911	Shizuishan	108	0.050	0.425
Lianyungang	54	0.008	0.908	Beihai	109	0.060	0.309
Anshan	55	0.008	0.908	Chifeng	110	0.087	0.000

Appendix VII Chronicle of Events about Low-Carbon Development Indicators in China (2011-2012)

*Collated and provided by graduate students Yu Li and Wang Ao from the School of Packaging Design & Art, Hunan University of Technology.

On January 1, 2011, *Circular Concerning Value-added Tax, Business Tax and Corporate Income Tax Policy Issues for Promoting Development of Energy Saving Service Industry* jointly released by the Ministry of Finance and State Administration of Taxation officially took effect, which increases support for energy saving industry.

On January 19, 2011, National Development and Reform Commission issued *Guidelines for Preparation of Circular Economy Development Plan* ("Guidelines") to promote implementation of *Circular Economy Promotion Law* and stimulate circular economy to form a relatively large scale as soon as possible. These Guidelines require various localities to plan overall layout for the development of local circular economy, and vigorously boost the development of circular agriculture, optimize industrial structure, build circular economy industry chain, and energetically cultivate and impel the development of emerging industries relating to circular economy, and develop social circular economy system including the tertiary industry; these Guidelines require construction of resource recycling network system and intensified efforts in waste recycling in relevant regions; these Guidelines state actions to reinforce publicity and education, popularize green consumption mode, give prominence to research, development and application of circular economy technologies and low-carbon technologies.

On February 1, 2011, the Ministry of Environmental Protection released *the 12th Five-year Plan for Implementation of Work on Environmental Protection Science Popularization*. According to this Plan, work on environmental protection science popularization will be taken as one basic, long-term work for enhancing citizens' awareness about environmental science quality and strengthening environmental protection capability during the 12th Five-year Plan period.

On February 5, 2011, National Development and Reform Commission and Certification and Accreditation Administration organized and convened launching meeting for "Special Subject for Addressing Climate Change——Research on Establishment of China's Low-carbon Certification System" and the first working meeting, which marks comprehensive launch of research on China's low-carbon certification system. This Project is designed to conduct comparative study of international low-carbon certification systems and proceed from China's national conditions and current development to analyze China's low-carbon certification policy and technical needs, and research and establish China's low-carbon certification system framework system, and gradually make improvement through pilot study so as to provide the basis for relevant departments to develop China's low-carbon certification management system.

On February 19, 2011, Zhou Shengxian, Minister of the Ministry of Environmental

Protection, said in recently convened video working conference concerning *the 12th Five-year Plan for Comprehensive Prevention and Control of Heavy Metal Pollution that the 12th Five-year Plan for Comprehensive Prevention and Control of Heavy Metal Pollution* had been officially approved by the State Council. The Ministry of Environmental Protection requires local governments at various levels to incorporate goals, tasks and projects specified in this Plan into economic and social development plans for their regions, and prepare plans for heavy metal prevention and control in various key regions. The Ministry of Environmental Protection and relevant departments have studied and developed measures for evaluating implementation of this Plan.

On March 2, 2011, the press conference for green paper *China's Urban Low-carbon Development (2011)* was held in Beijing, in which attempts are made to carry out innovations in theory, practice, general situation and prospect, etc; energy and carbon emission data concerning 110 cities above prefecture level in China are prepared and collated, and difficulties and problems in China's urban low-carbon development are investigated and analyzed; characteristics and framework system of China's urban low-carbon plan are clarified and summarized; attempt is made to develop evaluation indicator system for China's low-carbon cities and urban low-carbon development; basic route and main direction for China's low-carbon cities and urban low-carbon development are revealed; specific measures and suggestions for promoting China's urban low-carbon development are put forward.

On March 5, 2011, the opening ceremony for the 4th Session of the 11th National People's Congress was held in Great Hall of the People, in which deputies listened to and deliberated government work report delivered by Premier Wen Jiabao. Wen Jiabao stressed main goals and tasks for the 12th Five-year Plan period in this report, and vowed to accelerate strategic adjustment of economic structure. He emphasized that this was main direction for transforming economic development mode. He pointed out efforts to drive the economy on the track of endogenous growth and innovation-driven development as soon as possible while strengthening energy conservation, environmental protection and ecological construction and actively addressing climate change.

On March 10-11, 2011, Public Institution Energy Conservation Management Department of Government Offices Administration of the State Council organized and convened Public Institution Energy Conservation Liaison Meeting in Wuhan City, Hubei Province in 2011, in which the spirit of the 5th Plenary Session of the 17th Central Committee of CPC and the 4th session of the 11th National People's Congress was implemented; public institution energy conservation work progress nationwide in 2010 was summarized; this year's key work was arranged; evaluation plan for public institution energy conservation during the 11th Five-year Plan period was studied and discussed; opinions and suggestions about revision of *Public Institution Energy and Resource Consumption Statistics System and the 12th Five-year Plan for Public Institution Energy* Conservation were solicited; communication concerning promotion of public institution energy conservation was conducted.

On March 21, 2011, the State Council approved *Plan for Implementation of Control of Heavy Metal Pollution in Xiangjiang River Basin*, which was so far the first pilot plan for heavy metal pollution control nationwide approved by the State Council.

On March 24, 2011, 2011 International Conference on Low-carbon Energy & Climate Change organized by Tsinghua University, University of Cambridge and Massachusetts Institute of Technology was held in Beijing. This Conference was the first large academic conference organized by "Three-University Alliance". Tsinghua University—University of Cambridge—Massachusetts Institute of Technology Low-carbon Energy University Alliance ("Three-University Alliance") was established in November, 2009. As leading scientific and technological organizations in major energy consumption and carbon emission regions worldwide, these three prestigious universities have worked to establish low-carbon energy university alliance and made concerted efforts to build one platform so as to provide advanced energy technologies and policy choices for developing low-carbon economy and low-carbon society and addressing global climate change.

On March 28, 2011, the Ministry of Industry and Information Technology announced four large binding indicators for industrial energy conservation and emission reduction in China during this year and the 12th Five-year Plan period. It stressed that by 2015, in China, energy consumption per unit industrial added value, carbon dioxide emission and water consumption would decrease by 18%, more than 18% and 30% compared with that at the end of the 11th Five-year Plan period respectively, and rate of industrial solid wastes for comprehensive utilization would increase by about 72%; these four indicators would decrease by 4%, more than 4% and about 7% and increase by 2.2 percentage points this year.

On March 28, 2011, the 7th International Green Building & Building Energy Conservation Conference and New Technology and Product Expo were held. This Conference focused on the theme "Green Building: Make City Life More Low-carbon, Brighter" and showcased domestic and foreign latest achievements in green building and building energy conservation fields, development trends, successful cases and the latest technology and application developments in energy conservation and emission reduction, low carbon, ecology and environmental protection to the whole world.

On March 30, 2011, the Ministry of Environmental Protection, National Development and Reform Commission, the Ministry of Industry and Information Technology, the Ministry of Supervision, the Ministry of Justice, the Ministry of Housing and Urban-Rural Development, State Administration for Industry & Commerce, State Administration of Work Safety, State Electricity Regulatory Commission ("nine departments under the State Council") jointly convened video and telephone conference to arrange special environmental protection action for rectifying enterprises with illegal pollution discharge and protecting people's health nationwide in 2011. Zhou Shengxian, convener of inter-Ministerial joint conference on special environmental protection action, Minister of the Ministry of Environmental Protection, entrusted Zhang Lijun, Deputy Minister of the Ministry of Environmental Protection, to mobilize and arrange work on special environmental protection action for rectifying enterprises with illegal pollution discharge and protecting people's health nationwide in 2011.

On April 6, 2011, in order to quicken the pace of renewable energy development and utilization in rural areas, optimize rural energy structure, make rural energy clean and modernized, improve production and living conditions of farmers, National Energy Administration, the Ministry of Finance and the Ministry of Agriculture released *Interim Measures for Management of Subsidy Funds for Construction of Green Energy Demonstration County* to guarantee smooth construction of green energy demonstration county.

On April 7, 2011, "China Tour for Addressing Climate Change——Enter Jiangxi" launching ceremony organized by China Meteorological Administration was held in Nanchang. According to relevant information, pilot work on low-carbon development is officially carried out in ten counties (cities, regions) from this year. This is Jiangxi Province's first provincial-level pilot county project for low-carbon development in China.

On April 8. 2011, the Ministry of Environmental Protection, the Ministry of Commerce, National Development and Reform Commission, General Administration of Customs and General Administration of Quality Supervision jointly issued *Administrative Measures for Import of Solid Wastes*. According to relevant person in charge from the Ministry of Environmental Protection, implementation of these Measures would promote import and utilization of wastes, stimulate enterprises to further increase environmental protection awareness and level, standardize China's solid waste import administration, prevent illegal entry of overseas wastes and safeguard China's environmental safety.

On April 9, 2011, Civil Aviation Administration of China issued *Guiding Opinions Concerning Acceleration of Industrial Energy Conservation and Emission Reduction* ("Opinions") which specify guiding thought, main principles and working goals for quickening energy conservation and emission reduction in the whole industry and arrange work on civil aviation energy conservation and emission reduction prior to 2020. These Opinions are planning-oriented guiding opinions issued by Civil Aviation Administration of China, on the basis of *Plan for Energy Conservation and Emission Reduction in the Civil Aviation Industry* released during the 11th Five-year Plan period, according to changes of domestic and international situation, work arrangement made by the State Council, strategy for building great power through civil aviation, and requirements for building civil aviation as specified in the 12th Five-year Industry Plan. These Opinions constitute one programmatic document for guiding industrial energy conservation and emission reduction in a certain period of time.

On April 12, 2011, the Ministry of Environmental Protection required full implementation of various tasks specified in *the 12th Five-year Plan for Comprehensive Prevention and Control of Heavy Metal Pollution*. The Ministry of Environmental Protection released *Key Points for National Pollution Prevention and Control in 2011*. These Key Points stress crackdown efforts and severe punishment to ensure a good start for implementing *the 12th Five-year Plan for Comprehensive Prevention and Control of Heavy Metal Pollution.*

On April 19, 2011, the State Council issued *Circular Concerning Opinions About Further Strengthening Disposal of Urban Domestic Garbage* (GF (2011) No.9 Document), which states that with rapid urbanization, there is a dramatic increase in urban domestic garbage and garbage disposal capability is relatively insufficient, thus some cities are besieged by garbage. Various localities and relevant departments shall fully recognize the importance and necessity for strengthening disposal of urban domestic garbage, and comprehensively implement various policies and measures, promote disposal of urban domestic garbage to build low-carbon society.

On April 22, 2011, six departments including the Ministry of Environmental Protection, the Publicity Department of the Central Committee of the CPC, Central Civilization Office, the Ministry of Education, the Central Committee of the Communist Young League and All-China Women's Federation jointly issued *National Action Plan for Environmental Publicity Education* (2011-2015). This is one programmatic document for guiding national environmental publicity education jointly issued by six departments for the first time; it will play an important role in enhancing environmental awareness of the whole people, developing social action system with involvement from the whole people, boosting construction of energy-saving, environmentally friendly society and increasing ecological civilization level.

On April 28, 2011, National Development and Reform Commission issued *Administrative Measures for Orderly Utilization of Electricity* in order to reinforce power demand management and work on orderly utilization of electricity, which officially took effect on May 1.

On June 9, 2011, the Ministry of Environmental Protection printed and issued *the 12th Five-year National Scientific and Technological Development Plan for Environmental Protection.* This Plan states actions to establish industrialization platform through major special projects and support research and development of key technologies, equipment and products, research of environmental service industry supporting technologies, etc. to strengthen scientific and technological support for strategic emerging environmental protection industries. In the future, the Ministry of Environmental Protection will also implement "special water projects", other special scientific and technological projects, scientific and technological programs to vigorously research, develop and popularize advanced technologies, equipment and products for environmental protection, and guide and cultivate healthy development of strategic emerging environmental protection industries.

On June 22, 2011, the Ministry of Finance and National Development and Reform Commission printed and issued *Circular Concerning Comprehensive Demonstration Work on Fiscal Policy for Energy Conservation and Emission Reduction* and decided to conduct comprehensive demonstration of fiscal policy for energy conservation and emission reduction in part of cities during the 12th Five-year Plan period, and chose eight cities including Beijing, Shenzhen, Chongqing, Hangzhou City in Zhejiang Province, Changsha City in Hunan Province, Guiyang City in Guizhou Province, Jilin City in Jilin Province, Xinyu City in Jiangxi Province as the first batch of demonstration cities, and identified local governments as main body for assuming responsibility, and took cities as platforms to intensify integration of various fiscal policies for energy conservation and environmental protection, speed up system and mechanism innovations, actively push forward the adjustment of economic structure and transformation of development mode to realize energy conservation and emission reduction goals for the 12th Five-year Plan period.

On July 7, 2011, China International Low-carbon Industry Exhibition 2011 jointly organized by China Renewable Energy Industry Association, United Nations Industrial Development Organization and World Wildlife Fund, etc. was held in Beijing Exhibition Center. With the theme "low carbon", this Exhibition gave prominence to energy conservation, low carbon, emission reduction and intelligence and was designed to provide one professional, international cooperation communication platform for exhibitors and visitors.

On July 14, 2011, Chengdu Municipal Commission for Economy and Information Technology printed and issued *the 12th Five-year Development Plan of Chengdu City for New Energy Industry*, which underlines efforts to turn Chengdu into a new energy industry base being No.1 in Western China and nationally first-class within five years. According to relevant information, Chengdu is "National Hi-tech Industry Base for New Energy Industry" and "Chengdu (National) High and New Technology Industrialization Base for New Energy Equipment". Last year, 30km^2 new energy industry function zone was planned in Chengdu to guide concentrated, intensive and clustered development of new energy industry. In the same year, main business revenue was 28,675 million yuan, up 187% compared with 2008, and an industry system dominated by solar energy, nuclear energy, wind energy, power and energy storage battery, smart power grid preliminarily took shape.

On July 19, 2011, Premier Wen Jiabao, Leader of Leading Group for Energy Conservation and Emission Reduction, chaired and convened meeting of National Leading Group for Climate Change, Energy Conservation and Emission Reduction, in which *the 12th Five-year Comprehensive Work Plan for Energy Conservation and Emission Reduction* was deliberated and approved in principle and the following six energy conservation and emission reduction measures were deliberated and approved: carry forward energy conservation and emission reduction in key fields; further adjust and optimize industrial structure; implement key projects for energy conservation and emission reduction; popularize and apply advanced technologies; strengthen energy conservation and emission reduction management; improve long-term mechanism for energy conservation and emission reduction.

On July 19, 2011, the State Council announced *the 12th Five-year Comprehensive Work Plan for Energy Conservation and Emission Reduction* ("Plan"), which specifies overall arrangement of work on energy conservation and emission reduction in four large fields including industry, transportation, building and life during the 12th Five-year Plan period, and requires various localities and departments to implement specific requirements of this Plan at the next stage. It requires building energy conservation to be conducted by rationally constructing existing buildings, vigorously developing green buildings, intelligent buildings, maximizing saving of energy, land, water and materials.

On July 20, 2011, the Ministry of Finance and the Ministry of Environmental Protection issued *Pilot Administrative Measures for Protection of Lake Ecological Environment*, which is designed to protect lake ecological environment, establish high-quality ecological lake protection mechanism and incorporate eight lakes nationwide including "southern four lakes" in Shandong Province into pilot work on environmental ecological protection.

On July 21, 2011, General Office of the State Council issued *Circular Concerning Establishment of System for Compulsory Government Procurement of Energy-saving Products*, which requires the government to take the lead in the procurement of energy-saving products.

On July 28, 2011, the Ministry of Finance and the Ministry of Environmental Protection jointly issued *Circular Concerning Adjustment and Release of the 8th Government Procurement List for Environmental Label Products* (CK (2011) No.108) to further standardize types and categories of low-carbon products which the government takes the lead in purchasing.

On July 29, 2011, the State Council released *Circular Concerning Work on Public Institution Energy Conservation in 2011.*

On July 29, 2011, the Ministry of Environmental Protection and State Administration for Quality Supervision and Inspection and Quarantine jointly released newly revised *Air Pollutant Discharge Standard for Thermal Power Plant*, which would take effect from January 1, 2012. Tao Detian, news spokesman of the Ministry of Environmental Protection, stated that new standard will increase environmental protection access threshold for the thermal power industry, and promote decrease in emission intensity in the thermal power industry, lower pollutant discharge, accelerate transformation of development direction for the thermal power industry, optimize industrial structure and boost sustainable and healthy development of the power industry.

On August 2, 2011, the Ministry of Environmental Protection issued *Circular Concerning Strengthening Radiation Safety Supervision of Scrap and Old Metal Recovery Melting Enterprises*. This Circular provides that: (1)All melting enterprises must carry out radiation monitoring, and immediately report to local environmental protection department when discovering radioactive pollution. (2)Environmental protection departments at various localities should take effective measures for discovered and seized polluted metals, and timely urge owner to recover and properly dispose of polluted metals. (3)Costs incurred from handling discovered and seized out-of-control radioactive source or polluted metals should be borne by the original owner or supplier of out-of-control radioactive source or polluted metals. (4)When radiation monitoring data show that radioactive radiation level of monitored metals is higher than radiation background level in local natural environment, gamma spectrometer should be used for analyzing it and confirming whether radioactive pollution occurs. (5)The corresponding measures are developed for handling metals with different Co-60 activity concentrations. (6) Polluted metals which should be stored and handled may be delivered to radioactive waste repository or other qualified organizations in the city where such metals are located for temporary storage until their decay reaches the recyclable level.

On August 11, 2011, the Ministry of Finance and National Development and Reform Commission issued *Circular Concerning Adjustment and Release of the 10th Government Procurement List for Energy-saving Products.*

Main content: The 10th energy conservation list becomes effective from the day of its release; subsequent government procurement activities will be governed by the 10th energy conservation list and will be no longer governed by previously released energy conservation list. If products supplied under implemented centralized government procurement agreement are confirmed to involve product category in energy conservation list, centralized procurement organizations shall reorganize agreement supply activity or make adjustment according to current energy conservation list. Government procurement projects shall be strictly governed by the system for preferential procurement and compulsory procurement of energy-saving products by the government. Energy conservation list will be readjusted and released in January, 2012. Relevant enterprises shall ensure that products under models and series specified in energy conservation list are stably supplied within the effective period of the current energy conservation list.

On August 12, 2011, the Ministry of Finance and the Ministry of Housing and Urban-Rural Development issued *Circular Concerning Strengthening Management of Follow-up Work on Solar Photoelectric Building Application Demonstration.* Main content: Demonstration projects approved in 2009 must undergo completion acceptance prior to September 30, 2011; demonstration projects approved in 2010 must undergo completion acceptance prior to December 31, 2011; demonstration projects approved in 2011 must undergo completion acceptance prior to August 30, 2012. Action shall be taken to accelerate the organization of completion acceptance of demonstration projects; comprehensive acceptance check and evaluation of completed project application execution, project operation effect, system energy efficiency monitoring and fiscal fund utilization, etc. shall be conducted.

On August 12, 2011, the Ministry of Finance and State Forestry Administration issued *Circular Concerning Opinions Involving Pilot Work on Ecological Civilization Demonstration Project in Western Region.* Main content: pilot work mainly covers actions to strengthen ecological construction and environmental protection, quicken the transformation of economic development mode, optimize the consumption mode.

On August 17, 2011, Xie Zhenhua, Vice-minister of National Development and Reform Commission, chaired and convened the liaison meeting of the State Council's

Leading Group for Energy Conservation and Emission Reduction, in which the spirit of meeting of National Leading Group for Climate Change, Energy Conservation and Emission Reduction was conveyed and implemented, and work division among departments with respect to the 12th Five-year comprehensive work plan for energy conservation and emission reduction was studied and discussed. 32 relevant departments including the Ministry of Environmental Protection, the Ministry of Education and the Ministry of Science and Technology attended this meeting.

On August 18, 2011, the Ministry of Science and Technology, etc. strengthened safety management in new energy vehicle demonstration and popularization and issued *Circular Concerning Strengthening Safety Management in Energy-saving and New Energy Vehicle Demonstration and Popularization* (GKBHG: 2011 No.322).

On August 26, 2011, Low-carbon Emission Reduction Committee of China New Energy Chamber of Commerce, the first national organization in China's low-carbon emission reduction field, was established in Beijing. More than 100 personages including officials from competent departments in charge of climate change in ministers and commissions including National Development and Reform Commission and the Ministry of Foreign Affairs, famous entrepreneurs, famous experts and scholars from low-carbon emission reduction field, personages from domestic and foreign organizations attended the establishment meeting. Low-carbon Emission Reduction Committee was co-sponsored and co-founded by more than 30 organizations including Taimao Shengyuan Industrial Development Co., Ltd., Tianjin Climate Exchange, China Beijing Environment Exchange, First Climate, CDM Center of Tsinghua University, etc. among which Taimao Shengyuan Industrial Development Co., Ltd. tool the lead. It is one national non-profit industry organization consisting of project owners, project advisory bodies, carbon credit buyers, auditing bodies, low-carbon investment organizations, emission right exchanges and low-carbon research institutions, etc.

On October 19, 2011, Low Carbon Earth Summit was held in Dalian. 50 international low-carbon organizations, 100 government representatives, enterprise leaders and famous experts from nearly 50 countries and regions were invited from economic, scientific and technological perspectives to attend this world low-carbon economic forum. 10 large low-carbon sub-forums, 200 cutting-edge technology workshops, 1,000 leading low-carbon enterprises for display and 5 science popularization activities for the general public would be organized; it would be a technology-based, economy-focused, leading-edge magnificent meeting; representatives from more than 50 countries and regions would attend this meeting.

On October 29, 2011, World Economic and Environmental Conference 2011 (WEC2011) China International Climate Change Forum was held in Beijing during October 29-30, 2011; the theme of this Conference was "green development policy and action". This conference was sponsored by the United Nations Industrial Development Organization (UNIDO) in China, co-hosted by UNIDO and International Energy Conservation Environmental Protection Association (IEEPA) and co-organized by embassies of relevant countries in China and Chinese famous academic institution - Institute of Low Carbon Economy, Tsinghua University, etc.

On November 9, 2011, Premier Wen Jiabao chaired and convened executive meeting of the State Council to discuss and approve *the 12th Five-year Work Plan for Greenhouse Gas Emission Control*. This meeting specified overall requirements and key tasks for controlling greenhouse gas emissions in China. This meeting required various localities and departments to make active response to climate change as major strategy for economic and social economy and major opportunity for quickening transformation of economic development mode, adjusting economic structure and promoting new industrial change for carrying out various tasks according to the following goal specified in the 12th Five-year Plan: carbon dioxide emission per unit of GDP will decrease by 17% by 2015 compared with 2010.

On November 14, 2011, in order to implement *Environmental Protection Law of the People's Republic of China,* prevent and control pollution, guarantee people's health, the Ministry of Environmental Protection and State Administration for Quality Supervision and Inspection and Quarantine jointly released *Guidelines for Evaluation of Air Quality within Passenger Vehicle*. This standard specifies concentration requirements for benzene, methylbenzene, dimethylbenzene, ethylbenzene, styrene, formaldehyde, acetaldehyde and acraldehyde in air within vehicle. This standard is applied to evaluate air quality within passenger vehicle. This standard is mainly applicable to sold new vehicles; reference may also be made to this standard in the case of vehicles in use. This standard took effect on March 1, 2012.

On November 15, 2011, the Ministry of Housing and Urban-Rural Development issued *the 12th Five-year Plan for Urban Green Lighting*. This Plan is developed according to relevant requirements of *the 12th Five-year Plan of the People's Republic of China for National Economic and Social Development, the 12th Five-year Comprehensive Work Plan for Energy Conservation and Emission Reduction* and the 12th Five-year Plan for the housing and urban-rural development undertaking in order to promote urban green lighting nationwide and improve urban energy conservation management. This Plan mainly specifies guiding thought, basic principles, development goals, key tasks and safeguard measures for urban green lighting; this Plan serves as the basis for carrying out urban green lighting in various localities during the 12th Five-year Plan period.

On November 17, 2011, the person in charge of Department of Science, Technology and Standards, the Ministry of Environmental Protection stated that public opinions about closely-watched *Ambient Air Quality Standard* started being solicited for the second time from this day. The maximum adjustment in this exposure draft lies in incorporating fine particulate matter (PM2.5) and ozone (8-h concentration) into routine air quality evaluation and tightening standard limit values for fine particulate matter (PM10) and nitric oxide, etc; new standard proposes to be fully implemented in 2016. As formulation and implementation in 1982, China's first ambient air quality standard *Ambient Air Quality Standard* has been revised three times The Ministry of Environmental Protection, etc. completed *Ambient Air Quality Standard* (exposure draft) at the end of 2010.

On November 22, 2011, Information Office of the State Council released white paper *China's Policy and Actions for Addressing Climate Change* (2011). This white paper states that China will place emphasis on strengthening legal construction and strategic planning, accelerating adjustment of economic structure, optimizing energy structure and developing clean energy, continuing implementing key projects for energy conservation, vigorously developing circular economy, steadily pushing forward low-carbon pilot work, gradually developing carbon emission trading market, increasing carbon sink, enhancing the capability for addressing climate change, continuing reinforcing capacity building, engaging in international cooperation in an all-round way and propelling work on climate change in 11 aspects.

On December 9, 2011, the Ministry of Housing and Urban-Rural Development posted *Plan for Implementation of Circular of the State Council Concerning Printing and Distribution of the 12th Five-year Comprehensive Work Plan for Energy Conservation and Emission Reduction* on its official website. This Plan clarifies building energy conservation and emission reduction goals for the 12th Five-year Plan period. With respect to building energy conservation, building energy conservation capability equivalent to 116 million tons standard coal will be built by the end of the 12th Five-year Plan period; renewable energy and building integration application will be promoted to develop conventional energy replacement capability equivalent to 30 million tons standard coal. With regard to building emission reduction, by the end of the 12th Five-year Plan period, sewage treatment capacity will be basically built in all counties and key towns; daily sewage treatment capacity equivalent to 42 million tons will be newly added nationwide; about 160,000km supporting pipe network will be newly built; urban sewage treatment rate will reach 85%; chemical oxygen demand reduction capacity will reach 2.8 million tons; ammonia nitrogen reduction capacity will stand at 0.3 million tons; rate of innocent treatment of urban domestic garbage will exceed 80%.

On December 19, 2011, *the 12th Five-year National Plan for Energy Science and Technology (2011-2015)* ("this Plan") developed by National Energy Administration was officially released. It is the first plan released by National Energy Administration in three years since its establishment and also the first energy science and technology plan in China. According to relevant information, this Plan was completed by more than 200 experts from specialized departments of National Energy Administration and the energy field as organized by Energy Conservation & Technical Equipment Department of National Energy Administration in two and a half years. This Plan stresses efforts to build "four-in-one" national energy science and technology innovation system which integrates major science and technology research, major technical equipment, major demonstration projects and technical innovation platform. This Plan determines 19 major special projects for energy application technologies and project demonstration and their technical routes, and plans 37 major technical researches, 24 major technical equipments, 34 major demonstration projects and 36 technical innovation platforms.

On December 28, 2011, the opening ceremony for the 17th Conference of the Parties (COP17) to United Nations Framework Convention on Climate Change (UNFCCC) and the 7th Conference of the Parties (CMP7) to Kyoto Protocol was held in Durban, South Africa.

On January 31, 2012, 17 departments including National Development and Reform Commission, Publicity Department of the Central Committee of the CPC, the Ministry of Education, the Ministry of Science and Technology etc jointly released *the 12th Five-year All-people Action Plan for Energy Conservation and Emission Reduction* to organize and lead launch of ten special actions for energy conservation and emission reduction in household communities, youngsters, enterprises, schools, military camps, rural areas, government departments, science and technology, science popularization and media. According to this Plan, various ways including typical demonstration, special activity, exhibition and display, job creation and rational suggestions will be adopted to extensively mobilize people in the whole society to participate in energy conservation, emission reduction, and advocate civilized, saving, green and low-carbon production mode, consumption mode and living habit.

On February 24, 2012, the Ministry of Industry and Information Technology developed *the 12th Five-year Development Plan for the Solar Photovoltaic Industry in order to implement Industrial Transformation Upgrading Plan (2011-2015), the 12th Five-year Plan for the Information Industry and 12th Five-year Development Plan for the Electronic Information Manufacturing Industry* and promote sustainable development of the solar photovoltaic industry. This Plan mainly covers six parts including development retrospect, situation, guiding thought and development goals, major tasks, development priorities, policies and measures, involving various links of the photovoltaic industry including materials, equipment, batteries, photovoltaic power stations and supporting services.

On February 29, 2012, the Ministry of Environmental Protection and State Administration for Quality Supervision and Inspection and Quarantine jointly released national ambient quality standard *Ambient Air Quality Standard*. This standard is valid for compulsory execution according to relevant laws and will take effect nationwide from January 1, 2016. New standard adds fine particulate matter (PM2.5) and ozone 8-h concentration limit value monitoring indicators. This shows that China's environmental protection has entered a new stage at which not only there is high attention from the government but also environmental awareness of the general public has also been gradually enhanced. In addition, new standard still retains 150 mcg/m^3 as PM10 average daily concentration limit.

On March 26, 2012, the Ministry of Industry and Information Technology released *2012 Key Points for Work on Industrial Energy Conservation and Comprehensive Utilization*. These Key Points for Work include: this year action will be taken to strengthen energy conservation management in key energy-consuming enterprises, drive enterprise energy efficiency to reach benchmark and standard, and push forward phase-out and renovation of energy-consuming equipments, and vigorously apply new mechanisms including contract energy management. According to these Key Points for Work, in 2011, efforts will be made to further phase out backward production capacity, and break down and assign tasks for phase-out of backward production capacity during the 12th Five-year Plan period according to *Circular of the State Council Concerning Further Strengthening Phase-out of Backward Production Capacity* as soon as possible and restrict blind expansion of industries with high energy consumption.

On March 27, 2012, *Measures of Shenzhen City for Promotion of Green Buildings (Draft)* were announced; this suggests that Shenzhen is dedicated to building a low-carbon ecological city, while one priority for future low-carbon development of Shenzhen lies in green buildings. This Draft specifies specific provisions for construction of city-wide green buildings from the perspectives of green building materials, green municipal facilities, green construction, green decoration and green property management, etc; five categories of building include new, reconstructed and expanded buildings in which government fiscal funds are invested, new large public buildings with individual building area more than 50,000m^2 in which social investments are made, new residential quarters with building area more than 100,000m^2, and all new, reconstructed and expanded buildings within national, provincial and municipal green building demonstration areas designated in relevant agreement or document, and l new, reconstructed and expanded security housing, school, hospital, cultural sports tourism buildings, and all newly-added urban renewable buildings. Shenzhen will take a green path and develop green standards.

On March 28, 2012, the launching ceremony for pilot carbon emission right trading was held in Beijing, in which overall arrangement of national carbon emission trading work and work requirements for pilot areas for carbon emission right trading were introduced, and announcement was made to establish Beijing Municipal Climate Change Expert Committee, Beijing Climate Change Research and Talent Cultivation Base, Beijing Carbon Emission Right Trading Enterprise Alliance, Beijing Carbon Emission Right Trading Intermediary Consulting & Certification Body Alliance, Beijing Green Financial Institution Alliance. As Beijing was included in the first batch of pilot areas for carbon emission right trading in China, carbon emission trading was officially launched in Beijing in 2013. Currently, Beijing's plan for pilot work on carbon emission right trading has been finalized and approved by municipal government. Pilot work focuses on establishing regional carbon emission right trading market system and promoting greenhouse gas emission reduction to realize the goal for carbon dioxide emission reduction per unit of gross regional domestic product during the 12th Five-year Plan period.

On March 28, 2012, the Ministry of Industry and Information Technology printed and distributed *2012 Key Points for Work on Industrial Energy Conservation and Comprehensive Utilization*. The Ministry of Industry and Information Technology vowed to reduce energy consumption and carbon dioxide emission per unit of industrial added value by 5% and more than 5% respectively, and decrease water consumption per unit of industrial added value by 7%, and increase rate of comprehensive utilization of bulk industrial solid wastes by 2 percentage points this year so as to lay a solid foundation for realizing binding goals for energy conservation and emission reduction during the 12th Five-year Plan period and promoting industrial transformation upgrading.

On March 29, 2012, *2012 Key Points for Work on Industrial Energy Conservation and Comprehensive Utilization* printed and distributed by the Ministry of Industry and Information Technology call for reducing energy consumption and carbon dioxide emission per unit of industrial added value by 5% and more than 5% respectively, and decrease water consumption per unit of industrial added value by 7%, and increase rate of comprehensive utilization of bulk industrial solid wastes by 2 percentage points this year so as to lay a solid foundation for realizing binding goals for energy conservation and emission reduction during the 12th Five-year Plan period and promoting industrial transformation upgrading.

On March 29, 2012, the Ministry of Environmental Protection printed and issued *Opinions Concerning Strengthening Capacity Building for Ambient Air Quality Monitoring*. According to these Opinions, national ambient air quality monitoring network will be completed by 2015. Environmental protection departments at various localities shall actively coordinate with finance departments at the same levels to incorporate expenses for ambient air quality monitoring capacity building and operation guarantee into public financial budgets at various levels. The Ministry of Environmental Protection also issued circular to require various localities to apply for special funds from central finance for major pollutant emission reduction. According to these Opinions, overall goal for ambient air quality monitoring capacity building during the 12th Five-year Plan period is to integrate various information resources including national atmospheric background monitoring network, rural monitoring network, acidic precipitation monitoring network, monitoring network for impact of sand and dust weather on atmospheric environment, greenhouse gas test monitoring, and add monitoring indictors, and develop and improve unified quality management system and point location management system, and improve air quality evaluation techniques and methods and information release mechanism.

Postscript

Report of the 18th CPC National Congress elevates ecological civilization construction to an unprecedented level and stresses efforts to build on respecting and safeguarding nature, and proceed from harmonious coexistence between person and person, between man and nature, between man and society to promote urban transformation towards low-carbon economic mode by fostering low-carbon development, green development, circular development and intelligent development so as to achieve sustainable development.

Since 2010, the Institute for Urban and Environmental Studies Chinese Academy of Social Sciences has cooperated with Hunan University of Technology to establish Joint Research Center of Global Low-carbon City and choose 110 representative Chinese cities to conduct urban low-carbon development survey and research, build carbon emission database covering 110 Chinese cities and China's urban low-carbon development evaluation indicator system, and put forward specific measures and suggestions concerning promotion of China's urban low-carbon transformation from five perspectives including output, consumption, resource, society and environment.

China's Urban Intelligent Low-carbon Development Report is the very important basic work in ecological civilization construction project for building beautiful China and is masterpiece of the times jointly produced by Institute for Urban and Environmental Studies Chinese Academy of Social Sciences, Hunan University of Technology and Joint Research Center of Global Low-carbon City. Overall framework and guiding thought were finalized by researcher Pan Jiahua and Professor Wang Hanqing. "Theme" chapters were written by researcher Liang Benfan and Professor Zhou Yueyun, revised, supplemented and finalized by researcher Pan Jiahua and Professor Wang Hanqing. "General" and "special topics" chapters were separately written by authors after members of Research Group put forward writing outlines and group discussion was conducted. "Cases" chapters were written by members of Research Group according to their research practice in case cities. Indicator calculation with data concentration was completed by researcher Liang Benfan and Dr. Zhu Shouxian et al; data collection and processing was conducted students Chen Mengmei, Liang Tao and Li Minrui et al from Graduate School of Chinese Academy of Social Sciences, Beijing University of Posts and Telecommunications and Central Academy of Fine Arts. Memorabilia were complied and completed by graduate students Yu Li and Wang Ao from Hunan University of Technology. Researcher Liang Benfan and Professor Zhou Yueyun served as compilers in chief of the whole book, while researcher Zhang Taoxin and Yao Dihui assisted unified compilation.

Intelligent low-carbon development is the future for human city development and new hope for human civilization development. Mr. Wu Jichuan, former Minister of the Ministry of Information Industry, gladly prefaced this book with the hope that people from all walks of life can actively pay attention to and promote construction of low-carbon intelligent cities in China. This Report was funded by Innovation Project of Chinese Academy of Social Sciences. As this book will be immediately available for publication and readers, we hereby express sincere thanks to all members of Research Group, leaders and people from all walks of life who cared for and supported this project research! We also extend heartfelt thanks to Wang Yin from China Social Sciences Publishing House!

It should be noted that though all members of Research Group discussed some major theoretical and practical issues many times and absorbed diverse opinions and suggestions as far as possible during research and writing, flaws even errors inevitably exist in this book due to restrictions from various subjective and objective conditions, domestic and foreign colleagues are highly appreciated for making corrections.

Liang Benfan
December 24, 2012

This book is the result of a co-publication agreement between China Social Sciences Press (China) and Paths International Ltd (UK)

This book is published with the financial support of the Innovation Project of CASS

Smart Low-Carbon Development of Cities in China
Edited by Pan Jiahua, Wang Hanqing, Liang Benfan, Zhou Yueyun
ISBN: 978-1-84464-406-3
Ebook ISBN: 978-1-84464-407-0

Paths International Ltd
PO Box 4083
Reading
United Kingdom
RG8 8ZN
www.pathsinternational.com

Published in the United Kingdom

CPSIA information can be obtained
at www.ICGtesting.com
Printed in the USA
LVOW03*0908250916
505571LV00012BD/32/P

9 781844 644063